Contents

= Physical chemistry

= Inorganic chemistry

= Organic chemistry

Get the most from this book

Welcome to the **AQA A-level Chemistry 1 Student's Book**. This book covers Year 1 of the AQA A-level Chemistry specification and all content for the AQA AS Chemistry specification.

The following features have been included to help you get the most from this book.

Prior knowledge

This is a short list of topics that you should be familiar with before starting a chapter. The questions will help to test your understanding.

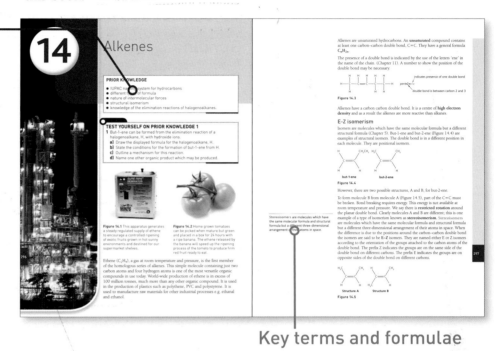

Key terms and formulae

These are highlighted in the text and definitions are given in the margin to help you pick out and learn these important concepts.

Tips

These highlight important facts, common misconceptions and signpost you towards other relevant topics.

Examples

Examples of questions or calculations are included to illustrate topics and feature full workings and sample answers.

Test yourself questions

These short questions, found throughout each chapter, are useful for checking your understanding as you progress through a topic.

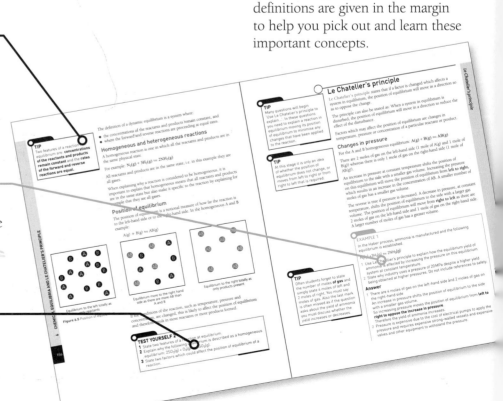

Activities

These practical-based activities will help consolidate your learning and test your practical skills.

In this edition the authors describe many important experimental procedures as "Activities" to conform to recent changes in the A-level curriculum. Teachers should be aware that, although there is enough information to inform students of techniques and many observations for exam-question purposes, there is not enough information for teachers to replicate the experiments themselves or with students without recourse to CLEAPSS Hazcards or Laboratory worksheets which have undergone a thorough risk assessment procedure.

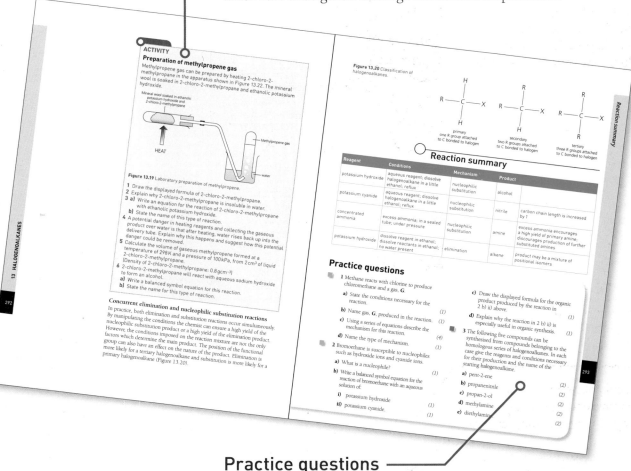

Practice questions

You will find practice questions, including multiple-choice questions, at the end of every chapter. These follow the style of the different types of questions with short and longer answers that you might see in your examination, and they are colour coded to highlight the level of difficulty.

- Green – Basic questions that everyone should be able to answer without difficulty.
- Orange – Questions that are a regular feature of exams and that all competent candidates should be able to handle.
- Purple – More demanding questions which the best candidates should be able to do.

A dedicated chapter for developing your **Maths** can be found at the back of this book.

1

Atomic structure

TEST YOURSELF ON PRIOR KNOWLEDGE 1

1 What is meant by the term atomic number?

2 State the relative mass and relative charge of: a) a proton; b) an electron and c) a neutron?

3 $^{35}_{17}Cl$ and $^{37}_{17}Cl$ are isotopes of chlorine.

   a) What is meant by the term isotopes?

   b) Calculate the number of protons, electrons and neutrons in an atom of $^{37}_{17}Cl$.

4 Write electron configurations for the following atoms:

   a) P **b)** Li **c)** O **d)** K **e)** Ar **f)** He **g)** Al

5 Identify the element which has atoms with 15 electrons and 16 neutrons.

The development of theories about atoms and their structure has spanned centuries. The model of the atom that will be familiar from previous study is of a heavy nucleus containing protons (and neutrons) surrounded by electrons orbiting in shells. Many scientists contributed to the development of theories of atomic structure.

The basics of atomic structure

An atom is the smallest particle of an element which has the characteristic properties of the element. The symbol for an element can represent an atom of that element. For example, Mg represents an atom of magnesium, C represents an atom of carbon.

Relative mass of subatomic particles

Atoms are composed of subatomic particles called electrons, protons and neutrons. The table below shows the relative mass, relative charge and location of the subatomic particles. The actual mass of a proton is $1.67262178 \times 10^{-27}$ kg. For ease of calculation the masses of the subatomic particles are measured relative to the mass of a proton, which is given a value of 1. The mass of a neutron is the same and the mass of an electron is 1840 times less. The same idea of a relative scale is used with charge. The charge on a proton is given a value of $+1$ and the charge on an electron has the same magnitude but it is oppositely charged.

Particle	Relative mass	Relative charge	Location in the atom
Proton	1	+1	nucleus
Neutron	1	0	nucleus
Electron	$\frac{1}{1840}$	−1	energy levels

As the relative mass of an electron is substantially smaller than the mass of a proton and a neutron, most of the mass of an atom is concentrated in the nucleus. The nucleus occupies only a small fraction of the total volume of the atom.

The incredibly high density within the nucleus suggests that the particles within it are drawn very close together by extremely powerful forces. These forces are obviously so powerful that they can overcome the repulsion which the protons have for each other as they are positively charged. Neutrons have no charge and are not involved in the repulsion in the nucleus.

Atomic number and mass number

- The **atomic number** is equal to the number of protons in the nucleus of an atom.
- The **mass number** is the total number of protons and neutrons in the nucleus of an atom.
- Atoms of the same element can have different masses. For example an atom of hydrogen can have a mass number of 1, 2 or 3.
- Atoms of the same element, which have different mass numbers, are called **isotopes**.
- All the isotopes of an element have the same atomic number as they have the same number of protons in the nucleus.

Isotopes

Isotopes should be written as shown below.

$$\text{Mass number} \rightarrow A \atop \text{Atomic number} \rightarrow Z} E$$

where E is the symbol for the element. A is often used to represent the mass number and Z is used to represent the atomic number. The atomic number can also be called the proton number.

There are three isotopes of hydrogen which can be written as:

^1_1H Mass Number 1, Atomic Number 1, called protium
^2_1H Mass Number 2, Atomic Number 1, called deuterium
^3_1H Mass Number 3, Atomic Number 1, called tritium

From the mass number and atomic number, it is possible to determine the quantities of the various subatomic particles in an atom of each isotope, using the formulae:

Mass Number = Number of protons + Number of neutrons

Atomic Number = Number of protons

So

Number of neutrons = Mass Number − Atomic Number

Number of electrons = Number of protons [in a neutral atom]

Figure 1.1 The explosive power of a hydrogen bomb results from an uncontrolled, chain reaction in which isotopes of hydrogen – namely deuterium and tritium combine under extremely high temperatures. What element do you think is produced?

> **TIP**
> The number of protons defines the element. Any particles with 17 protons are particles of the element chlorine – they may be ions or atoms.

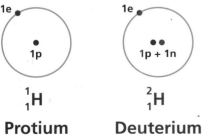
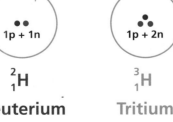

Figure 1.2 The three isotopes of hydrogen, showing their subatomic particles.

	Number of protons	Number of electrons	Number of neutrons
Protium ^1_1H	1	1	0
Deuterium ^2_1H	1	1	1
Tritium ^3_1H	1	1	2

Not all isotopes of elements have individual names, for example, with the two isotopes of chlorine:

$^{35}_{17}Cl$ (often called chlorine-35): Mass Number 35, Atomic Number 17

$^{37}_{17}Cl$ (often called chlorine-37): Mass Number 37, Atomic Number 17

The table below shows the numbers of subatomic particles for the two isotopes of chlorine.

	Number of protons	Number of electrons	Number of neutrons
Chlorine-35	17	17	18
Chlorine-37	17	17	20

Isotopes have the same number and arrangement of electrons and so the atoms of each isotope will have the same chemical properties. However due to the different numbers of neutrons, the atoms have different masses and hence different physical properties. For example pure $^{37}_{17}Cl_2$ will have a higher density, higher melting point and higher boiling point than pure $^{35}_{17}Cl_2$.

Two important isotopes of carbon are ^{12}C and ^{13}C.

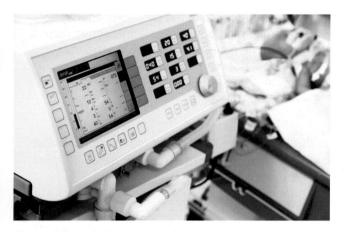

Figure 1.3 Analysis of carbon isotope ratios ($^{13}CO_2$:$^{12}CO_2$) in the breath of paediatric patients on ventilators can lead to early detection of sepsis. Sepsis is a whole-body inflammation caused by severe infection, which can lead to death.

TIP
Often the atomic number is not included when identifying isotopes but the mass number should always be written. Chlorine-35 may be written as $^{35}_{17}Cl$ or ^{35}Cl.

The carbon-12 standard

The masses of all atoms are measured relative to the mass of an atom of carbon-12, which is given a value of 12.0000. In all the following definitions the term relative means that the mass is measured against $\frac{1}{12}$ the mass of an atom of carbon-12.

The reason for the choice of carbon as the standard is an historical one and a practical one. Originally hydrogen was chosen as the standard against which the masses of all other elements was measured, however when the accuracy of atomic mass measurement reached 0.00001, it was realised that elements were a mixture of atoms of different masses. So it was decided to choose one isotope of an element as the standard. $^{12}_{6}C$ was chosen for the practicality of carrying a solid around as opposed to a gas.

The **relative isotopic mass** is the mass of a single isotope of an element relative to $\frac{1}{12}$ the mass of an atom of carbon-12. This is most often derived from mass spectrometry where the mass of the individual isotopes can be determined. This will be explained in the next section. For all purposes the value of the relative isotopic mass is the same as the mass number for a particular isotope of an element.

The **relative atomic mass** (A_r) is the average mass of an atom of an element relative to $\frac{1}{12}$ the mass of an atom of carbon-12. A_r is often used as shorthand for relative atomic mass.

Calculating relative atomic mass

Naturally occurring elements are composed of a mixture of different isotopes. These isotopes occur in different proportions.

If the proportion of each isotope were to be changed, the relative atomic mass of the element would change. The following hypothetical example shows what would happen if the proportion of the two isotopes of chlorine were different.

% chlorine-35	100.0	75.0	50.0	25.0	0.0
% chlorine-37	0.0	25.0	50.0	75.0	100.0
Relative atomic mass	35.0	35.5	36.0	36.5	37.0

EXAMPLE 1

In a naturally occurring sample of chlorine, 75% is ^{35}Cl and the remainder (25%) is ^{37}Cl. Calculate the relative atomic mass of the sample of chlorine.

Assuming 100 atoms of chlorine.

Answer

Total mass of ^{35}Cl atoms = 75 × 35 = 2625

Total mass of ^{37}Cl atoms = 25 × 37 = 925

Total mass of 100 atoms = (75 × 35) + (25 × 37) = **3550**

Average mass $= \frac{3550}{100} = 35.5$

Relative atomic mass of chlorine = 35.5.

The relative abundance of the two isotopes may not be given as a percentage. The relative atomic mass is calculated in the same way. Multiply the mass by the relative abundance for each isotope. Add these values together and divide by the total of all the relative abundances.

EXAMPLE 2

This table shows the relative abundances of the three different isotopes of magnesium. Calculate the relative atomic mass of magnesium to one decimal place.

Isotope	Relative abundance
^{24}Mg	15.8
^{25}Mg	2.0
^{26}Mg	2.2

For each isotope multiply the mass by the relative abundance and add these together. Finally divide this number by the sum of the relative abundances.

Answer

$$\text{Relative atomic mass} = \frac{(15.8 \times 24) + (2.0 \times 25) + (2.2 \times 26)}{15.8 + 2.0 + 2.2}$$

$$= \frac{379.2 + 50 + 57.2}{20}$$

$$= \frac{486.4}{20} = 24.32$$

The answer to one decimal place is 24.3.

TIP

Often you may be asked to quote the answer to a specific number of decimal places. The answer on the right to 1 decimal place is 24.3. 24.48 to 1 decimal place becomes 24.5. 48.775 to 2 decimal places becomes 48.78. If you are unsure about decimal places see Chapter 17.

TEST YOURSELF 2

1 State the number of protons, electrons and neutrons present in the following atoms:

a) $^{39}_{19}K$ **b)** $^{19}_{9}F$ **c)** $^{137}_{56}Ba$ **d)** $^{226}_{88}Ra$

2 Bromine has two isotopes which are detailed in this table.

Isotope	Atomic number	Mass number	% abundance
$^{79}_{35}Br$			
	35	81	49.5

a) Copy and complete the table above.

b) Calculate the relative atomic mass of bromine.

3 Copper exists as two isotopes, ^{65}Cu and ^{63}Cu. ^{65}Cu has an abundance of 30.8%.

a) Calculate the relative abundance of ^{63}Cu.

b) Calculate the relative atomic mass of copper to two decimal places.

4 Silver has two isotopes, ^{107}Ag and ^{109}Ag. 52% of silver is ^{107}Ag. Determine the relative atomic mass of silver.

5 92.0% of lithium is ^{7}Li and the remainder is ^{8}Li. Determine the relative atomic mass of lithium to 1 decimal place.

6 Boron has two isotopes, ^{10}B and ^{11}B. The relative atomic mass of boron is 10.8. Calculate the percentage abundance of each of the two isotopes.

Mass spectrometry

Mass spectrometry can be used to determine information about elements and compounds. It can determine the relative isotopic masses of the isotopes of elements and their relative abundance. It shows the different isotopes of an element. This information is used to calculate the relative atomic mass (A_r) of an element. For compounds, mass spectrometry can identify unknown purified compounds by comparing the mass spectrum obtained to those in a database. The mass spectrum of a compound also gives its relative molecular mass (M_r).

Time-of-flight (TOF) mass spectrometer

A TOF mass spectrometer is used to analyse elements and compounds. The sample is dissolved in a polar, volatile solvent and pumped through a narrow capillary tube to create droplets of the solution. A polar volatile solvent is used to ensure that it evaporates.

The five processes which occur in a TOF mass spectrometer are:

1 Electrospray ionisation

A high voltage is applied to the tip of the capillary to produce highly charged droplets. The solvent evaporates from these droplets to produce gaseous charged ions. All the ions in this simple treatment of TOF mass spectrometry are considered to be mononuclear ions (with a single positive charge).

2 Acceleration

An electric field is applied to give all the ions with the same charge a constant kinetic energy. As kinetic energy $= \frac{1}{2}mv^2$, it depends on the mass of the particles (m) and their velocity (v, or speed). As all particles are given the same kinetic energy, heavier particles (larger M_r) move more slowly than lighter particles.

3 Ion drift

The ions enter a region with no electric field called the flight tube. Here the ions are separated based on their different velocities. The smaller fast ions travel though the flight tube much more rapidly and arrive at the detector first.

4 Ion detection

The detector records the different flight times of the ions. The positively charged ions arrive at the detector and cause a small electric current because of their charge.

5 Data analysis

The flight times are analysed and recorded as a mass spectrum by the data analyser. The mass spectrum obtained is a plot of relative abundance against mass to charge ratio (m/z).

General information about a mass spectrum

The trace from the mass spectrometer (the mass spectrum) is a series of peaks on a graph where the vertical axis is relative abundance (which is the same as the electric current from the detector) and the horizontal axis is the mass to charge ratio.

The relative abundance is a measure of how many of each ion is present. Often the highest peak is given a value of 100 and the other peaks are worked out relative to this value.

The mass/charge ratio for single charge ions is equivalent to the mass of the ion, as mass divided by 1 = mass. The horizontal axis should be labelled 'm/z' but it may also be labelled 'm/e' or 'mass to charge ratio'. The m/z values for isotopes of an element are the relative isotopic masses for these isotopes.

Mass spectrum of an element

Figure 1.4 shows is an example of a mass spectrum of naturally occurring magnesium (an element). The numbers above the peaks indicate the relative abundance of each ion detected. Sometimes it is given like this and for other examples the relative abundance may have to be read from the scale on the vertical axis.

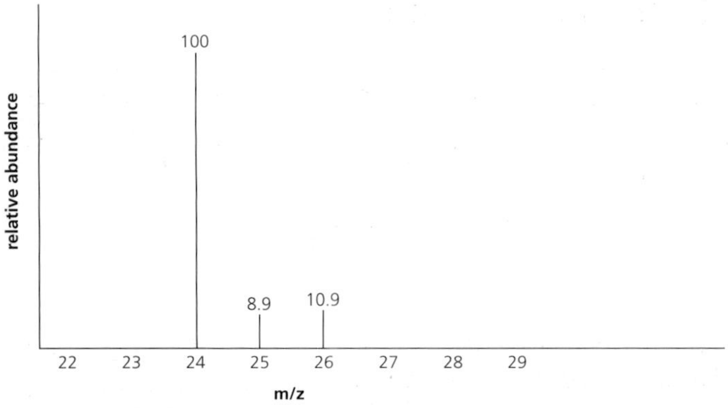

Figure 1.4 Mass spectrum of magnesium.

The peaks in the spectrum are caused by the isotopes of magnesium.

The mass spectrum tells us several things:

1 Magnesium has three isotopes as there are three peaks on the spectrum corresponding to each of the isotopes.

2 The relative isotopic masses of these isotopes are 24, 25 and 26 as these are the m/z values for each of the isotopes.

3 The most abundant isotope of magnesium has a relative isotopic mass of 24. This is the m/z value for the peak with the highest relative abundance.

4 The ion responsible for the peak at 24 is $^{24}Mg^+$; the ion responsible for the peak at 25 is $^{25}Mg^+$; the ion responsible for the peak at 26 is $^{26}Mg^+$.

5 The relative atomic mass (A_r) of magnesium can be calculated from the information in the mass spectrum.

The relative atomic mass of magnesium is simply the average mass of all the atoms of the different isotopes. This calculation is carried out by multiplying the relative isotopic mass by the relative abundance for each peak. These are then added together and the total is divided by the total of all the relative abundances.

Working this out

$$\text{relative atomic mass} = \frac{(100 \times 24) + (8.9 \times 25) + (10.9 \times 26)}{119.8} = 24.26$$

To 1 decimal place this is 24.3 as given on the Periodic Table of the Elements supplied in the Data Booklet with examinations.

EXAMPLE 3

Cubic zirconia, the cubic crystalline form of zirconium dioxide is extensively used in gems as a cheap alternative to diamond. It is artificially manufactured and contains the element zirconium, a transition metal which has four isotopes.

The mass spectrum on page 9 is for zirconium. Determine the relative atomic mass of zirconium to one decimal place.

Figure 1.5 Cubic zirconia resembles diamond, and sparkles with brilliance. Can you compare its chemical composition and properties with that of diamond?

Answer

There are four isotopes of zirconium with relative isotopic masses of 90, 91, 92 and 94.

The relative abundances of the isotopes are: $^{90}Zr = 9.0$; $^{91}Zr = 2.0$; $^{92}Zr = 3.0$; $^{94}Zr = 3.0$.

relative atomic mass

$$= \frac{(90 \times 9.0) + (91 \times 2.0) + (92 \times 3.0) + (94 \times 3.0)}{9 + 2 + 3 + 3}$$

$$= \frac{1550}{17} = 91.2 \text{ (to 1 decimal place)}$$

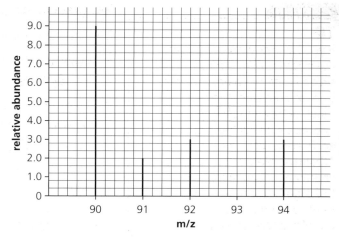

Figure 1.6 Mass spectrum of zirconium.

TIP

The element in Example 3 could have been an unknown element and you may have had to identify it from its relative atomic mass. In the Periodic Table provided with your examination, zirconium has a relative atomic mass of 91.2.

TIP

The ability to read data from a graph or in this case a spectrum and to translate these data into numerical form is an important skill throughout Chemistry. Use the level of precision given on the graph/spectrum – in this case both the m/z values and the relative abundance values are given to two significant figures. If you were asked for an appropriate level of precision in the answer, it should also be to two significant figures.

Mass spectrum of a molecular element

Elements like chlorine exist as diatomic molecules. In the mass spectrometer the molecules of an element like chlorine can form ions with a single positive charge but also the mass spectrometer can break up the molecule into atoms, which can also form single positively charged ions.

Chlorine has two different isotopes: ^{35}Cl and ^{37}Cl

There are five different possible ions which should be detected in the mass spectrometer:

- $^{35}Cl^+$
- $^{37}Cl^+$
- $(^{35}Cl\text{—}^{35}Cl)^+$
- $(^{35}Cl\text{—}^{37}Cl)^+$
- $(^{37}Cl\text{—}^{37}Cl)^+$

There should be peaks seen at m/z values of 35, 37, 70, 72 and 74 on a mass spectrum of molecular chlorine.

The mass spectrum of molecular chlorine is shown in Figure 1.7.

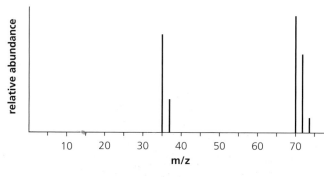

Figure 1.7 Mass spectrum of molecular chlorine.

- As predicted there are peaks at 35, 37, 70, 72 and 74.
- As the ratio of ^{35}Cl:^{37}Cl is 3:1 the relative abundance of the peaks at m/z values of 35 and 37 are in a 3:1 ratio.
- The ratio of the peaks at m/z values of 70:72:74 are 9:6:1.

Mass spectrum of a compound

For a compound, the last major peak at the highest m/z value is the **molecular ion**. This means it is caused by the molecular ion formed from the whole molecule. **The m/z value of this peak is the relative molecular mass (M_r) of the compound.**

The **relative molecular mass** is the mass of a molecule relative to $\frac{1}{12}$ the mass of an atom of carbon-12. The relative molecular mass is often written as M_r. It can be calculated by adding up the relative atomic masses of all the atoms in a compound.

When a molecule is put through a mass spectrometer, the molecule breaks up. This process is called fragmentation. The fragments of the molecule form ions and these are detected. The pattern of peaks caused by these fragments is called the fragmentation pattern.

Figure 1.8 shows the spectrum of ethanol, C_2H_5OH. The molecular ion peak (often written as M^+) is at an m/z value of 46. This is the last major peak in the spectrum of the highest m/z value. There is a small peak at 47 and this is caused by the presence of one ^{13}C atom.

The peak with the greatest abundance is usually not the molecular ion peak. The peak with the greatest abundance is called the base peak (in this case the one at an m/z value of 31) and in computer generated mass spectra, the height of this peak is usually taken as 100 and all other peaks are measured relative to it.

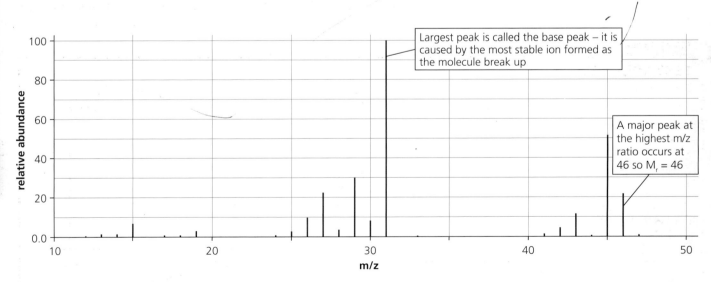

Largest peak is called the base peak – it is caused by the most stable ion formed as the molecule break up

A major peak at the highest m/z ratio occurs at 46 so M_r = 46

Figure 1.8 Mass spectrum of ethanol, CH_3CH_2OH.

EXAMPLE 4

A sample of an unknown compound was analysed in a time-of-flight (TOF) mass spectrometer. The spectrum obtained is shown in Figure 1.9. What is the M_r of this compound and what is the m/z value of the base peak?

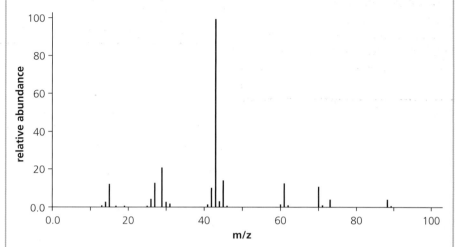

Figure 1.9 Spectrum of unknown compound.

Answer

The M_r of the compound is 88. This is the highest m/z of a peak in the spectrum and corresponds to the molecular ion.

The peak at an m/z value of 43 is the base peak and corresponds to the most stable ion as the molecule breaks up in the mass spectrometer.

Identification of elements (and compounds)

A mass spectrum is obtained, and information about the peak heights and m/z values are fed into a computer. This computer compares the spectrum of the unknown element (or compound) with those in its data banks and can identify the element (or compound). The fragmentation pattern of a compound is the same and acts like a fingerprint for that compound. An exact match is required for identification and this requires a pure sample.

TEST YOURSELF 3

1 The mass spectrum of an element is shown below. Calculate the relative atomic mass of the element to one decimal place and identify the element.

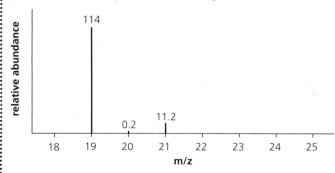

Figure 1.10

2 The mass spectrum of lead is shown below. Calculate the relative atomic mass of the lead.

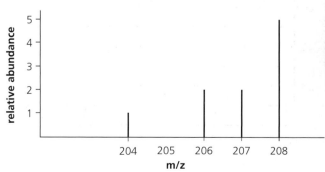

Figure 1.11

3 The spectrum below is for an unknown alkane. All alkanes have the general formula C_nH_{2n+2}.

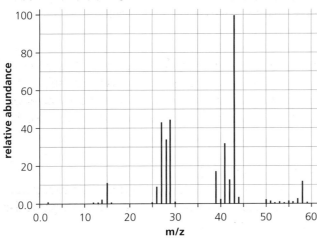

Figure 1.12 Spectrum for unknown alkane.

a) State the m/z value of the molecular ion peak.
b) State the m/z value of the peak with the highest relative abundance.
c) i) What is the relative molecular mass (M_r) of this alkane?
   ii) Suggest a formula for the alkane using the M_r.

TIP
For some work in bonding and structure, GCSE notation of electrons in shells is still used, particularly in dot and cross diagrams and shapes of molecules.

Electron configuration

The shells which may be familiar from GCSE are now called **energy levels**. Each energy level may be divided into sub-shells.

The evidence for the existence of **sub-shells** comes from evidence of ionisation energies of elements. Ionisation energies will be examined in the next part of this topic but before this can be done we need to understand electron configuration.

Basic information

● The first energy level (moving out from the nucleus) is called $n=1$; the second $n=2$ and so on. This number is called the principal quantum number.
● The energy levels get closer together as you move further from the nucleus. The difference between the first energy level ($n=1$) and the second energy level ($n=2$) is larger than the distance between the second energy level ($n=2$) and the third energy level ($n=3$). This continues further from the nucleus.
● Each energy level is divided into sub-shells.
● A sub-shell is an orbital or a combination of orbitals.
● An orbital is a three-dimensional space and each orbital can hold up to two electrons.
● Two electrons in the same orbital spin in opposite directions to minimise repulsions.
● There are four main types of orbitals: s, p, d and f, but at this level only s, p and d orbitals are studied.
● There is only one s orbital at each energy level.
● There are three p orbitals at each energy level starting at $n=2$ making up a p sub-shell.
● There are five d orbitals at each energy level starting at $n=3$ making up a d sub-shell.

Sub-shells

A sub-shell is always written as the energy level (principal quantum number) and then the type of orbital which makes up that sub-shell.

For example:

- The s orbital at energy level $n=1$ is written as 1s; this is the 1s sub-shell
- The s orbital at energy level $n=2$ is written as 2s; this is the 2s sub-shell
- The p orbital at energy level $n=2$ is written as 2p; this is the 2p sub-shell
- The s orbital at energy level $n=3$ is written as 3s; this is the 3s sub-shell and so on

Writing notation for electrons in sub-shells

When electrons are placed in a sub-shell, a number (written as a superscript) is written after the sub-shell notation. If there are no electrons in a particular sub-shell, it does not have to be written again.

For example:

- One electron in the 1s sub-shell is written as $1s^1$
- Two electrons in the 1s sub-shell is written as $1s^2$
- Three electrons in the 2p sub-shell is written as $2p^3$
- Six electrons in the 2p sub-shell is written as $2p^6$

The 3d sub-shell is at a slightly higher energy level than the 4s sub-shell as energy levels overlap slightly but by convention the 3d should be written before the 4s.

Sub-shells available at each energy level

- The $n=1$ energy level can have two electrons in the same sub-shell (1s).
- The $n=2$ energy level can have two electrons in one sub-shell (2s) and six electrons in a slightly higher sub-shell (2p).
- The $n=3$ energy level can have two electrons in one sub-shell (3s), six electrons in a slightly higher sub-shell (3p) and ten electrons in a sub-shell slightly higher again (3d).
- The $n=4$ energy level can have two electrons in one sub-shell (4s), six electrons in a slightly higher sub-shell (4p), ten electrons in the next sub-shell (4d) and 14 electrons in a sub-shell slightly higher again (4f).

Filling order of sub-shells

The sub-shells fill in the following order shown in Figure 1.13.

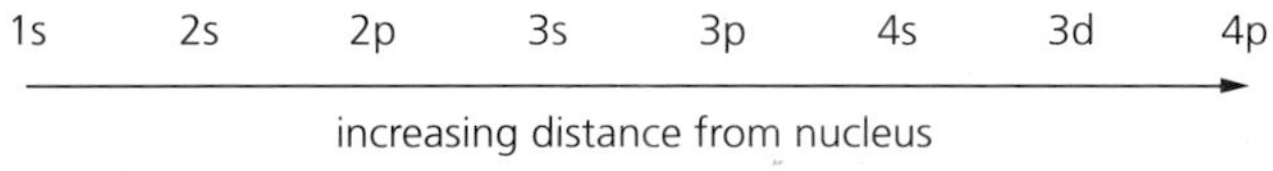

A typical sub-shell diagram

Figure 1.15 shows the electron configuration of a potassium atom.

At this stage it is important to note the following from the diagram:

- the distance between the energy levels decreases further from the nucleus
- the $n=3$ and $n=4$ energy levels overlap so that the 4s sub-shell is at a lower level than the 3d sub-shell
- electrons fill the sub-shells closest to the nucleus first
- in a potassium atom there is one electron in the outer 4s sub-shell and all other sub-shells closer to the nucleus are full

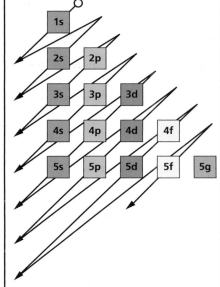

Figure 1.14 Diagram to remember the order in which sub-shells are filled. Follow the arrow!

| 1s | 2s | 2p | 3s | 3p | 4s | 3d | 4p |

increasing distance from nucleus

Figure 1.13 Sub-shells filling order.

13

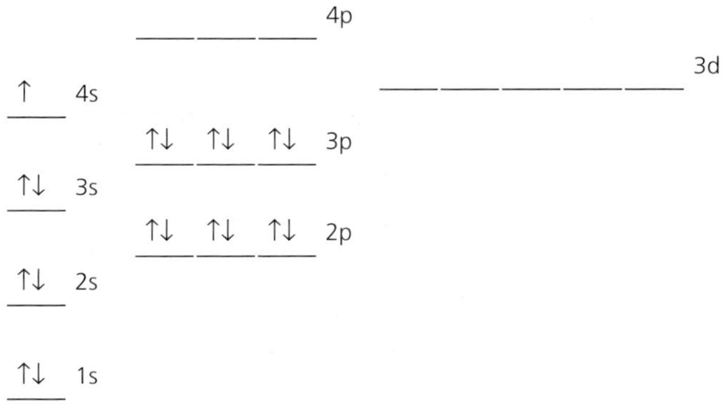

$$\begin{array}{ll}
\underline{}\,\underline{}\,\underline{} & 4p \\
\end{array}$$

↑ 4s   ___ ___ ___ ___ ___ 3d

  ↑↓ ↑↓ ↑↓ 3p

↑↓ 3s

  ↑↓ ↑↓ ↑↓ 2p

↑↓ 2s

↑↓ 1s

Figure 1.15 The electron configuration of a potassium atom.

- the electrons are indicated by an up arrow (↑) and a down arrow (↓); the up and down arrows represent the different directions of spin of the electrons in an orbital to minimise repulsions
- the 3d and 4p sub-shells are not occupied by electrons and so are not written in the electron configuration
- from the diagram below the electron configuration of a potassium atom is written as: $1s^2\ 2s^2\ 2p^6\ 3s^2\ 3p^6\ 4s^1$.
- when electrons are in their lowest possible energy levels, the atom is said to be in the ground state.

Writing electron configurations

Electrons repel each other and so when forming pairs in an orbital, they will only do this when they must.

An s sub-shell has only one orbital so two electrons will occupy this (spinning in opposite directions).

A p sub-shell has three p orbitals. If two electrons are placed in a p sub-shell they will go into different orbitals which make up the sub-shell.

Figure 1.16 The three electrons are shown to be spinning in the same direction (all as ↑) in the electron in box notation diagram.

EXAMPLE 5

Write the electron configuration of a nitrogen atom.

Answer

An electron in box notation diagram helps to show this as it shows the electrons at each of the energy levels. Nitrogen atoms (atomic number = 7) have full 1s and 2s sub-shells and there are three more electrons to place in the 2p sub-shell. There is one electron in each of the three 2p orbitals that make up the 2p sub-shell. The three electrons are shown to be spinning in the same direction (all as ↑) in the electron in box notation diagram (Figure 1.16).

The electron configuration of nitrogen is written $1s^2\ 2s^2\ 2p^3$.

Nitrogen atoms have three unpaired electrons.

EXAMPLE 6

Write the electron configuration of an oxygen atom.

Answer

The atomic number of oxygen is 8.

The 1s and 2s sub-shells are full and this leaves 4 electrons to place in the 2p sub-shell. There is one orbital with 2 electrons (spinning in opposite directions shown as ↑↓) and the other two 2p orbitals in this sub-shell have one electron (both shown as ↑)

Oxygen atoms have only two unpaired electrons as one of the 2p orbitals has a pair of electrons which spin in opposite directions to minimise their repulsion for each other.

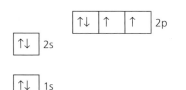

Figure 1.17 The electron configuration of oxygen is written $1s^2\,2s^2\,2p^4$.

EXAMPLE 7

Write the electron configuration of a vanadium atom.

Answer

The atomic number of vanadium is 23. For 23 electrons once again the $1s^2\,2s^2\,2p^6\,3s^2\,3p^6$ sub-shells are full (18 electrons in total). This leaves five electrons to place in the 4s and 3d sub-shells. The 4s fills first taking two electrons and then the remaining three electrons are placed in the 3d (all spinning in the same direction in different orbitals).

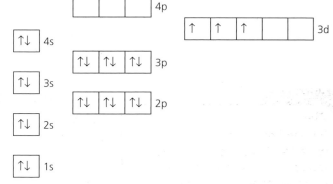

Figure 1.18 The electron configuration of a vanadium atom is $1s^2\,2s^2\,2p^6\,3s^2\,3p^6\,3d^3\,4s^2$.

EXAMPLE 8

Write the electron configuration of a chromium atom.

Answer

The atomic number of chromium is 24.

24 electrons means the $1s^2\,2s^2\,2p^6\,3s^2\,3p^6$ are full as before (18 electrons in total). This leaves six electrons to place in the 4s and 3d sub-shells. By moving one electron from the 4s to the 3d the chromium atom can have a half-filled 3d sub-shell ($3d^5$). A half-filled or filled sub-shell is more stable so this is a more stable electron configuration for the chromium atom.

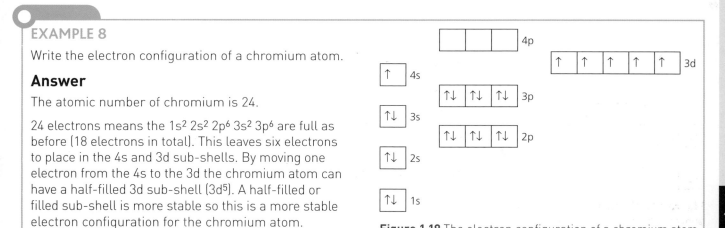

Figure 1.19 The electron configuration of a chromium atom is $1s^2\,2s^2\,2p^6\,3s^2\,3p^6\,3d^5\,4s^1$.

EXAMPLE 9

Write the electron configuration of a copper atom.

Answer

The atomic number of copper is 29.

29 electrons means the $1s^2 2s^2 2p^6 3s^2 3p^6$ are full (18 electrons in total). This leaves 11 electrons to place in the 4s and 3d sub-shells. As with chromium, copper has an unusual electron configuration as one electron moves from the 4s to the 3d sub-shell giving a more stable full 3d sub-shell.

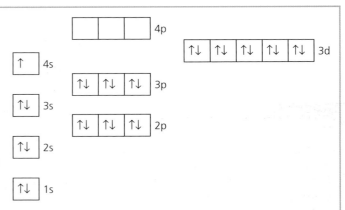

Figure 1.20 The electron configuration of a copper atom is $1s^2 2s^2 2p^6 3s^2 3p^6 3d^{10} 4s^1$.

TIP

When you go into any Chemistry AS or A2 exam you should put a star at chromium and copper on the Periodic Table as their atoms have unusual electron configurations.

The atoms of chromium and copper have an unusual electron configuration because of the stability of filled and half-filled sub-shells. Chromium is $1s^2 2s^2 2p^6 3s^2 3p^6 3d^5 4s^1$ instead of $3d^4 4s^2$. This is because one electron in each of the orbitals of the 3d sub-shell makes it more stable – they are symmetrical around the nucleus. Copper is $1s^2 2s^2 2p^6 3s^2 3p^6 3d^{10} 4s^1$ instead of $3d^9 4s^2$ as would have been expected. Again the ten electrons in the 3d sub-shell make it more stable due to symmetry around the nucleus.

Electron configuration of simple ions

When atoms form simple ions they can either lose or gain electrons.

Ions formed from metal atoms

- Metal atoms tend to lose electrons to become positive ions.
- The number of electrons they lose is the same as the positive charge on the ion.
- The name of a positive ion is the same as the atom, e.g. Na is a sodium atom and Na^+ is a sodium ion; Al is an aluminium atom and Al^{3+} is an aluminium ion.
- When metal atoms lose electrons they lose them from the outermost level except atoms of d block elements.
- Atoms of d block elements lose their 4s electrons first then their 3d.

Ions formed from non-metal atoms

- Non-metal atoms tend to gain electrons to become negative ions.
- The number of electrons they gain is the same as the negative charge on the ion.
- The name of a negative ion is the atom stem with '-ide' on the end, e.g. O is an oxygen atom and O^{2-} is an oxide ion; Br is a bromine atom and Br^- is a bromide ion.

Hydrogen

- A hydrogen atom has only one 1s electron.
- A hydrogen atom can either lose this electron to become a hydrogen ion, H^+ or it can gain an electron to become a hydride ion, H^-.

EXAMPLE 10

Write the electron configuration of an iron(III) ion.

Answer

- Atoms of iron have 26 electrons.
- An iron(III) ion is Fe^{3+}.
- The electrons in an atom of iron are arranged $1s^2\ 2s^2\ 2p^6\ 3s^2\ 3p^6\ 3d^6\ 4s^2$.
- Iron is a d block element.
- Atoms of d block elements lose their 4s electrons first.
- An iron atom loses three electrons to become Fe^{3+} so it loses two electrons from the 4s and one electron from the 3d.
- The electron configuration of an iron(III) ion is $1s^2\ 2s^2\ 2p^6\ 3s^2\ 3p^6\ 3d^5$.

TIP

Atoms elements in the main groups (not transition metals) form ions which have the electron configuration of the nearest Noble gas. For example nitride ions, N^{3-}, have the same electron configuration as Ne atoms.

EXAMPLE 11

Write the electron configuration of a sulfide ion, S^{2-}.

Answer

Atoms of sulfur have 16 electrons, arranged $1s^2\ 2s^2\ 2p^6\ 3s^2\ 3p^4$.

When sulfur atoms form sulfide ions, they gain two electrons to give them the same electron configuration as argon.

The electron configuration of a sulfide ion is $1s^2\ 2s^2\ 2p^6\ 3s^2\ 3p^6$.

Isoelectronic

- Particles which are isoelectronic have the same electron configuration.
- An atom of neon has an electron configuration of $1s^2\ 2s^2\ 2p^6$.
- An oxide ion, O^{2-}, also has an electron configuration of $1s^2\ 2s^2\ 2p^6$.
- A neon atom has 10 protons and an oxide ion has 8 protons. This is what makes them different despite the fact they have the same electron configuration (are isoelectronic).

TEST YOURSELF 4

1 Write electron configurations in spd format for the following ions:

  i) Li^+ **ii)** Cl^- **iii)** O^{2-} **iv)** Na^+ **v)** Fe^{2+}
  vi) Fe^{3+} **vii)** Ni^{2+} **viii)** Cu^{2+} **ix)** Cr^{3+} **x)** Br^-

2 Which of the following ionic compounds contains ions that are isoelectronic (have the same electron configuration)?

  A sodium chloride
  B aluminium oxide
  C potassium fluoride
  D zinc oxide

3 Using the following electron configurations:

  A $1s^2\ 2s^2\ 2p^6$
  B $1s^2$
  C $1s^2\ 2s^2\ 2p^6\ 3s^2\ 3p^6\ 3d^{10}\ 4s^2$
  D $1s^2\ 2s^2\ 2p^6\ 3s^2\ 3p^6\ 3d^{10}\ 4s^2\ 4p^6$
  E $1s^2\ 2s^2\ 2p^6\ 3s^2\ 3p^6\ 3d^5$

Give the letter (A to E) which represents the electron configuration of the following atoms and ions.

  i) H^- **i)** N^{3-} **iii)** Zn **iv)** Ga^+ **v)** Mg^{2+} **vi)** Kr
  vii) Se^{2-} **viii)** Rb^+ **ix)** Li^+ **x)** He **xi)** O^{2-} **xii)** Sr^{2+}

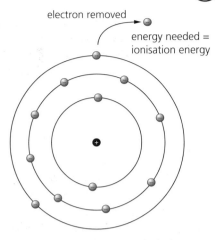

electron removed

energy needed = ionisation energy

Figure 1.21 What element is being ionised in this diagram?

Ionisation energies

The electrons in atoms and ions are attracted to the positive nucleus. Energy is required to overcome this attraction and remove electrons. The process of removing electrons from atoms and ions is called ionisation. The ionisation energy is the energy required to remove electrons.

The **first ionisation energy** is the energy required to remove one mole of electrons from one mole of gaseous atoms to form one mole of gaseous 1+ ions.

Successive ionisations give the first, second, third, fourth, etc. ionisation energies. Only one mole of electrons is removed with each ionisation.

For example for sodium

The first ionisation of sodium is represented by the equation:

$$Na(g) \rightarrow Na^+(g) + e^-$$

The second ionisation of sodium is represented by the equation:

$$Na^+(g) \rightarrow Na^{2+}(g) + e^-$$

The third ionisation of sodium is represented by the equation:

$$Na^{2+}(g) \rightarrow Na^{3+}(g) + e^-$$

TIP
Equations for ionisation energies are often asked for. You must include state symbols. The atoms and ions must be in the gaseous state.

Figure 1.22 Ionisation in plasma televisions.

Ionisation occurs in plasma screen televisions. The screen consists of two glass panels with millions of tiny cells containing xenon and neon gas, sandwiched in between. When energy in the form of an electrical voltage is supplied, the atoms become ionised into a mixture of positive ions and negative electrons, which is called a plasma.

$$Ne(g) \rightarrow Ne^+(g) + e^-$$

$$Xe(g) \rightarrow Xe^+(g) + e^-$$

In a plasma with an electrical current running through it, positively charged ions collide with electrons, exciting the gas atoms in the plasma and causing them to release ultraviolet photons that interact with the phosphor material coated on the inside of the cell, and gives off visible light.

Values for ionisation energies

Ionisation energies are measured in kJ mol^{-1} (kilojoules per mole). The units include per mole (mol^{-1}) as it is the energy per mole of gaseous atoms. The first ionisation energy of sodium is $+496$ kJ mol^{-1}. 496 kJ of energy are required to convert 1 mole of Na(g) to 1 mole of Na$^+$(g) by removing 1 mole of electrons.

Often this is written as:

$$Na(g) \rightarrow Na^+(g) + e^- \qquad \Delta H = +496 \text{ kJ mol}^{-1}$$

All ionisation energy values will be positive as they are endothermic as energy is required to remove an electron from the attractive power of the nucleus. The higher the value, the more energy is required to remove 1 mole of electrons. The first ionisation energy for magnesium is $+738$ kJ mol^{-1}. 738 kJ of energy are required to convert 1 mole of Mg(g) to 1 mole of Mg$^+$(g) by removing 1 mole of electrons. There are trends in ionisation energy values in the Periodic Table that provide evidence for the existence of electron arrangement in energy levels and in sub-shells.

Ionisation energies as evidence for energy levels

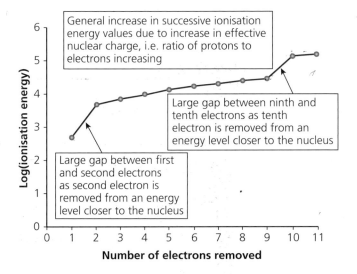

Figure 1.23 The log(ionisation energy) against the number of electrons removed from a sodium atom.

The graph shows the successive ionisation energies of a sodium atom, it becomes clear that there is a distinct set of energy levels. The diagram shows the log(ionisation energy) against the number of electrons removed from a sodium atom.

The log of the ionisation energy is used to condense the diagram as the ionisation energies vary across a wide range of values.

The existence of energy levels is proven by the large gaps in the successive ionisation energies as these correspond to the removal of electrons from energy levels closer to the nucleus and so more energy is required to remove the electron.

The general increase in successive ionisation energies is caused by the increase in the ratio of protons to electrons as successive electrons are removed. This is often called effective nuclear charge.

TIP
Only 1 mole of electrons is removed with each ionisation, even though the ions formed may look odd, for example Cl$^+$(g) or Al^{2+}(g).

Using successive ionisation energies

Successive ionisation energies are an indicator of the group to which an element belongs.

EXAMPLE 12

The first five successive ionisation energies for four different elements are given in the table.

Element	Ionisation energy (kJ mol⁻¹)				
	First	Second	Third	Fourth	Fifth
W	+496	+4562	+6912	+9543	+13353
X	+1087	+2353	+4621	+6223	+37831
Y	+578	+1817	+2745	+11577	+14842
Z	+738	+1451	+7733	+10543	+13630

1 State which element belongs to Group 4.
2 Which element would form a simple ion with a charge of 2+?
3 Which element would have one electron in its outer energy level?
4 Which element would form an oxide with the formula M_2O_3 where M represents the element?

Answers

1 Element X has a large increase in ionisation energy after the fourth electron has been removed (+6223 to remove the fourth electron and +37831 to remove the fifth electron). This would suggest four electrons in the outer energy level and the fifth electron in an energy level closer to the nucleus. So, element X has four electrons in the outer energy level, which is characteristic of an element in Group 4. Element X is actually carbon.

2 An element in Group 2 would form a simple ion with a charge of 2+. Element Z has a large increase in ionisation energy after the second electron has been removed (+1451 to remove the second electron and +7733 to remove the third electron). This would suggest there are two electrons in the outer energy level and the third electron is in an energy level closer to the nucleus. So, element Z has two electrons in its outer energy level, which is characteristic of an element in Group 2. Element Z is actually magnesium.

3 One electron in the outer energy level would suggest a Group 1 element. Element W has a large increase in ionisation after the first electron has been removed (+496 to remove the first electron and +4562 to remove the second electron). This would suggest one electron in the outer energy level and the second electron in an energy level closer to the nucleus. So, element W has one electron in the outer energy level, which is characteristic of an element in Group 1. Element W is actually sodium.

4 An element which forms an oxide with the formula M_2O_3 would suggest an element in Group 3. Element Y has a large increase in ionisation energy after the third electron has been removed (+2745 to remove the third electron and +11577 to remove the fourth electron). This would suggest three electrons in the outer energy level and the fourth electron in an energy level closer to the nucleus. So, element X has three electrons in the outer energy level, which is characteristic of an element in Group 3. Element Y is actually aluminium.

TIP

All of the above questions are different ways of getting you to place an element in a particular group.

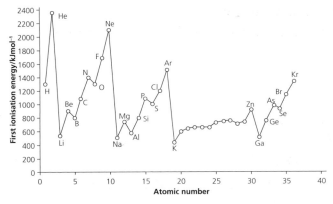

Figure 1.24 First ionisation energies for the elements, hydrogen (atomic number 1) to krypton (atomic number 36).

Ionisation energy pattern for elements 1 to 36

The graph (Figure 1.24) shows the pattern in first ionisation energies for the elements, hydrogen (atomic number 1) to krypton (atomic number 36).

There are three general patterns which should be apparent from this diagram:

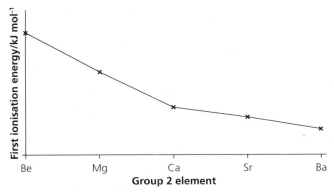

Figure 1.25 Graph of first ionisation energy of the Group 2 elements.

1 Ionisation energy decreases down a group.

Look at the decrease from helium to neon to argon and then krypton. The same pattern is clear for Group 2.

This is seen more clearly by examining a graph of first ionisation energies of the Group 2 elements (Figure 1.25).

2 Ionisation energy shows a general increase across a period.

From sodium to argon there is a general increase in first ionisation energy.

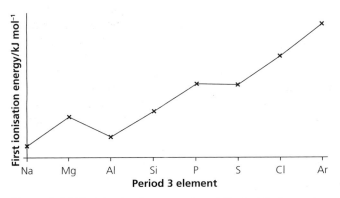

Figure 1.26 First ionisation energies of Period 3 elements.

The graph in Figure 1.26 shows the first ionisation energies of the Period 3 elements from sodium to argon.

3 Within the short periods (periods 2 and 3), there is a zig-zag pattern.

The first ionisation drops below the general increase for elements in Group 3 (boron and aluminium) and Group 6 (oxygen and sulfur) (Figure 1.26).

Explaining trends in ionisation energies

The patterns in ionisation energies can be explained by the electronic structure of the atoms. There are three main factors which can be used to explain these patterns:

1 Atomic radius (how far the outer electrons are from the attractive power of the nucleus).

- The atomic radius of beryllium is less than the atomic radius of magnesium.
- Atomic radius increases down a group and decreases across a period.
- The further an outer electron is from the attractive power of the nucleus the less energy is required to ionise it.

2 Nuclear charge (how many protons are attracting the outer electron).

- The greater the number of protons, the greater the nuclear charge.
- A greater nuclear charge leads to a stronger attraction to the outer electron so more energy is required to ionise it.

3 Shielding by inner electrons (how many electrons are between the nucleus and the outer electron so shielding the attractive power of the nucleus).

- The attractive power of the nucleus can be shielded by inner electrons.
- The more inner electrons there are, the more the nucleus is shielded and the less energy is required to ionise the outer electron.

First ionisation energy decreases down a group

This pattern is due to the existence of energy levels within the atom. As a group is descended, the outer electrons are further away from the nucleus (at higher energy levels) and so are more easily removed as the positive nucleus has less of a hold on them. This factor is called atomic radius.

Also as the outer electron is at a higher energy level, there are more electrons between it and the positive nucleus so this shields the attractive power of the nucleus from the outer electrons. This factor is called shielding by inner electrons. The charge of the nucleus also increases as a group is descended and so works against the other two factors to hold the electron more firmly. This factor is called nuclear charge. However, the combination of the three factors causes the ionisation energy to decrease as the group is descended.

First ionisation energy increases across a period

The pattern of a general increase in ionisation energy as the period is crossed from left to right is caused by the increase in nuclear charge.

Also the atomic radius decreases across the period. The outer electron is closer to the nucleus with a greater nuclear charge holding it so more energy is required to remove it.

There is no increased shielding (as the electron being removed is in the same energy level).

Group 1 elements, such as lithium, have low ionisation energies and are likely to form positive ions. Lithium's low ionisation energy, for example, is important for its use in lithium-ion computer backup batteries where the ability to lose electrons easily makes a battery that can quickly provide a large amount of energy.

Group 3 and Group 6 first ionisation energies

The zig-zag pattern gives us evidence for the existence of sub-shells in energy levels.

The atoms of Group 3 elements, such as aluminium, show a lower than expected first ionisation energy. This is due to the division of the energy level into sub-shells. Aluminium has the electron configuration $1s^2\ 2s^2\ 2p^6\ 3s^2\ 3p^1$. The $3p^1$ electron is further from the nucleus and has additional shielding from the $3s^2$ inner electrons so it requires less energy to ionise it.

The atoms of Group 6 elements, such as sulfur, show a lower than expected first ionisation energy. This is due to the pairing of electrons in the p sub-shell. The $3p^4$ electron configuration of sulfur ($1s^2\ 2s^2\ 2p^6\ 3s^2\ 3p^4$) means that two electrons are paired in a p orbital in this sub-shell. The repulsion between these two electrons lowers the energy required to remove one of the electrons and this decreases first ionisation energy.

Figure 1.27 Lithium is used in batteries. Why is a low ionisation energy important?

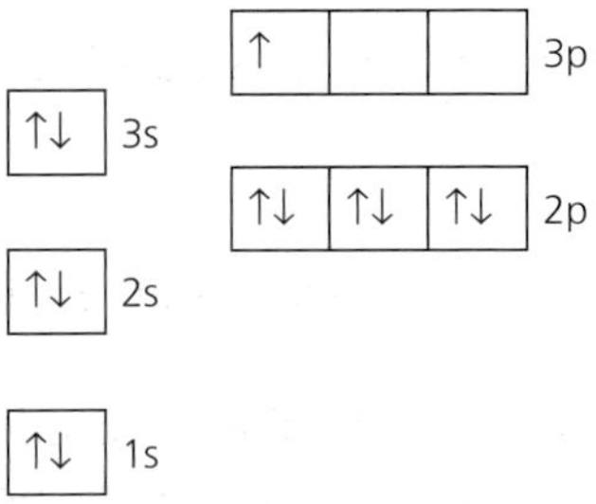

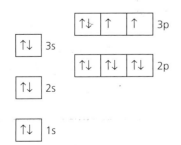

Figure 1.28 Electron configuration of aluminium.

Figure 1.29 Electron configuration of sulfur.

The fact that the atoms of elements in Group 3 and Group 6 show a lower first ionisation energy than would be expected provides evidence that the second and third energy levels are divided into two sub-shells, the first taking two electrons and the second taking six. This would explain the decrease in first ionisation energy from Group 2 to Group 3 and also the decrease from Group 5 to Group 6.

Group 1 and Group 0 elements

Atoms of Group 1 elements have the lowest first ionisation energy in every period as they have the greatest atomic radius and the lowest nuclear charge in a particular period.

Atoms of Group 0 elements have the highest first ionisation energy in every period for the opposite reasons – they have the smallest atomic radius and the highest nuclear charge in a period.

The increased pressure that scuba divers experience far below the water's surface can cause too much oxygen to enter their blood, which would result in confusion and nausea. To avoid this, divers sometimes use a gas mixture called heliox – oxygen diluted with helium. Helium's high ionisation energy ensures that it will not react chemically in the bloodstream.

Figure 1.30 Why is helium safe to use in a diver's tank?

Summary of patterns in ionisation energy and their explanation

1 First ionisation energy decreases down a group

- Atomic radius increases
- Shielding by inner electron increases
- So less energy required to remove the electron.

2 First ionisation energy increases across a period

- Atomic radius decreases
- Nuclear charge increases
- Shielding by inner electrons is the same
- More energy required to remove the electrons.

3 Lower first ionisation energy than expected for elements in Group 3 and 6

- Group 3 atoms have an $s^2\,p^1$ arrangement
- Outer p^1 electron is further from the nucleus
- Inner s^2 electrons increase shielding so less energy is required to ionise the outer p^1 electron
- Group 6 atoms have a p^4 arrangement – the repulsion of two electrons in the same p orbital leads to less energy being required to ionise the outer electron.

Patterns in second ionisation energies

The patterns in first ionisation energy are shifted one to the left when the patterns of second ionisation energy are considered. Where a Group 1 element would have the lowest first ionisation energy, it would have the highest second ionisation energy.

TIP

The pattern in ionisation energies for Group 2 can be applied to any group in the Periodic Table. The patterns in Period 3 can be applied to Period 2.

23

Table 1.1 shows the first and second ionisation energies for the Period 3 elements.

Group 1 elements have the highest second ionisation energy in a particular period as the second electron is being removed from an energy level closer to the nucleus. Group 2 elements have the lowest second ionisation energy in a particular period.

Table 1.1 First and second ionisation energies for the Period 3 elements.

Element	First ionisation energy/kJ mol^{-1}	Second ionisation energy/kJ mol^{-1}
Na	496	4560
Mg	738	1450
Al	577	1816
Si	786	1577
P	1060	1890
S	1000	2260
Cl	1256	2295
Ar	1520	2665

TEST YOURSELF 5

1 What is the definition of first ionisation energy?
2 What are the units of ionisation energy?
3 Write equations for the following ionisations including state symbols:
   a) first ionisation of silicon
   b) second ionisation of potassium
   c) third ionisation of carbon.
4 The first six successive ionisation energies, in kJ mol^{-1}, of an element M are:
   578, 1817, 2745, 11 578, 14 831, 18 378
   What is the formula of the oxide of M?

Practice questions

1 Which of the following represents the second ionisation of magnesium?

   A $Mg(s) \rightarrow Mg^+(s) + e^-$

   B $Mg(s) \rightarrow Mg^{2+}(g) + 2e^-$

   C $Mg^+(g) \rightarrow Mg^{2+}(g) + e^-$

   D $Mg^+(g) \rightarrow Mg^{2+}(s) + e^-$ (1)

2 Which of the following is the electron configuration of an iron(II) ion?

   A $1s^2\ 2s^2\ 2p^6\ 3s^2\ 3p^6\ 3d^6\ 4s^2$

   B $1s^2\ 2s^2\ 2p^6\ 3s^2\ 3p^6\ 3d^6$

   C $1s^2\ 2s^2\ 2p^6\ 3s^2\ 3p^6\ 3d^5\ 4s^1$

   D $1s^2\ 2s^2\ 2p^6\ 3s^2\ 3p^6\ 3d^5$ (1)

3 An unknown element in Period 3 has the following successive ionisation energies in $kJ\,mol^{-1}$.

First ionisation energy	+1000
Second ionisation energy	+2252
Third ionisation energy	+3357
Fourth ionisation energy	+4456
Fifth ionisation energy	+7004
Sixth ionisation energy	+8496
Seventh ionisation energy	+27 107
Eighth ionisation energy	+31 719

To which group of the Periodic Table does the element belong?

   A Group 3   **B** Group 4

   C Group 5   **D** Group 6   (1)

4 Write equations to represent the following ionisations:

   a) First ionisation of aluminium   (1)

   b) Third ionisation energy of lithium   (1)

   c) Explain why there is no fourth ionisation of lithium   (1)

5 The graph below shows the first ionisation energies of the elements hydrogen to sodium.

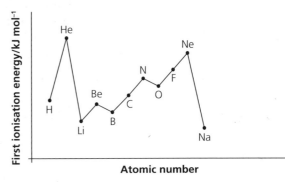

Figure 1.31

   a) Write an equation for the first ionisation of boron including state symbols.   (1)

   b) Explain why the first ionisation energy of sodium is less than that of lithium.   (3)

   c) Explain why the first ionisation energy of oxygen is less than the first ionisation energy of nitrogen.   (2)

   d) Continue the sketch above for the next three elements after sodium.   (3)

   e) Explain why the Noble gases have the highest ionisation energy in each period.   (3)

6 The graph below shows the first ionisation energies of the elements lithium (atomic number 3) to neon (atomic number 10).

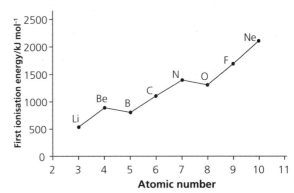

Figure 1.32

   a) Explain why there is a general increase in first ionisation energy across the period.   (3)

   b) Explain why the value for the first ionisation energy of boron is lower than the value for beryllium.   (2)

   c) Write the electron configuration of an atom of nitrogen.   (1)

7 An element was analysed using a TOF (time-of-flight) mass spectrometer. The spectrum showed that there were four isotopes. The relative isotopic masses and relative abundances are given in the table below.

Relative isotopic mass	Relative abundance
50	9.4
52	72.5
53	14.5
54	3.6

a) Calculate the relative atomic mass of the element to one decimal place. (2)

b) Identify the element. (1)

c) Identify the species responsible for the peak at 54. (1)

8 The spectrum below is for ethanoic acid.

a) State the m/z value of the peak with the highest relative abundance. (1)

b) What is the relative molecular mass (M_r) of this compound? (1)

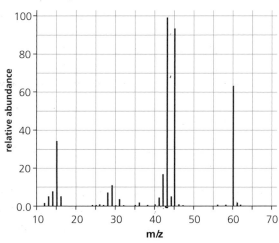

Figure 1.33

9 The second ionisation of sodium is represented by the equation:

$$Na^+(g) \rightarrow Na^{2+}(g) + e^-$$

a) Explain why the second ionisation energy is greater than the first ionisation energy of sodium. (2)

b) Write an equation for the first ionisation of sodium. (1)

c) Which element in Period 3 would be expected to have the lowest second ionisation energy? (1)

d) The electron configuration of an ion is shown below:

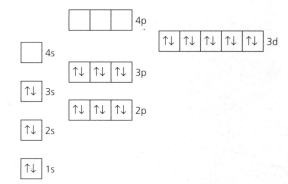

Figure 1.34

i) Write the electron configuration. (1)

ii) An ion of zinc and an ion of copper have this electron configuration. Write the formulae of these two ions. (2)

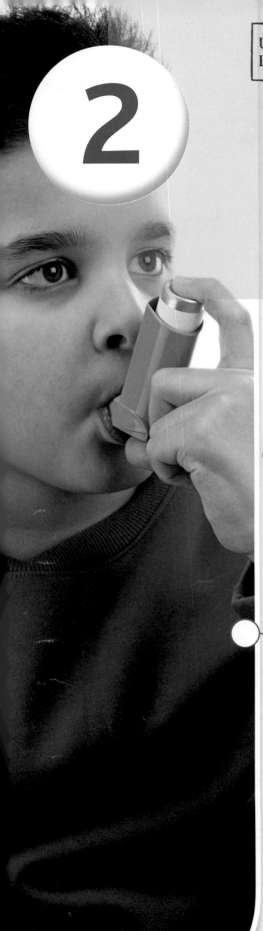

2

Amount of substance

TEST YOURSELF ON PRIOR KNOWLEDGE 1

1 Write the symbols for the following elements including state symbols.
   a) solid sodium   **b)** liquid mercury
   c) gaseous helium   **d)** chlorine water (aqueous chlorine)
2 Write the formula for the following compounds including state symbols.
   a) liquid water   **b)** gaseous carbon dioxide
   c) ice   **d)** sodium chloride solution

Chemical formulae

A chemical formula shows the atoms/ions of each element that are present in a compound or molecule, as well as the number of each atom/ion.

For example H_2O is the chemical formula for water and it shows that a water molecule contains two hydrogen atoms and one oxygen atom; $MgCl_2$ is the chemical formula for magnesium chloride and it contains one magnesium ion and two chloride ions.

The formulae of familiar and unfamiliar ionic compounds can be determined using the charges on the ions.

- The formula and charge of molecular ions need to be learned as you encounter them, for example, sulfate is SO_4^{2-}, nitrate is NO_3^-, carbonate is CO_3^{2-}, hydroxide is OH^-, ammonium is NH_4^+.
- Where more than one of a molecular ion is needed, brackets are used around the molecular ion and the number of these needed is placed outside the bracket.
- The formulae of some compounds just need to be learned. For example: water is H_2O, methane is CH_4, ammonia is NH_3, hydrogen peroxide is H_2O_2.

Using charges on ions

In an ionic compound, the number of each ion required is determined by using the charge on the ion. The compound should have no overall charge as the total positive charge and the total negative charge on the ions should cancel out.

For example:

lithium oxide

lithium ion is Li^+; oxide ion is O^{2-}

$2Li^+$ and $1\ O^{2-}$ are required to cancel out the charge

Formula for lithium oxide is Li_2O

iron(III) chloride

iron(III) ion is Fe^{3+}; chloride ion is Cl^-

$1\ Fe^{3+}$ and $3\ Cl^-$ are required to cancel out the charge

Formula for iron(III) chloride is $FeCl_3$

aluminium sulfate

aluminium ion is Al^{3+}; sulfate ion is SO_4^{2-}

$2Al^{3+}$ and $3SO_4^{2-}$ are required to cancel out the charge

Formula of aluminium sulfate is $Al_2(SO_4)_3$

Types of formulae

An **empirical formula** shows the simplest whole number ratio of the atoms of each element in a compound.

This type of formula is used for ionic compounds and macromolecules (giant covalent molecules).

Examples: NaCl (ionic); $MgCl_2$... MgO (ionic); $CaCl_2$ (ionic); SiO_2 (macromolecular).

Glucose and ethanoic acid are both found in foods. Glucose tastes sweet and ethanoic acid tastes sour. Glucose is found in sweets, cake and chocolate. Ethanoic acid is the main ingredient in vinegar. These are different chemicals,

> **TIP**
> If an unusual ion is given, the same rules apply. For example the ethanedioate ion is $C_2O_4^{2-}$. Potassium ethanedioate is $K_2C_2O_4$.

Figure 2.1 Sweets containing glucose $C_6H_{12}O_6$ and vinegar containing ethanoic acid CH_3COOH. Both have the same empirical formula – what is it?

yet both have the same empirical formula. Formaldehyde, a pungent smelling substance used as a preservative in medical laboratories also has the same empirical formula. All three chemicals have different molecular formulae. What is the difference between molecular and empirical formula?

A **molecular formula** shows the actual number of atoms of each element in one molecule of the substance.

This is used for all molecular (simple) covalent substances.

Examples: H_2O; CO_2; O_2; CH_4; NH_3; H_2O_2; I_2; S_8 (all molecular covalent).

Some elements exist as simple molecules. The following exist as diatomic molecules (H_2, N_2, O_2, F_2, Cl_2, Br_2, I_2) whereas sulfur exists as S_8 molecules and phosphorus as P_4 molecules.

An empirical formula can be written for molecular covalent substances and this may be the same as the molecular formula or it may be different, e.g. the molecular formula of hydrogen peroxide is H_2O_2 but its empirical formula is written as HO (simplest ratio).

Figure 2.2 The energy drink Red Bull contains about 80 mg of caffeine in 50 ml. This is similar to the amount found in a cup of freshly brewed coffee. Caffeine has the molecular formula $C_8H_{10}N_4O_2$. What is its empirical formula?

TEST YOURSELF 2

1 Write formulae for the following compounds:
   a) sodium fluoride   **b)** magnesium oxide
   c) potassium oxide   **d)** barium chloride
2 Name the following ions: **a)** OH^-; **b)** O^{2-}; **c)** Cl^-; **d)** Al^{3+}; **e)** SO_4^{2-}
3 Write formulae for the following transition metal compounds:
   a) copper(II) chloride   **b)** zinc oxide
   c) copper sulfate   **d)** iron(III) hydroxide
4 Name the following compounds:
   a) CO_2 **b)** KNO_3 **c)** $CuCO_3$ **d)** HF   **e)** $MgSO_4$
5 Write formulae for the following compounds:
   a) ammonium sulfate   **b)** sulfur dioxide
   c) calcium carbonate   **d)** aluminium sulfate

Balanced equations

A balanced equation shows the rearrangement of atoms in a chemical reaction.

TIP

Remember you cannot change a formula to balance an equation

$$2Mg \quad + \quad O_2 \quad \rightarrow \quad 2MgO$$

STEP 2: With the oxygen now balanced, you need a **2** here in front of the Mg to give 2 Mg atoms and balance the equation

STEP 1: A **2** here balances the oxygen but creates another issue and now there are 2 Mg on the right hand side – this now has to be balanced as well

Figure 2.3

The equation is now balanced; often you have to balance one element which may create another issue with another element.

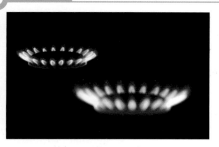

Figure 2.4 When natural gas burns it produces a clear, steady, blue flame. What is produced?

TIP
It is common to be given the names of reactants and products and have to write a balanced equation for an unfamiliar reaction. Follow the pattern shown in previous examples balancing each type of atom on both sides of the equation.

EXAMPLE 1

Natural gas is used as a fuel. It is made up mainly of methane, with about 2.5% ethane.

Ethane burns completely in air to form carbon dioxide and water. Write a balanced equation for this reaction.

Answer

$$C_2H_6 \quad + \quad O_2 \quad \rightarrow \quad CO_2 \quad + \quad H_2O$$

ethane is C_2H_6 oxygen is diatomic   carbon dioxide   water is H_2O
so is O_2 is CO_2

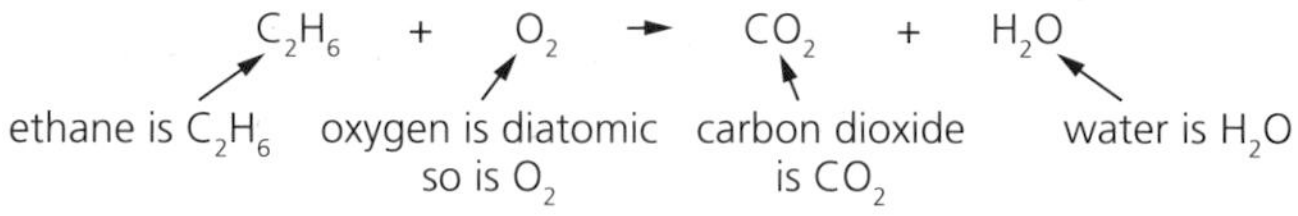

Figure 2.5

This equation is not balanced as there are:

2 C atoms on the left (in C_2H_6) and only 1 C atom on the right (in CO_2);

also there are 6 H atoms on the left (in C_2H_6) and only 2 H atoms on the right (in H_2O); 2 O atoms on the left (in O_2) but 3 O atoms on the right (2 O in CO_2 and 1 O in H_2O).

Sort out the carbon first:

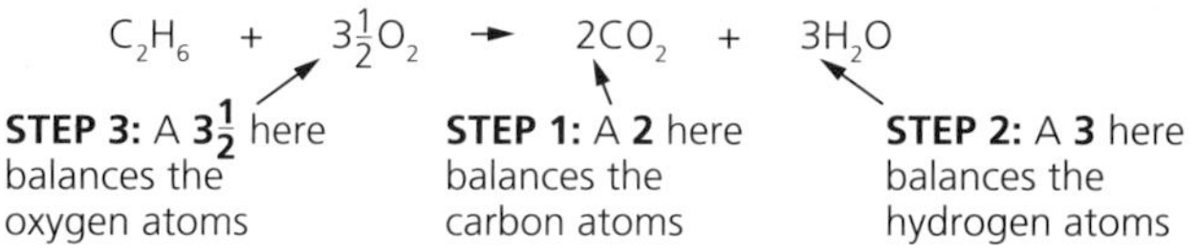

$$C_2H_6 \quad + \quad 3\tfrac{1}{2}O_2 \quad \rightarrow \quad 2CO_2 \quad + \quad 3H_2O$$

STEP 3: A $3\tfrac{1}{2}$ here balances the oxygen atoms   **STEP 1:** A **2** here balances the carbon atoms   **STEP 2:** A **3** here balances the hydrogen atoms

Figure 2.6

It may seem unusual to balance an equation using decimals and the balancing numbers in the equation can be doubled to avoid the decimal giving:

$$2C_2H_6 + 7O_2 \rightarrow 4CO_2 + 6H_2O$$

TIP
At AS level it is fine to use decimals to balance equations particularly combustion equations and this will even be encouraged in the energetics section of AS level.

EXAMPLE 2

Copper(II) oxide reacts with nitric acid to form copper(II) nitrate and water. Write a balanced equation for this reaction.

Answer

The unbalanced equation is written as:

$$CuO + HNO_3 \rightarrow Cu(NO_3)_2 + H_2O$$

In this type of equation it is easy to get lost with the oxygen atoms as there are three in HNO_3 and one in CuO on the left, and six in $Cu(NO_3)_2$ and one in H_2O on the right of the equation. As the NO_3 is not broken up, it is best to treat it as a unit. As there is only one NO_3 on the left and two on the right, so two $2HNO_3$ are required; this will also balance the hydrogen. The oxygen in CuO is balanced with the oxygen in H_2O.

$$CuO + 2HNO_3 \rightarrow Cu(NO_3)_2 + H_2O$$

Ionic equations

Some chemical reactions which involve ionic compounds are actually reactions between only some of the ions involved in the reaction. The balanced equation can be rewritten as an **ionic equation** leaving out the ions which do not take part in the reaction. Ions which do not take part in the reaction are known as **spectator ions**.

For each **ionic** substance in the reaction write the ions present below it and how many of each ion are present. Covalent substances should be left as they are. If an ion appears on both sides of the equation (in the same state), it should not be included in the ionic equation.

Sometimes state symbols need to be included to show exactly what ions have changed and which have not changed.

EXAMPLE 3

Write the ionic equation for the reaction between hydrochloric acid and sodium hydroxide.

Answer

The balanced equation is:

$$HCl + NaOH \rightarrow NaCl + H_2O$$

Ions:

$$H^+ + Cl^- + Na^+ + OH^- \rightarrow Na^+ + Cl^- + H_2O$$

The Na^+ and Cl^- ions are on both sides of the equation so are not part of the ionic reaction. These are the spectator ions.

The ionic equation is:

$$H^+ + OH^- \rightarrow H_2O$$

EXAMPLE 4

Write an ionic equation for the reaction between copper(II) sulfate solution and sodium hydroxide solution.

Answer

The balanced equation is:

$$CuSO_4(aq) + 2NaOH(aq) \rightarrow Cu(OH)_2(s) + Na_2SO_4(aq)$$

Ions:

$$Cu^{2+}(aq) + SO_4^{2-}(aq) + 2Na^+(aq) + 2OH^-(aq) \rightarrow Cu^{2+}(s) + 2OH^-(s) + 2Na^+(aq) + SO_4^{2-}(aq)$$

The $2Na^+(aq)$ and $SO_4^{2-}(aq)$ are on both sides of the equation so are not part of the ionic equation.

The ionic equation is:

$$Cu^{2+}(aq) + 2OH^-(aq) \rightarrow Cu(OH)_2(s)$$

This reaction is a precipitation reaction. Copper hydroxide is insoluble and is formed as a blue precipitate, as shown in Figure 2.7.

Figure 2.7 Copper hydroxide is a blue precipitate.

TIP

State symbols are most often not required but they may help you work out the spectator ions in the equation.

EXAMPLE 5

Write an ionic equation for the reaction between potassium chloride solution and silver nitrate solution.

Answer

The balanced equation is:

$KCl(aq) + AgNO_3(aq) \rightarrow AgCl(s) + KNO_3(aq)$

Ions:

$K^+(aq) + Cl^-(aq) + Ag^+(aq) + NO_3^-(aq) \rightarrow Ag^+(s) + Cl^-(s) + K^+(aq) + NO_3^-(aq)$

The $K^+(aq)$ and $NO_3^-(aq)$ are on both sides of the equation so are not part of the ionic equation.

The ionic equation is:

$Cl^-(aq) + Ag^+(aq) \rightarrow AgCl(s)$

TIP

These equations are useful in the tests for cations and anions which will be looked at in Group 2 and Group 7 on pages 238–241.

TEST YOURSELF 3

1 Write balanced equations for the following reactions.

a) potassium hydroxide + sulfuric acid → potassium sulfate + water

b) calcium + oxygen → calcium oxide

c) aluminium + chlorine → aluminium chloride

2 Barium chloride solution reacts with potassium sulfate solution to produce solid barium sulfate and potassium chloride remains in solution.

a) Write a balanced symbol equation for the reaction.

b) Write an ionic equation for the reaction.

3 Write a balanced symbol equation for the reactions:

a) calcium hydroxide + hydrochloric acid → calcium chloride + water

b) aluminium oxide + sulfuric acid → aluminium sulfate + water

c) zinc + hydrochloric acid → zinc chloride + hydrogen

4 Write an ionic equation for the reaction between zinc(II) ions and hydroxide ions to produce zinc(II) hydroxide.

5 Convert the following balanced symbol equations into ionic equations by removing spectator ions:

a) $Mg(s) + CuSO_4(aq) \rightarrow MgSO_4(aq) + Cu(s)$

b) $Zn(s) + 2HCl(aq) \rightarrow ZnCl_2(aq) + H_2(g)$

The mole and the Avogadro constant

A balanced equation for a reaction gives the rearrangement of the atoms within a chemical reaction.

Using the equation: $C + O_2 \rightarrow CO_2$

We can read this as 1 C atom reacts with 1 O_2 molecule to form 1 CO_2 molecule.

However, the mass of 1 atom of carbon is approximately 2.0×10^{-23} g. This sort of measurement is not practically possible and so the number of particles measured must be scaled up but they will still react in the same ratio.

For example 1 million C atoms will react with 1 million oxygen molecules to form 1 million carbon dioxide molecules. However, the mass of 1 million carbon atoms is $1\,000\,000 \times 2.0 \times 10^{\times 23} = 2 \times 10^{-17}$ g. This is still too small to be measured practically. We need to multiply by 6.02×10^{23}. This number is called the Avogadro constant. The Avogadro constant is often represented by the capital letter L and may be defined as the number of atoms in 12.000 g of carbon-12. The Avogadro constant is used to get a measureable mass.

- 1 carbon atom has a mass of 2×10^{-23} g so 6.02×10^{23} carbon atoms have a mass of approximately 12 g.
- The amount of a substance that contains the number of particles (atoms, ions, molecules or electrons) equal to the Avogadro constant (6.02×10^{23}) is called a <u>mole</u> of the substance.
- The term 'amount' is the quantity which is measured in moles. Mole is written as mol for unit purposes.

Reading equations in moles

This equation: $C + O_2 \rightarrow CO_2$ can be read as:

1 mole of carbon atoms reacts with 1 mole of oxygen molecules to form 1 mole of carbon dioxide.

For any substance the mass of 1 mole is simply the total of the relative atomic masses (A_r) of all the atoms which make up the substance:

1 mole of Mg = 24.3 g   (A_r of Mg = 24.3)

1 mole of O_2 = 32.0 g   (A_r of O = 16.0)

1 mole of H_2O = 18.0 g   (A_r of H = 1.0; A_r of O = 16.0)

1 mole of $Ca(OH)_2$ = 74.1 g   (A_r of H = 1.0; A_r of O = 16.0; A_r of Ca = 40.1)

1 mole of $Fe_2(SO_4)_3$ = 399.9 g   (A_r of O = 16.0; S = 32.1; Fe = 55.8)

There are 2 moles of Fe atoms, 3 moles of S atoms and 12 moles of O atoms in 1 mole of $Fe_2(SO_4)_3$.

Often the total of the relative atomic masses in any substance is referred to as the relative molecular mass (represented by M_r). Relative formula mass (RFM) may be used for ionic compounds but M_r is accepted for all compounds.

The M_r of H_2O = 18.0 (no units required)

The M_r of $Fe_2(SO_4)_3$ = 399.9

Figure 2.8 Just how big is the Avogadro constant? The Shard near London Bridge is 304 000 m high. Currently it is the tallest building in the European Union. If there was a building 6×10^{23} m tall, it would be 1 980 000 000 000 000 000 000 times taller than the Shard.

Figure 2.9 Each beaker contains 1 mole of a different coloured compound and hence each contains the Avogadro constant number of particles. The compounds shown are copper carbonate, hydrated copper sulfate, sodium chloride, potassium manganate(VII) and potassium dichromate(VI).

TIP

The A_r values should be used to 1 decimal place as given on the Periodic Table supplied with the AQA examinations. It is available to download from the past paper section of the AQA website.

The term molar mass is also used which means the mass of 1 mole

The units of molar mass are g/mol or $g\,mol^{-1}$

Molar mass of $Ca(OH)_2 = 74.1\,g\,mol^{-1}$

Molar mass of $O_2 = 32.0\,g\,mol^{-1}$

The terms molar mass, M_r and RFM have the same numerical value. M_r and RFM do not have units but molar mass has units of $g\,mol^{-1}$.

TIP

During calculations it is acceptable to follow through with the answer on your calculator. If writing down a number from your calculator during a calculation (e.g. 0.02927400468) you would be advised to write down a number less than 1 to 4 significant figures (i.e. 0.02927) or a number greater than 1 to 3 decimal places (e.g. 20.482). Most calculations in this chapter will ask for an answer to 3 significant figures. If an 'appropriate level of precision' is asked for in an answer, the answer must be given to the limits of the least accurate measurement used (fewest number of significant figures).

✓Calculating moles from mass

The amount, in moles, of a substance can be calculated from the mass using the expression:

$$\text{amount (in moles)} = \frac{\text{mass (g)}}{M_r}$$

The mass must be in grams to calculate a correct amount, in moles.

Amount, in moles, is often written as n so the expression is often written as $n = \dfrac{m}{M_r}$.

TIP

The mass of water is given to 3 significant figures and the M_r is also calculated to 3 significant figures. It is then common to be asked for an answer to the same level of precision.

EXAMPLE 6

Calculate the amount, in moles, present in 0.274 g of water. Give your answer to 3 significant figures.

Answer

The M_r of water is $(2 \times 1.0) + 16.0 = 18.0$.

$n = \dfrac{m}{M_r}$ so $n = \dfrac{0.274}{18.0} = 0.0152\,mol$

TIP

As the mass is given in kg, it must be converted to grams by multiplying by 1000. Many mistakes are made in converting a measurement from one unit to another. Think carefully when you are doing this.
250 kg = 250 000 g

EXAMPLE 7

Calculate the amount, in moles, present in 250 kg of iron(III) oxide. Give your answer to 2 decimal places.

Answer

The M_r of iron(III) oxide (Fe_2O_3) = $2 \times 55.8 + 3 \times 16.0 = 159.6$

$n = \dfrac{m}{M_r}$, so $n = \dfrac{250000}{159.6} = 1566.42\,mol$

TIP

In this question decimal places are used as the mass and amount, in moles, are large numbers and the precision of the answer would be limited if 3 or 4 significant figures were used.

TIP

If a mass in tonnes is given multiply the number by 10^6 to change to grams. See Chapter 17 for more information.

Using the Avogadro constant

Sometimes calculations are set which require the calculation of the mass of one atom or molecule or a comparison between masses of atoms. Or there may be a comparison between the number of particles (atoms, molecules, ions) in a certain mass of substances.

The Avogadro constant is the number of particles in one mole.

The Avogadro constant is equal to 6.02×10^{23}.

All these calculations rely on these expressions:

$$\frac{\text{mass (g)}}{M_\text{r}} = \text{amount in moles} = \frac{\text{number of particles}}{\text{Avogadro constant}}$$

This is often written:

$$\frac{m}{M_\text{r}} = n = \frac{\text{number of particles}}{L}$$

where m is the mass in grams, M_r is the relative molecular mass, n is the amount (in moles) of a substance and L is the Avogadro constant.

The expression can be rearranged to:

$$m = n \times M_\text{r}$$

and

$$\text{number of particles} = n \times L$$

TIP

Rearranging an expression to change the subject is an important mathematical skill. You should rearrange the expression before substituting values into it. See Chapter 17 for more information.

EXAMPLE 8

Calculate the mass of 1 atom of Fe. Give your answer to 3 significant figures.

Answer

Amount (in moles) of Fe for 1 atom $= \dfrac{1}{6.02 \times 10^{23}} = 1.66 \times 10^{-24}\,\text{mol}$

Mass of 1 iron atom $= 1.66 \times 10^{-24} \times 55.8 = 9.27 \times 10^{-23}\,\text{g}$

EXAMPLE 9

Calculate the number of oxygen atoms present in 4.40 g of carbon dioxide.

Answer

M_r of CO_2 = 12.0 + (2 × 16.0) = 44.0. Give your answer to 3 significant figures.

$$n = \frac{m}{M_r} = \frac{4.40}{44.0} = 0.100 \text{ mol} \times L \ (6.02 \times 10^{23}) = 6.02 \times 10^{22} \text{ molecules of } CO_2$$

Each CO_2 contains two oxygen atoms

so the number of oxygen atoms = 2 × 6.02 × 10²²

= 1.204 × 10²³ atoms of oxygen

To 3 significant figures the answer is 1.20 × 10²³.

TEST YOURSELF 4

1 Calculate the number of **atoms** present in the following:
   a) 0.243 g of magnesium
   b) 14.2 g of neon
   c) 0.482 g of water (H_2O)
2 Calculate the number of oxygen atoms present in 1.10 g of carbon dioxide (CO_2)
3 Calculate the number of **molecules** present in the following:
   a) 9.05 g of water
   b) 11.0 g of propane, C_3H_8
4 What mass of magnesium contains the same number of atoms as the number of atoms in 3.807 g of iodine, I_2?

Using balanced equations quantitatively

Quantitatively means measuring and calculating quantities.

As no atoms are gained or lost in a chemical reaction, the equation can be read quantitatively. This allows mole calculations to be carried out.

A balanced equation such as the one below gives the mole ratios of the reactants to products in the chemical reaction.

$$2Pb(NO_3)_2(s) \rightarrow 2PbO(s) + 4NO_2(g) + O_2(g)$$

This equation shows that 2 moles of $Pb(NO_3)_2$ when heated to constant mass, break down to produce 2 moles of PbO, 4 moles of NO_2 and 1 mole of O_2.

If the $Pb(NO_3)_2$ is heated to constant mass, this should ensure that it all decomposes.

TIP

For some calculations it is important to realise that 2 moles of solid $Pb(NO_3)_2$ produce a total of 5 moles of gas ($4NO_2$ and O_2). State symbols (s) = solid; (l) = liquid; (g) = gas and (aq) = aqueous are often used to show the states of the reactants and products. Do not put them in unless they are specifically asked for in a question.

If there is a different amount, in moles, of $Pb(NO_3)_2$ to start with, the balancing numbers in the equation still give the ratio of how many moles react or are produced.

There are three steps to follow:

Step 1 Using the mass of one of the reactants, which will be given to you, calculate the amount, in moles, of this substance by dividing by the M_r.

Step 2 Using the balancing numbers in the equation, calculate the amount, in moles, of the substance asked for in the question.

Step 3 Change the amount, in moles, of this substance to mass by multiplying by the M_r.

EXAMPLE 10

$Pb(NO_3)_2$ undergoes thermal decomposition according to the equation:

$2Pb(NO_3)_2 \rightarrow 2PbO + 4NO_2 + O_2$

82.8 g of $Pb(NO_3)_2$ were heated to constant mass. Calculate the mass of PbO formed.

Answer

1 M_r of $Pb(NO_3)_2 = 207.2 + 2 \times (14.0 + (3 \times 16.0)) = 331.2$

As it is a solid, $n = \frac{m}{M_r} = \frac{82.8}{331.2} = 0.25$ moles of $Pb(NO_3)_2$

2 In the balanced equation, 2 moles of $Pb(NO_3)_2$ forms 2 moles of PbO. So 0.25 moles of $Pb(NO_3)_2$ forms 0.25 moles of PbO.

3 0.25 moles of PbO can be converted to mass by multiplying by its M_r.
M_r of PbO = 207.2 + 16.0 = 223.2.
Mass of PbO formed = $n \times M_r = 0.25 \times 223.2 = 55.8$ g

This type of calculation can be set out in a table below the balanced equation

	$2Pb(NO_3)_2$	$\rightarrow$	2PbO	+	$4NO_2$	+	O_2
Mass	82.8 g		***55.8 g				
M_r	331.2		**223.2				
Moles	0.25		*0.25				

1 Put in the mass you have been given and calculate the M_r value of that substance. Divide the mass by the M_r to calculate the amount, in moles. This is shown in the $Pb(NO_3)_2$ column.

2 Then calculate the other moles using the balancing numbers. 0.25 moles of $Pb(NO_3)_2$ produces 0.25 moles of PbO (* in the table).

3 Calculate the M_r of PbO (** in the table) and multiply it by the amount, in moles, to determine the mass of PbO (*** in the table).

EXAMPLE 11

27.5 kg of aluminium were heated in a stream of oxygen until constant mass was achieved. Determine the mass of aluminium oxide formed. Give your answer to 2 decimal places.

Answer

$4Al + 3O_2 \rightarrow 2Al_2O_3$

1 27.5 kg of aluminium is 27 500 g.

As it is a solid, moles $= \dfrac{m}{A_r} = \dfrac{27500}{27.0} = 1018.52$ moles of Al.

2 In the balanced equation 4 moles of Al forms 2 moles of Al_2O_3.

So 1018.52 moles of Al forms $\dfrac{1018.52}{2} = 509.26$ moles of Al_2O_3.

3 M_r of $Al_2O_3 = (2 \times 27.0) + (3 \times 16.0) = 102.0$.

Mass of Al_2O_3 formed = mass × M_r = 509.26 × 102.0 = 51 944.52 g = 51.94 kg.

The table shows step 1 in the first column, step 2 working out other moles* and step 3 working out M_r** and mass***.

	$4Al$	+	$3O_2$	→	$2Al_2O_3$
mass	27 500 g				51 944.52 g ***
M_r	27.0				102.0**
moles	1018.52				509.26*

EXAMPLE 12

Calcium oxide reacts with carbon to form calcium carbide, CaC_2, and carbon monoxide.

Answer

$CaO + 3C \rightarrow CaC_2 + CO$

Calculate the mass of carbon required to react completely with 25.2 g of calcium oxide. Give your answer to 3 significant figures.

	CaO	+	3C	$\rightarrow$	CaC_2	+	CO
mass	25.2 g		16.1712 g				
M_r	56.1		12.0				
moles	0.4492		1.3476				

The answer to 3 significant figures is 16.2 g.

TEST YOURSELF 5

1 1.28 g of sodium hydrogen carbonate, $NaHCO_3$, are heated to constant mass. Calculate the mass of sodium carbonate formed. Give your answers to 3 significant figures.

$2NaHCO_3(s) \rightarrow Na_2CO_3(s) + H_2O(l) + CO_2(g)$

2 0.900 g of copper(II) nitrate, $Cu(NO_3)_2$, are heated to constant mass releasing nitrogen dioxide and oxygen gas.

$2Cu(NO_3)_2(s) \rightarrow 2CuO(s) + 4NO_2(g) + O_2(g)$

Calculate the mass of copper(II) oxide formed. Give your answer to 3 significant figures.

3 Methane reacts with steam according to the equation:

$CH_4(g) + 2H_2O(g) \rightarrow CO_2(g) + 4H_2(g)$

Calculate the mass of carbon dioxide formed when 0.0124 g of methane reacts. Give your answer to 3 significant figures.

4 Calculate the mass of PCl_5 required to form 0.124 moles of HCl. Give your answer to 3 significant figures.

$PCl_5(s) + 4H_2O(l) \rightarrow H_3PO_4(aq) + 5HCl(aq)$

Percentage yield

- During a chemical reaction the calculated amount, in moles, of the product formed or the calculated mass of the product formed is called the theoretical yield. It is the moles or mass you would expect to be produced if the reaction goes to completion.
- However many chemical reactions do not give the moles or mass you would expect and the moles or mass you obtain is called the actual yield. This is what you obtain experimentally.

- The percentage yield is the percentage of the theoretical yield which is achieved in the reaction. It is calculated using the expression:

$$\text{Percentage yield} = \frac{\text{actual yield}}{\text{theoretical yield}} \times 100$$

- The actual yield and theoretical yield may be in moles or as a mass measurement, usually in grams. The calculation will be correct as long as both are moles or both are masses with the same units.
- The reasons why the percentage yield is not 100% are often asked and the main reasons are:
  - loss by mechanical transfer (transferring from one container to another)
  - loss during a separating technique, for example filtration, separating funnel
  - side reactions occurring
  - reaction not being complete.

EXAMPLE 13

A sample of 3.72 g of magnesium was heated with a Bunsen burner. The magnesium reacts with the oxygen in the air forming magnesium oxide according to the equation:

$2Mg + O_2 \rightarrow 2MgO$

1 Calculate the theoretical yield of magnesium oxide obtained in the reaction.
2 4.44 g of magnesium oxide were obtained, calculate the percentage yield of this reaction using your answer to 1.
3 Suggest one reason why the percentage yield was not 100%.

Answers

	2Mg	+	O$_2$	→	2MgO
Mass	3.72 g				6.17 g
A$_r$ or M$_r$	24.3				40.3
Moles	0.1531				0.1531

1 The theoretical yield is 6.17 g of magnesium oxide, MgO.

2 Percentage yield $= \dfrac{\text{actual yield}}{\text{theoretical yield}} \times 100$

$= \dfrac{4.44}{6.17} \times 100 = 71.96\%$

3 Reaction is not complete.

EXAMPLE 14

Phosphorus(V) oxide, P_4O_{10} reacts with water according to the equation:

$P_4O_{10} + 6H_2O \rightarrow 4H_3PO_4$

72.0 g of phosphorus(V) oxide reacts with an excess of water to form H_3PO_4. The percentage yield is 70.0%. Calculate the mass of H_3PO_4 formed to 3 significant figures.

Answer

	P$_4$O$_{10}$	+	6H$_2$O	→	4H$_3$PO$_4$
mass	72.0 g				99.37 g
M$_r$	284.0				98.0
moles	0.2535				1.014

The theoretical yield of H_3PO_4 is 99.37 g.

Percentage yield $= \dfrac{\text{actual yield}}{\text{theoretical yield}} \times 100$

$70.0 = \dfrac{\text{actual yield}}{99.37} \times 100$

actual yield $= \dfrac{70.0}{100} \times 99.37 = 0.700 \times 99.37 = 69.56\,g$

The mass of H_3PO_4 formed to 3 significant figures is 69.6 g.

EXAMPLE 15

Ammonia reacts with oxygen to form nitrogen(II) oxide, NO, and steam.

$$4NH_3 + 5O_2 \rightarrow 4NO + 6H_2O$$

Calculate the mass of ammonia required to react with excess oxygen to form 5.00 g of nitrogen(II) oxide given a 40.0% yield. Give your answer to 3 significant figures.

Answer

In this question the actual yield is 5.00 g and the theoretical yield is determined from the percentage yield.

$$\text{Percentage yield} = \frac{\text{actual yield}}{\text{theoretical yield}} \times 100$$

$$40.0 = \frac{5.00}{\text{theoretical yield}} \times 100 \quad \text{so} \quad \text{theoretical yield} = \frac{5.00 \times 100}{40.0} = 12.5\,g$$

The reacting mass calculation is carried out in reverse to determine the mass of ammonia required to form 12.5 g of nitrogen(II) oxide.

STEP 3: Calculate the mass of NH_3 from the amount, in moles, and the M_r.

	$4NH_3$	+	$5O_2$	→	$4NO$	+	$6H_2O$
mass	7.0833 g				12.5 g		
M_r	17.0				30.0		
moles	0.4167				0.4167		

STEP 2: Determine the amount, in moles, of NH_3 using the ratio in the balanced equation

STEP 1: Determine the amount, in moles, of NO

Figure 2.10

The answer to 3 significant figures is 7.08 g of ammonia are required.

Drugs are produced in industry using different chemical steps, each of which typically occurs with less than 100% yield. The overall percentage yield for the production of the drug is the sum of the product of the percentage yields of the individual steps. For a drug synthesis with many steps the overall percentage yield can be very small and this is one factor contributing to the huge cost of some drugs.

Salbutamol is an inhaled drug used to treat asthma. It is purified in a five step synthesis. Table 2.1 below shows the percentage yield for each step, and the overall percentage yield. Only about one fourteenth of the original material was turned into the purified drug. Scientists are continually researching to find new ways to improve percentage yields of the steps in the synthesis of high purity drugs in order to decrease costs and improve profits.

Table 2.1 The percentage yield for each step, and the overall percentage yield for the drug salbutamol.

Step 1	impure salbutamol → intermediate A	percentage yield = 70%
Step 2	intermediate A → intermediate B	percentage yield = 100%
Step 3	intermediate B → intermediate C	percentage yield = 40%
Step 4	intermediate C → intermediate D	percentage yield = 72%
Step 5	intermediate D → purified salbutamol	percentage yield = 35%
	overall percentage yield = 70% × 100% × 40% × 72% × 35% = 7.06%	

Figure 2.11 Many people use salbutamol to relieve the symptoms of asthma. It is administered by inhaler.

TEST YOURSELF 6

1 Calculate the percentage yield of magnesium sulfate if 2.10 g of magnesium reacts with excess sulfuric acid to form 8.30 g of magnesium sulfate. Give your answer to 3 significant figures.
$Mg(s) + H_2SO_4(aq) \rightarrow MgSO_4(aq) + H_2(g)$

2 Calculate the mass of NO_2 formed from 5.00 kg of ammonia assuming a 75.0% yield. Give your answer in kg to 3 significant figures.
$4NH_3(g) + 7O_2(g) \rightarrow 4NO_2(g) + 6H_2O(g)$

3 Calculate the percentage yield of sulfur when 42.1 g of hydrogen iodide reacts completely with concentrated sulfuric acid. 1.20 g of sulfur were formed. Give your answer to 3 significant figures.
$6HI(g) + H_2SO_4(l) \rightarrow 3I_2(s) + S(s) + 4H_2O(g)$

4 Iron(III) hydroxide undergoes thermal decomposition to form iron(III) oxide and water.
Calculate the mass of Fe_2O_3 formed from the decomposition of 17.2 g of $Fe(OH)_3$, assuming a 40.0% yield. Give your answer to 3 significant figures.
$2Fe(OH)_3(s) \rightarrow Fe_2O_3(s) + 3H_2O(g)$

Calculations involving solutions

Solutions contain a solute dissolved in a solvent. The amount, in moles, of the solute can be determined from the solution volume used (most often in cm^3) and the concentration of the solution (most often in $mol\,dm^{-3}$).

TIP
1 cm^3 (cubic centimetre) is the same as 1 ml (millilitre). 1 dm^3 (cubic decimetre) is the same as 1 l (litre) of volume. 1000 cm^3 is equal to 1 dm^3. The concentration of a solution in $mol\,dm^{-3}$ is the amount, in moles, of solute which would be dissolved in 1 dm^3 to form the same concentration of solution.

TIP
$mol\,dm^{-3}$ can be written as mol/dm^3 or sometimes as molar (M).
A 1 molar solution is written as 1M and is described as having a molarity of 1M.
1M = 1 $mol\,dm^{-3}$. For a 1 $mol\,dm^{-3}$ solution, 1 mole of solute is dissolved in deionised water and the volume made up to form 1 dm^3 of solution.

TIP

For a solution volume in dm^3 the expression becomes $n = v \times c$.

The amount, in moles, (n) of solute dissolved in a solution is calculated from the volume of the solution and the concentration based on the expression:

$$\text{amount (in moles)} = \frac{\text{solution volume (cm}^3) \times \text{concentration (mol dm}^{-3})}{1000}$$

or more simply as $n = \dfrac{v \times c}{1000}$

EXAMPLE 16

Calculate the amount, in moles, of calcium hydroxide present in 25.0 cm^3 of a solution of concentration 0.127 mol dm^{-3}. Give your answer to 3 significant figures.

$$n = \frac{v \times c}{1000} = \frac{25.0 \times 0.127}{1000} = 0.003175 \text{ mol}$$

0.003175 may be written in standard form as 3.175×10^{-3}.

To 3 significant figures the answer is 0.00318 mol.

TIP

Rearranging a mathematical expression to make some other quantity the subject is an essential skill. Make sure you understand how this process works.

Calculating volumes and concentrations

The expression can be rearranged to allow the calculation of solution volume or concentration.

$$v = \frac{n \times 1000}{c} \text{ and } c = \frac{n \times 1000}{v}$$

For volumes (v) in cm^3 and concentrations (c) in mol dm^{-3} and where n = amount, in moles.

EXAMPLE 17

Calculate the concentration of a solution of sodium hydroxide formed by dissolving 6 g of sodium hydroxide in deionised water and the volume of the solution is made up to 250 cm^3.

M_r of NaOH = 23.0 + 16.0 + 1.0 = 40.0

$$n = \frac{m}{RFM} = \frac{6}{40.0} = 0.15 \text{ mol}$$

There are several ways to approach the next part of this calculation and you may move from one method to the other with practice at these types of calculations.

Method 1: Using the expression

As you want to find the concentration of the solution the expression $c = \dfrac{n \times 1000}{v}$ may be used. v = 250 cm^3; n = 0.15 mol and c is the quantity you want to find.

$$c = \frac{0.15 \times 1000}{250} = 0.6 \text{ mol dm}^{-3}.$$

Method 2: Using the volumes

- From the initial calculation you know that there are 0.15 mol of NaOH dissolved in 250 cm³ of the solution made.

- The concentration in mol dm⁻³ is the amount, in moles, which would be required to make a solution of the same concentration if the volume was 1 dm³ (1000 cm³).

- To scale up from 250 cm³ to 1 dm³ there is a factor of four so $0.12 \times 4 = 0.6$ mol dm⁻³.

- 0.6 mol of NaOH would be required to be dissolved in 1 dm³ to produce a solution of the same concentration as when 0.15 mol of NaOH were dissolved in 250 cm³.

EXAMPLE 18

Calculate the concentration of a solution formed when 1.25 g of calcium chloride are dissolved in deionised water and the volume of the solution is made up to 25.0 cm³. Give your answer to 3 significant figures.

M_r of $CaCl_2 = 40.1 + (2 \times 35.5) = 111.1$

$$n = \frac{m}{M_r} = \frac{1.25}{111.1} = 0.01125 \text{ mol}$$

Method 1: Using the expression

As you want to find the concentration of the solution the expression $c = \dfrac{n \times 1000}{v}$ may be used. $v = 25.0$ cm³; $n = 0.1125$ mol and c is the quantity you want to find.

$$c = \frac{0.01125 \times 1000}{25.0} = 0.450 \text{ mol dm}^{-3}.$$

Method 2: Using the volumes

- From the initial calculation you know that there are 0.01125 mol of $CaCl_2$ dissolved in 25.0 cm³ of the solution made.

- The concentration in mol dm⁻³ is the amount, in moles, which would be required to make a solution of the same concentration if the volume was 1 dm³ (1000 cm³).

- To scale up from 25.0 cm³ to 1 dm³ there is a factor of 40 so $0.01125 \times 40 = 0.450$ mol dm⁻³ (to 3 significant figures).

- 0.450 mol of $CaCl_2$ would be required to be dissolved in 1 dm³ to produce a solution of the same concentration as when 0.01125 mol of $CaCl_2$ were dissolved in 25.0 cm³.

TIP

×40 is a very common calculation feature when using solutions as most pipettes measure 25.0 cm³ of solution and concentration is often asked.

TIP

Throughout this calculation you should work to at least 3 significant figures. 0.01125 mol is to 4 significant figures. The final answer is stated to 3 significant figures as 0.450 mol dm⁻³.

Calculating concentrations from reactions forming a solution

Some chemical reactions where a substance reacts with water or other solutions such as acids can form the basis of calculations. The concentration of the solution formed can be determined from the mass of the substance which reacts, the ratios in the balanced equation and the final volume of the solution.

EXAMPLE 19

Solid phosphorus pentasulfide, P_2S_5, reacts with water to form a solution of phosphoric(V) acid, H_3PO_4, and hydrogen sulfide gas.

$$P_2S_5(s) + 8H_2O(l) \rightarrow 2H_3PO_4(aq) + 5H_2S(g)$$

Calculate the concentration of the solution of phosphoric(V) acid formed when 3.82 g of phosphorus pentasulfide reacts with water and the solution volume is made up to 100 cm^3. Give your answer to 3 significant figures.

Answer

M_r of P_2S_5 = $(2 \times 31.0) + (5 \times 32.1)$ = 222.5

Calculating moles of P_2S_5:

$$n = \frac{m}{M_r} = \frac{3.82}{222.5} = 0.01717 \text{ mol (to 4 significant figures)}$$

Moles of H_3PO_4 formed = 0.01717×2 = 0.03434 mol

Method 1: Using the expression

$$c = \frac{n \times 1000}{v}$$

v = 100 cm^3; n = 0.03434 mol and c is the quantity you want to find.

$$c = \frac{0.03434 \times 1000}{100} = 0.3434 \text{ mol dm}^{-3}.$$

The answer to 3 significant figures is 0.343 mol dm^{-3}

Method 2: Using the volumes

From 100 cm^3 to 1 dm^3 (1000 cm^3) the factor is 10 so $0.03434 \times 10 = 0.3434$ mol dm^{-3} and the answer to 3 significant figures is 0.343 mol dm^{-3}.

You may not have to calculate the amount, in moles, from the mass of the substance which reacts with water. The amount, in moles, may be given to you.

EXAMPLE 20

Calculate the concentration of nitric acid produced when 0.253 moles of nitrogen dioxide (NO_2) reacts with water and the solution volume is made up to 200 cm^3. Give your answer to an appropriate level of precision.

Answer

$3NO_2(g) + H_2O(l) \rightarrow 2HNO_3(aq) + NO(g)$

0.253 moles of NO_2 forms $\frac{0.253}{3} \times 2 = 0.1687$ mol of nitric acid (working to 4 significant figures during the calculation)

Method 1: Using the expression

$$c = \frac{n \times 1000}{v}$$

$v = 200$ cm³; $n = 0.1687$ mol and c is the quantity you want to find.

$$c = \frac{0.1687 \times 1000}{200} = 0.8435 \, mol \, dm^{-3}.$$

The concentration of the nitric acid formed is 0.844 mol dm⁻³ (to 3 significant figures)

Method 2: Using the volumes

From 200 cm³ to 1 dm³ (1000 cm³) the factor is 5 so
$0.1687 \times 5 = 0.8435 \, mol \, dm^{-3}$.

The answer to 3 significant figures is 0.844 mol dm⁻³.

Calculating volumes and concentrations required to react

When a solution reacts with a solid or another solution, calculations can be carried out to determine the concentration of solution required or the volume of a particular concentration which is required.

All of these calculations are again carried out using the ratio in the balanced equation and the expressions linking n, v and c shown below:

$$n = \frac{v \times c}{1000}; \, c = \frac{n \times 1000}{v}; \, v = \frac{n \times 1000}{c}$$

EXAMPLE 21

Calculate the volume of 0.140 mol dm⁻³ hydrochloric acid required to react with 1.00 g of magnesium. Give your answer to 3 significant figures.

$Mg + 2HCl \rightarrow MgCl_2 + H_2$

Answer

● Determine the amount, in moles, of magnesium

$$\text{moles of Mg} = \frac{m}{A_r} = \frac{1.00}{24.3} = 0.04115 \, mol \, (\text{to 4 significant figures})$$

● Moles of HCl that react with 0.04115 moles of Mg = $0.04167 \times 2 = 0.08230$ mol (using 1:2 ratio in the balanced equation).

● Calculating volume of hydrochloric acid required

$$v = \frac{n \times 1000}{c}; \, c = 0.140 \, mol \, dm^{-3}; \, n = 0.08230$$

$$v = \frac{0.08230 \times 1000}{0.14} = 588 \, cm^3 \, (\text{to 3 significant figures}).$$

EXAMPLE 22

$25.0\,cm^3$ of nitric acid reacts completely with $5.00\,g$ of copper. The balanced equation for the reaction is:

$$Cu + 4HNO_3 \rightarrow Cu(NO_3)_2 + 2NO_2 + 2H_2O$$

Calculate the concentration of the nitric acid to 3 significant figures.

Answer

● Determine the amount, in moles, of copper

$$\text{moles of Cu} = \frac{m}{A_r} = \frac{5.00}{63.5} = 0.07874\,mol \text{ (to 4 significant figures)}$$

● Moles of HNO_3 which reacts with 0.07874 moles of Cu = 0.07874×4 = 0.31496 mol (using 1:4 ratio in the balanced equation).

● Calculating concentration of nitric acid required

● $c = \dfrac{n \times 1000}{v}$; $v = 25.0\,cm^3$; $n = 0.31496$

$$c = \frac{0.31496 \times 1000}{25.0} = 12.5984\,mol\,dm^{-3}$$

Remember this last calculation can be carried out using the volumes method where you simple multiply by the factor which is 25.

Answer to 3 significant figures is $12.6\,mol\,dm^{-3}$.

Figure 2.12 The dramatic reaction of a penny with nitric acid. The copper in the penny reacts and a green solution and brown nitrogen dioxide are produced.

TIP

This last calculation could be reversed. You could easily be asked to calculate the mass of copper required to react with $25.0\,cm^3$ of $12.6\,mol\,dm^{-3}$ nitric acid. $12.6\,mol\,dm^{-3}$ is to 3 significant figures but your answer should be very close to 5 g. This is similar to the questions in the next section.

Calculating the mass or identity of a solid reacting with a solution

In this style of question, you will be given information on the solution and asked to calculate the amount, in moles, of the solute present in the solution. From here you can calculate the amount, in moles, of solid which reacted with the solute in the solution. From this point, the mass of a solid can be calculated from the amount in moles and its M_r, or the M_r of the solid, can then be calculated using its mass and moles. The solid can be easily identified from its M_r.

The identity of an unknown compound can be determined by determining its M_r from the reaction of a given mass of solid reacting with a solution (usually an acid).

EXAMPLE 23

Copper(II) chloride can be formed from the **neutralisation** reaction of copper(II) carbonate with hydrochloric acid.

$CuCO_3(s) + 2HCl(aq) \rightarrow CuCl_2(aq) + CO_2(g) + H_2O(l)$

15.2 cm³ of 0.174 mol dm⁻³ hydrochloric acid is reacted with copper(II) carbonate.

a) Calculate the amount, in moles, of HCl in 15.2 cm³ of 0.174 mol dm⁻³ nitric acid. Give your answer to 3 significant figures.
b) Calculate the amount, in moles, of $CuCO_3$ which reacted with the nitric acid. Give your answer to 3 significant figures.
c) Calculate the minimum mass of powdered $CuCO_3$ which should be added to react with all of the nitric acid. Give your answer to 3 significant figures.

Answers

a) $n = \dfrac{v \times c}{1000} = \dfrac{15.2 \times 0.174}{1000} = 0.00264 \ (2.64 \times 10^{-3}) \text{ mol}$

b) Using the 2:1 ratio in the balanced equation for $HCl:CuCO_3$

$\text{moles of } CuCO_3 = \dfrac{0.00264}{2} = 0.00132 \text{ mol}$

c) M_r of $CuCO_3 = 63.5 + 12.0 + 3(16.0) = 123.5$
mass of $CuCO_3 = n \times M_r = 0.00132 \times 123.5 = 0.163 \text{ g}$ (to 3 significant figures)

EXAMPLE 24

An unknown metal carbonate may be written X_2CO_3. X_2CO_3 reacts with hydrochloric acid according to the balanced equation:

$X_2CO_3(aq) + 2HCl(aq) \rightarrow 2XCl(aq) + CO_2(g) + H_2O(l)$

A 3.69 g sample of solid X_2CO_3 was dissolved in deionised water and the solution volume made up to 1000 cm³. A 25.0 cm³ portion of this solution required 20.0 cm³ of 0.125 mol dm⁻³ hydrochloric acid for complete reaction.

1 Calculate the amount, in moles, of HCl in 20.0 cm³ of 0.125 mol dm⁻³.
2 Calculate the amount, in moles, of X_2CO_3 which reacted with this amount of HCl.
3 Calculate the amount, in moles, of X_2CO_3 present in the 3.69 g sample.
4 Calculate the M_r of X_2CO_3.
5 Calculate the relative atomic mass (A_r) of the metal X and deduce its identity.

Answers

1 $n = \dfrac{v \times c}{1000} = \dfrac{20.0 \times 0.125}{1000} = 0.0025 \ (2.5 \times 10^{-3}) \text{ mol}$

2 Using the 2:1 ratio in the balanced equation for $2HCl:X_2CO_3$

$\text{moles of } X_2CO_3 = \dfrac{0.0025}{2} = 0.00125 \text{ mol}$

3 All of 3.69 g of X_2CO_3 dissolved in 1000 cm^3
   moles in 1000 cm^3 = 0.00125 × 40 = 0.05 mol
4 Rearranging the expression: amount, in moles, $\frac{\text{mass (g)}}{M_r}$, allows us to

   calculate the M_r or relative formula mass of X_2CO_3

   $$n = \frac{m}{M_r} \text{ so } M_r = \frac{m}{n}$$

   M_r of $X_2CO_3 = \frac{3.69}{0.05} = 73.8$

5 M_r of 'CO_3' = 12.0 + 3(16.0) = 60.0
   M_r of 'X_2' = 73.8 − 60.0 = 13.8

   A_r of 'X' $= \frac{13.8}{2} = 6.9$

From the Periodic Table, X is lithium.

TIP

Remember you are asked for the identity of X in X_2CO_3 so identify X and not the compound X_2CO_3. This question provided a very definite answer but remember that rounding of solution volumes may give an answer close to the A_r of the element. It is always a good idea to carry out the calculation in reverse to see if lithium carbonate would react with the volume of acid stated in the question.

TIP

The calculation could also be an unknown hydrogen carbonate, $XHCO_3$ or $MHCO_3$ where M represents a metal such as a Group 1 metal. The calculation follows the same pattern except there is a 1:1 ratio for the reaction between $XHCO_3$ and HCl.

EXAMPLE 25

One indigestion table (0.850 g) containing calcium carbonate reacts with exactly 40.0 cm^3 of 0.220 mol dm^{-3} hydrochloric acid.

$CaCO_3$ (s) + 2HCl (aq) → $CaCl_2$(aq) + CO_2(g) + H_2O(l)

Calculate the percentage of calcium carbonate in one tablet. Give your answer to 3 significant figures.

Moles of HCl $= \frac{v \times c}{1000} = \frac{40.0 \times 0.220}{1000} = 0.0088$ mol

Moles of $CaCO_3$ in the tablet $= \frac{0.0088}{2} = 0.0044$ mol

Answer

M_r of $CaCO_3$ = 40.1 + 12.0 + 3(16.0) = 100.1

Mass of $CaCO_3$ in the tablet = 0.0044 × 100.1 = 0.440 g

Percentage of calcium carbonate in the tablet $= \frac{0.440}{0.850} \times 100 = 51.8\%$.

TIP

A back titration may also be used to determine the mass of calcium carbonate or any other insoluble solid using its reaction with an acid. See page 58.

TEST YOURSELF 7

1 What is the concentration, in $mol\,dm^{-3}$, of a solution prepared by dissolving 0.737 g of phosphoric acid, H_3PO_4 in 25.0 cm^3 of water? Give your answer to 3 significant figures.

2 What mass of potassium sulfate, K_2SO_4, should be dissolved in 100 cm^3 of water to make a solution of concentration $0.5\,mol\,dm^{-3}$?

3 0.137 g of an unknown metal hydroxide $M(OH)_2$ reacts completely with 15.4 cm^3 of $0.24\,mol\,dm^{-3}$ hydrochloric acid.

$$M(OH)_2(s) + 2HCl(aq) \rightarrow MCl_2(aq) + 2H_2O(l)$$

  a) Determine the relative atomic mass (A_r) of M to 1 decimal place.
  b) Identify M

4 Zinc reacts with hydrochloric acid:

$$Zn(s) + 2HCl(aq) \rightarrow ZnCl_2(aq) + H_2(g)$$

Calculate the volume in dm^3 of $0.12\,mol\,dm^{-3}$ hydrochloric acid required to react with 0.5 g of zinc powder. Give your answer to 2 decimal places.

Solutions reacting together

Solutions can react together. This is true in neutralisation reactions where an acid neutralises an alkali. Both an acid and an alkali are solutions. They are most often colourless solutions and an indicator can be used to determine the point (called the end-point) where the solutions have neutralised each other.

Carrying out a titration

A titration is a method of volumetric analysis. One solution is placed in a burette and the other is placed in a conical flask. An indicator is added to the solution in the conical flask. The solution in the burette is added to the solution in the conical flask. The indicator will show the end-point of the titration (when the indicator changes colour). This is the point when the reaction is complete.

Apparatus and practical techniques

The main pieces of apparatus used in a titration are a burette, a pipette with safety filler, a volumetric flask and several conical flasks. One conical flask can be used and it can be rinsed out between titrations.

Preparing a burette for use

1 rinse the burette with deionised water

2 ensure the water flows through the jet

3 discard the water

4 rinse the burette with the solution you will be filling it with

5 ensure the solution flows through the jet

6 discard the solution

7 charge (fill) the burette with the solution you will be using in it.

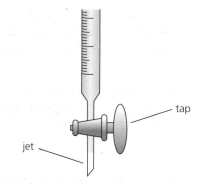

Figure 2.13 Burette.

jet

tap

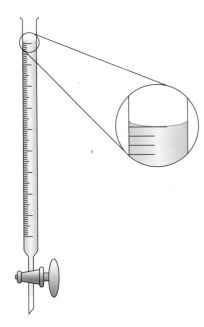

Figure 2.14 Using a burette.

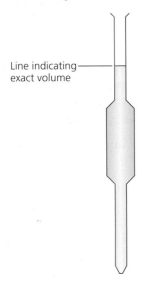

Line indicating exact volume

Figure 2.15 A typical pipette.

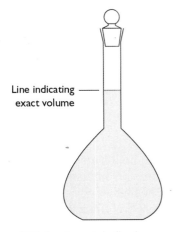

Line indicating exact volume

Figure 2.16 A volumetric flask.

Using a burette

When using a burette the volume of solution it contains is read at the bottom of the meniscus as shown in the diagram.

If you are right handed the tap of the burette is operated with the left hand to allow the right hand to be used to swirl to mix the contents of the conical flask.

Preparing a pipette for use in a titration

1 Using a pipette filler rinse the pipette using deionised water.

2 Discard the water.

3 Rinse the pipette with the solution you will be filling it with.

4 Discard this solution.

Using a pipette

A pipette accurately measures an exact volume of a solution and should be used in the following way:

1 A pipette filler is attached to the top of a pipette.

2 The pipette is placed in the solution and suction applied to draw the solution up.

3 The solution is drawn up above the line on the pipette.

4 The solution is released until the meniscus sits on the line.

5 The solution in the pipette is released into a conical flask.

6 Pipettes measure out exactly $25.0\,cm^3$ or $10.0\,cm^3$. (Remember that $1\,cm^3 = 1\,ml$ but chemists prefer cm^3 as their unit of volume).

7 An exact volume of a solution is vital in volumetric work as taking exactly $25.0\,cm^3$ of a known concentration of a solution means that we know exactly how many moles of the dissolved substance are present in the conical flask.

Conical flasks

Conical flasks are used in titrations as they can be swirled easily to mix the reactants. Also the sloped sides prevent any of the solution spitting out when it has been added. The conical flask should be rinsed out with deionised water before use.

The conical flask does not have to be completely dry before use as the exact volume of solution added contains an exact number of moles of solute. Extra deionised water does not add to the number of moles of solute.

Volumetric flasks

Volumetric flasks are used when diluting one of the solutions before the titration is carried out. They can also be used when preparing a solution of a solid.

The **dilution factor** is the amount the original solution is diluted by. It is calculated by dividing the new total volume by the volume of original solution put into the mixture.

Carrying out a dilution of a solution

1 pipette 25.0 cm³ of the original solution into a clean volumetric flask

2 add deionised water to the flask until the water is just below the line

3 using a disposable pipette add deionised water very slowly until the bottom of the meniscus is on the line

4 stopper the flask and invert to mix thoroughly.

If a 25.0 cm³ sample of a solution is made up to a total volume of 250 cm³ using deionised water then the dilution factor is 10.

If a 10.0 cm³ sample of a solution is made up to a total volume of 250 cm³ using deionised water then the dilution factor is 25.

Preparing a solution from a mass of solid

When preparing a solution from a solid it is important not to lose any of the solid or solution before it is placed in the volumetric flask.

1 Weigh out an accurate mass of a solid in a weighing boat and dissolve in a suitable volume (100 cm³, if the final volume in the volumetric flask is to be 250 cm³, to allow for rinsings) of deionised water in a beaker, stir with a glass rod and rinse the weighing boat into the beaker with deionised water.

2 Once dissolved, hold the glass rod above the beaker and rinse it with deionised water before removing it.

3 Place a glass funnel into the top of a clean volumetric flask and pour the prepared solution down a glass rod into the funnel.

4 Rinse the glass rod with deionised water into the funnel.

5 Rinse the funnel with deionised water.

6 Remove the funnel and add deionised water to the volumetric flask until the water is just below the line.

7 Using a disposable pipette add deionised water very slowly until the bottom of the meniscus is on the line.

8 Stopper the flask and invert to mix thoroughly.

> **TIP**
> You should calculate the mass required to make the solution and weigh this out accurately. Calculate the amount, in moles, of the solid from the solution volume required and the concentration. Calculate the mass required by multiplying the amount by the M_r of the solid.

> **TIP**
> Glass pipettes may be used in preference to disposable plastic pipettes.

> **EXAMPLE 26**
>
> Describe how you would prepare 250 cm³ of solution of 0.170 mol dm³ potassium hydroxide.
>
> Moles of KOH required $= \dfrac{250 \times 0.170}{1000} = 0.0425$ mol
>
> **Answer**
> 1 Mr of KOH = 39.1 + 16.0 + 1.0 = 56.1.
> 2 Mass of KOH required = 0.0425 × 56.1 = 2.38 g (to 3 significant figures)
> 3 Weigh out 2.38 g of KOH to 3 significant figures on a balance.
> 4 Dissolve in 100 cm³ of deionised water in a beaker.
> 5 Add the solution to a 250 cm³ volumetric flask and rinse the beaker into the flask.
> 6 Make the solution up to the mark with deionised water until the bottom of the meniscus is on the line.
> 7 Stopper the flask and invert to mix.

Carrying out a titration

The major points in carrying out a titration are shown in the diagram:

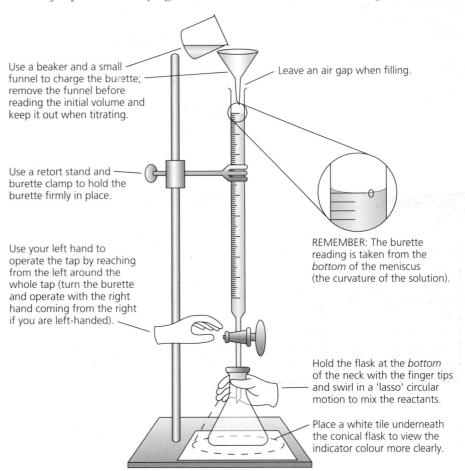

Use a beaker and a small funnel to charge the burette; remove the funnel before reading the initial volume and keep it out when titrating.

Leave an air gap when filling.

Use a retort stand and burette clamp to hold the burette firmly in place.

Use your left hand to operate the tap by reaching from the left around the whole tap (turn the burette and operate with the right hand coming from the right if you are left-handed).

REMEMBER: The burette reading is taken from the *bottom* of the meniscus (the curvature of the solution).

Hold the flask at the *bottom* of the neck with the finger tips and swirl in a 'lasso' circular motion to mix the reactants.

Place a white tile underneath the conical flask to view the indicator colour more clearly.

Figure 2.17 Carrying out a titration.

- In one volumetric analysis, several titrations should be carried out.
- The first titration should be rough and should be an overshoot and therefore the titre should be greater in value than the two accurate titrations.
- The subsequent titrations should be accurate (within $0.10\,cm^3$ of each other) with drop-wise addition as the end-point is reached. Titrations which are not within $0.10\,cm^3$ of each other are not used to calculate the average titre.
- Standard solutions are used in volumetric analysis. A **standard solution** is a solution of known concentration.
- A burette has a total graduated volume of $50.0\,cm^3$. You can perform two titrations using a burette if the titres are well below $25\,cm^3$. However if the rough is close to or above $25\,cm^3$, it is essential you refill the burette before starting the first accurate titration.
- The funnel is removed to ensure any drops of the solution which remains in the funnel do not interfere with the titration results.

Recording titration results

A typical table used to record titration results is shown below:

	Rough titration	1st titration	2nd titration
Initial burette reading/cm³			
Final burette reading/cm³			
Titre/cm³			

The titre is the volume delivered from the burette into the conical flask until the indicator changes colour (end-point).

Please note the following points about recording titration results.

- You should put units with the headings.
- All values should be recorded to 0.05 cm³, for example 0 cm³ is written as 0.00; 24.5 cm³ is written as 24.50. 24.35 is acceptable but 24.31 and 24.38 are not. The units should be with the headings and not in the main body of the table.
- The rough titration titre should be greater than the two accurate titration titres but not more than 1 cm³ greater.
- The two accurate titration titre values should be within 0.10 cm³ of each other. Titrations should be carried out until two values within 0.10 cm³ of each other are obtained.
- The average titre must be stated with units to 2 decimal places.
- Titres within 0.10 cm³ of each other should only be used to calculate the average titre. Any other titres should be disregarded in the calculation. You should indicate which titration results are being used to calculate the average titre.

Calculating the average titre

- When calculating the average titre, ignore the rough titration and any result, which is clearly not within 0.10 cm³ of the other accurate titration values.
- Write the average titre below the table and include the units. The average titre should be written to 2 decimal places.

The table below shows a sample set of titration results.

	Rough titration	1st titration	2nd titration
Initial burette reading/cm³	0.00	0.00	24.00
Final burette reading/cm³	24.50	24.00	47.90
Titre/cm³	24.50	24.00	23.90
		Average titre = 23.95 cm³	

Units

- Units of volume are in cm^3 ($1\,cm^3 = 1\,ml$).
- Concentration units are $mol\,dm^{-3}$. '$mol\,dm^{-3}$' can also be written as 'mol/dm^3'.
- Remember that $1\,dm^3$ is the same as 1 litre.
- However M (molar) may also be used and is the same as $mol\,dm^{-3}$.
- A 1M solution is $1\,mol\,dm^{-3}$. The molarity of a solution has units M but concentration is usually quoted as $mol\,dm^{-3}$ but also can be given in $g\,dm^{-3}$.
- Units of $g\,dm^{-3}$ (grams per dm^3) may be calculated by multiplying the concentration (or molarity) by the M_r of the solute.

Percentage error

The percentage error in the measurements obtained from different pieces of apparatus can be calculated.

A volume of $25\,cm^3$ measured with a measuring cylinder has an error of $\pm0.5\,cm^3$. The percentage error for this piece of apparatus is

$$\frac{0.5}{25} \times 100 = 2\%$$

A volume of $25\,cm^3$ measured using a class A pipette has an error of $\pm0.03\,cm^3$. The percentage error is

$$\frac{0.03}{25} \times 100 = 0.12\%.$$

A mass of $0.120\,g$ measured using a balance with an error of $\pm0.001\,g$. The percentage error is

$$\frac{0.001}{0.120} \times 100 = 0.833\%.$$

Indicators

The two main indicators used for acid-base titrations are phenolphthalein and methyl orange.

Indicator	Phenolphthalein	Methyl orange
Colour in acidic solutions	colourless	red
Colour in neutral solutions	colourless	orange
Colour in alkaline solutions	pink	yellow
Titrations suitable for:	strong acid-strong base weak acid-strong base	strong acid-strong base strong acid-weak base

The choice of indicator is based on the type of acid and base in the titration. It is important to be able to choose the correct indicator for a particular titration. If ethanoic acid (a weak acid) and sodium hydroxide solution (a strong alkali) are being used then phenolphthalein is the indicator of choice. Both methyl orange and phenolphthalein can be used for strong acid-strong base titrations.

TIP
Changes in temperature or titres measured from a burette will have double the ± error as they are calculated from two readings and each reading will have an error. For more information see pages 358-9.

Figure 2.18 Phenolphthalein and methyl orange indicator. Which beakers contain acid and which alkali?

The colour change of the indicator at the end-point of a titration is given in the table below:

Titration	Methyl orange	Phenolphthalein
Acid in conical flask Alkali in burette	red to yellow	colourless to pink
Alkali in conical flask Acid in burette	yellow to red	pink to colourless

Types of titrations at AS

For AS the titrations which are carried out are mainly monoprotic acids with bases.

A monoprotic acid is an acid which can release 1 mole of H^+ ions per mole of acid.

The main monoprotic acids which may be used are hydrochloric acid (HCl), nitric acid (HNO_3) and ethanoic acid (CH_3COOH). Hydrochloric acid and nitric acid are strong acids. Ethanoic acid is a weak acid.

Sodium hydroxide (NaOH) and potassium hydroxide (KOH) are strong bases. Ammonia (NH_3) is a weak base.

Diprotic acids such as sulfuric acid can be used in titrations. The only difference in using a diprotic acid is the ratios in the equation for the reaction, for example

$$2NaOH + H_2SO_4 \rightarrow Na_2SO_4 + 2H_2O.$$

There is a 2:1 ratio of $NaOH:H_2SO_4$

TIP

In this calculation you will calculate the amount, in moles, of sodium hydroxide (as you have a volume and concentration of sodium hydroxide). Then using the equation you can calculate the amount, in moles, of hydrochloric acid which will react with the sodium hydroxide. From this answer using the concentration you can calculate the volume of hydrochloric acid required to react.

EXAMPLE 27

Calculate the volume of $0.750\,mol\,dm^{-3}$ hydrochloric acid required to react with $18.2\,cm^3$ of $0.840\,mol\,dm^{-3}$ sodium hydroxide solution. Give your answer to 3 significant figures.

Answer

$NaOH(aq) + HCl(aq) \rightarrow NaCl(aq) + H_2O(l)$

Moles of NaOH $= \dfrac{v \times c}{1000} = \dfrac{18.2 \times 0.84}{1000} = 0.01529\,mol$ (to 4 significant figures)

Using the ratio in the balanced equation: NaOH + HCl; ratio = 1:1

Moles of HCl $= 0.01529\,mol$

Using the expression: $v = \dfrac{n \times 1000}{c}$

so $v = \dfrac{0.015229 \times 1000}{0.75} = 20.4\,cm^3$ (to 3 significant figures)

EXAMPLE 28

A solution of nitric acid was diluted by placing 25.0 cm³ of the solution in a volumetric flask and the volume was made up to 250 cm³ using deionised water. 25.0 cm³ of this diluted solution were placed in a conical flask and titrated against 0.0120 mol dm⁻³ potassium hydroxide solution. 24.25 cm³ of the potassium hydroxide solution were required for complete neutralisation. Calculate the concentration of the nitric acid in g dm⁻³. Give your answer to 3 significant figures.

Answer

$KOH(aq) + HNO_3(aq) \rightarrow KNO_3(aq) + H_2O(l)$

Moles of KOH $= \dfrac{v \times c}{1000} = \dfrac{24.25 \times 0.0120}{1000} = 0.000291$ mol

Ratio of KOH:HNO₃ in equation is 1:1

Moles of HNO_3 in 25.0 cm³ = 0.000291 mol

Using the expression: $c = \dfrac{n \times 1000}{v}$

so $c = \dfrac{0.00291 \times 1000}{25.0} = 0.01164$ mol dm⁻³

This is the concentration of the diluted solution. The dilution factor is 10 so the concentration of the original solution is
0.01164 × 10 = 0.1164 mol dm⁻³

M_r of HNO_3 = 1.0 + 14.0 + 3(16.0) = 63.0

Concentration in g dm⁻³ = concentration in mol dm⁻³ × M_r

= 0.1164 × 63.0 = 7.33 g dm⁻³ (to 3 significant figures)

TIP

This step is the same as multiplying by 40.

REQUIRED PRACTICAL

Finding the concentration of a sample of commercial vinegar

Commercial vinegar contains ethanoic acid. 10.0 cm³ of commercial vinegar was diluted to 1 dm³ with water before being used in a restaurant. 25.0 cm³ of the diluted solution was titrated against 0.10 mol dm⁻³ sodium hydroxide using phenolphthalein indicator. The results were recorded in the table below.

	Rough titration	1st titration	2nd titration
Initial burette reading (cm³)	0.00	0.00	0.00
Final burette reading (cm³)	18.10	17.55	17.45
Titre (cm³)	18.10	17.55	17.45

1 Describe how you would prepare the diluted solution of vinegar.
2 Describe how you would accurately transfer 25.0 cm³ of the diluted solution of vinegar into a conical flask.
3 State the colour change of the indicator at the end-point.

4 Calculate the average titre.
5 Write the equation for the reaction of ethanoic acid with sodium hydroxide.
6 Calculate the number of moles of sodium hydroxide used in the titration.
7 Calculate the number of moles of ethanoic acid in 25.0 cm³ of the diluted vinegar.
8 Calculate the concentration of the diluted vinegar solution.
9 Calculate the concentration of the commercial vinegar solution in g dm⁻³.

Figure 2.19 Vinegar contains ethanoic acid, and can be titrated with any alkali.

Back titration

The mass of calcium carbonate in an indigestion tablet can be determined using a back titration. A back titration involves reacting the tablet (or any insoluble solid) with a known excess of dilute hydrochloric acid. The excess acid is then titrated using a standard solution of an alkali such as sodium hydroxide solution. The basic method of any back titration is shown below though the total volume and concentration of the sodium hydroxide solution may vary.

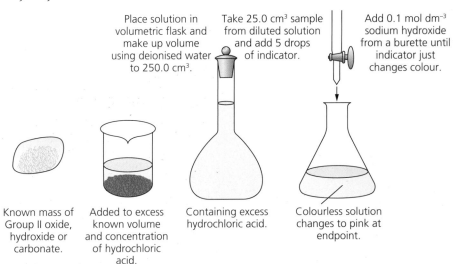

Known mass of Group II oxide, hydroxide or carbonate.

Added to excess known volume and concentration of hydrochloric acid.

Place solution in volumetric flask and make up volume using deionised water to 250.0 cm³.

Containing excess hydrochloric acid.

Take 25.0 cm³ sample from diluted solution and add 5 drops of indicator.

Add 0.1 mol dm⁻³ sodium hydroxide from a burette until indicator just changes colour.

Colourless solution changes to pink at endpoint.

Figure 2.20

EXAMPLE 29 ~~tablets~~

Two indigestion ~~tablets~~ were reacted with 25.0 cm³ of 2.00 mol dm⁻³ hydrochloric acid. The solution was transferred to a volumetric flask and the volume made up to 250 cm³ using deionised water. 25.0 cm³ of this solution were titrated against 0.100 mol dm⁻³ sodium hydroxide solution. The average titre was found to be 24.7 cm³.

Answer

$CaCO_3(s) + 2HCl(aq) \rightarrow CaCl_2(aq) + CO_2(g) + H_2O(l)$

$NaOH(aq) + HCl(aq) \rightarrow NaCl(aq) + H_2O(l)$

Calculate the mass of calcium carbonate, in mg, present in one tablet.

Moles of NaOH $= \dfrac{v \times c}{1000} = \dfrac{24.7 \times 0.100}{1000} = 0.00247 \, mol$

Moles of HCl in 25.0 cm³ = 0.00247 mol

Moles of HCl in 250 cm³ = 0.00247 × 10 = 0.0247 mol

Moles of HCl added initially to the tablets $= \dfrac{v \times c}{1000} = \dfrac{25.0 \times 2.00}{1000} = 0.0500 \, mol$

Moles of HCl which reacted with $CaCO_3$ in tablets = 0.0500 − 0.0247 = 0.0253 mol

Moles of $CaCO_3$ in two tablets $= \dfrac{0.0253}{2} = 0.01265 \, mol$

Mass of $CaCO_3$ in two tablets = 0.01265 × 100.1 = 1.266 g

Mass of $CaCO_3$ in one tablet $= \dfrac{1.266}{2} = 0.633 \, g$

Mass of $CaCO_3$ in one tablet in mg = 0.633 × 1000 = 633 mg

TEST YOURSELF 8

1 Calculate the volume of 0.220 mol dm⁻³ sodium hydroxide solution required to react with 20.5 cm³ of 0.400 mol dm⁻³ hydrochloric acid. Give your answer to 3 significant figures.

$NaOH(aq) + HCl(aq) \rightarrow NaCl(aq) + H_2O(l)$

2 Calculate the concentration of nitric acid in g dm⁻³ if 11.4 cm³ of nitric acid reacts completely with 25.0 cm³ of 0.350 mol dm⁻³ potassium hydroxide solution.

$KOH(aq) + HNO_3(aq) \rightarrow KNO_3(aq) + H_2O(l)$

3 A solution of nitric acid was diluted by placing 25.0 cm³ in a volumetric flask and making up the volume to 250 cm³ using deionised water. 25.0 cm³ of this solution required 17.7 cm³ of 0.200 mol dm⁻³ sodium hydroxide solution for neutralisation.
   a) Calculate the amount, in moles, of NaOH in 17.7 cm³ of 0.200 mol dm⁻³ solution.
   b) Calculate the concentration of the diluted solution of HNO_3.
   c) Calculate the concentration of the original solution of HNO_3.

Calculations involving gases

1 mole of any gas occupies a volume of $24\,dm^3$ ($24\,000\,cm^3$) at room temperature and pressure. This is called the molar volume and is often written as V_m.

The amount, in moles, of a gas (n) is related to the volume of the gas (V) by the following expression:

$$n = \frac{\text{gas volume}}{V_m} \quad \text{or this can be simplified to } n = \frac{V}{V_m}$$

V_m is 24 000 where the gas volume (V) is measured in cm^3. V_m is 24 where the gas volume (V) is measured in dm^3.

The expression can be rearranged to:

$$\text{gas volume } (V) = n \times V_m$$

EXAMPLE 30

4 g of dry ice (solid carbon dioxide) is allowed to expand to room temperature. Calculate the volume of carbon dioxide formed in dm^3. Give your answer to 3 decimal places.

Answer

M_r of CO_2 = 12.0 + 2(16.0) = 44.0

Amount of CO_2 (in moles) $= \dfrac{m}{M_r} = \dfrac{4}{44.0} = 0.09091\,mol$ (to 4 significant figures)

As units of volume in dm^3, $V_m = 24\,dm^3$.

gas volume (V) $= n \times V_m = 0.09091 \times 24 = 2.182\,cm^3$ (to 3 decimal places)

EXAMPLE 31

0.1 g of magnesium ribbon reacts completely with hydrochloric acid according to the equation below:

$Mg(s) + 2HCl(aq) \rightarrow MgCl_2(aq) + H_2(g)$

Calculate the volume of hydrogen produced in cm^3. Give your answer to 1 decimal place.

Answer

This calculation required the calculation of the amount of Mg in moles, using the ratio in the equation of Mg:H_2 and finally calculating the gas volume of hydrogen in cm^3 using the $V_m = 24\,000\,cm^3$.

A_r of Mg = 24.3

Amount of Mg (in moles) $= \dfrac{m}{A_r} = \dfrac{0.1}{24.3} = 0.004115\,mol$ (to 4 significant figures)

Amount of H_2 (in moles) = 0.004115 mol (1:1 ratio in equation)

$V_m = 24\,000\,cm^3$ (as answer required in cm^3)

Volume of H_2 = $n \times V_m$ = 0.004115 × 24 000 = 98.76 cm^3

Answer to 1 decimal place = 98.8 cm^3.

The Ideal Gas equation

As pressure increases at constant temperature, the volume of a gas decreases.

As temperature increases at constant pressure, the volume of a gas increases.

The amount of a gas in moles is directly proportional to its volume.

The ideal gas equation relates all these factors and includes the gas constant, R.

The ideal gas equation is $pV = nRT$ where p is pressure (measured in Pascals, Pa) V is gas volume (measured in cubic metres, m^3), n is the number of moles of the gas, R is the gas constant ($R = 8.31 \, J \, K^{-1} \, mol^{-1}$) and T is the temperature measured in kelvin (K).

Units in the Ideal Gas equation

Temperature

The kelvin temperature scale is an absolute scale. Temperatures in kelvin are written as 273 K. There is no need for a degree symbol when using kelvin. Zero kelvin (0 K) is called absolute zero and is the temperature at which all movement of particles stops. Temperatures in kelvin are either zero or a positive number. A rise of 1 °C is the same as a rise of 1 K. Absolute zero (0 K) is equivalent to −273.15 °C. This is usually taken as −273 °C. To convert temperatures from °C to kelvin, add 273. To change from kelvin to °C, subtract 273.

For example:

Convert 0 °C to kelvin

Add 273; 0 °C is 273 K

or

Convert 1000 K to °C.

Subtract 273; 1000 K is 727 °C.

Pressure

There are many different units of pressure. The units which are usually used in the ideal gas equation are Pascals. The symbol for a Pascal is Pa. 1 Pa = 1 Nm^{-2} (Newton per square metre). 101325 Pa is equal to 1 atmosphere pressure. Often kilopascals are used (kPa). When the gas constant is quoted as 8.31 J K^{-1} mol^{-1}, the units of pressure must be Pa.

Volume

As seen previously with the solution calculations, cm^3 and dm^3 are common units of volume and gas volumes may also be measured in these units. In the ideal gas equation, the gas volume (V) must be measured in m^3. $1 \, m^3 = 1000 \, dm^3 = 1\,000\,000 \, cm^3$.

> **TIP**
> Converting between different units is very important. 100 cm = 1 m so when dealing with cubic quantities (100 cm)^3 = (1 m)^3 so 1 000 000 cm^3 = 1 m^3 as 100^3 = 1 000 000. Also 10 dm = 1 m so when dealing with cubic quantities (10 dm)^3 = (1 m)^3 so 1000 dm^3 = 1 m^3.

Moles

The units of n are mol. This may be calculated from a mass of a gas or be given as an amount, in moles.

R, the gas constant

The gas constant can have various values depending on the units used for pressure and volume but $8.31\,J\,K^{-1}\,mol^{-1}$ is most often used where volume is measured in m^3 and pressure is measured in Pa.

Using the Ideal Gas equation

The ideal gas equation allows conversion between moles of gas and volume at a particular temperature and pressure.

> **TIP**
>
> If $n = 1$ (1 mole of a gas) and $T = 298\,K$ (room temperature of 25 °C), $p = 101\,325\,Pa$ (atmospheric pressure),
>
> $$V = \frac{nRT}{p} = \frac{1 \times 8.31 \times 298}{101325} = 0.02445\,m^3.$$
>
> $0.02445\,m^3 = 24.45\,dm^3$, which is why $24\,dm^3$ is used as the molar gas volume at room temperature and pressure.

> **EXAMPLE 32**
>
> Calculate the volume, in dm^3, of a sample of chlorine gas if there are 0.217 moles of chlorine gas at temperature of 350 K and a pressure of 200 kPa. (The gas constant $R = 8.31\,J\,K^{-1}\,mol^{-1}$.) Give your answer to 3 significant figures.
>
> **Answer**
>
> All of the above quantities are in the correct units to give a volume in m^3.
>
> p = 200 000 Pa
>
> T = 350 K
>
> n = 0.217 mol
>
> R = $8.31\,J\,K^{-1}\,mol^{-1}$
>
> $pV = nRT; V = \dfrac{nRT}{p}$
>
> $V = \dfrac{0.217 \times 8.31 \times 350}{200\,000}$
>
> $V = \dfrac{631.1445}{200\,000} = 0.003156\,m^3$ (to 4 significant figures)
>
> V = 0.003156 × 1000 = 3.16 dm^3

TIP

As mentioned earlier this type of reaction where two different gases are produced is common and you just need to combine the total number of moles of gas produced in the balanced equation.

TIP

The temperature is quoted in °C but +273 will convert to K so the temperature used in the calculation is 300 K. See Chapter 17 for more information.

EXAMPLE 33

Calculate the total volume of gas produced in m^3 at 27 °C and 100 kPa when 0.254 moles of calcium nitrate are heated to a constant mass.

(The gas constant R = 8.31 J K^{-1} mol^{-1}.) Give your answer to 3 significant figures.

Answer

$2Ca(NO_3)_2(s) \rightarrow 2CaO(s) + 4NO_2(g) + O_2(g)$

From the balanced equation: 2 moles of $Ca(NO_3)_2$ produces 5 moles of gas $(4NO_2 + O_2)$.

0.254 mol of $Ca(NO_3)_2 = \frac{0.254}{2} \times 5 = 0.635$ mol of gas

$pV = nRT$

$p = 100\text{ kPa} = 100\,000\text{ Pa}$

$R = 8.31\text{ J K}^{-1}\text{ mol}^{-1}$

$n = 0.635\text{ mol}$

$T = 300\text{ K}$

$V = \frac{nRT}{p}$

$V = \frac{0.635 \times 8.31 \times 300}{100\,000}$

$V = \frac{1583.055}{100\,000} = 0.0158\text{ m}^3$ (to 3 significant figures)

EXAMPLE 34

By using gas volumes, this type of calculation may be reversed as shown in the following example.

Boron trichloride (BCl_3) may be prepared as shown in the equation below.

$B_2O_3(s) + 3C(s) + 3Cl_2(g) \rightarrow 2BCl_3(g) + 3CO(g)$

A sample of boron oxide (B_2O_3) was reacted completely with carbon and chlorine.

The two gases produced occupied a total volume of 1250 cm^3 at a pressure of 110 kPa and a temperature of 375 K.

Calculate the mass of boron oxide that reacted. Give your answer to 3 significant figures. The gas constant R = 8.31 J K^{-1} mol^{-1}

Answer

The first step is to calculate the total number of moles of gas that were produced using the ideal gas equation.

$pV = nRT$

$p = 110\text{ kPa} = 110\,000\text{ Pa}$

$V = 1250\text{ cm}^3 = 0.00125\text{ m}^3$ (1250/1 000 000)

$R = 8.31\text{ J K}^{-1}\text{ mol}^{-1}$

$T = 375\,K$

$pV = nRT$

$n = \dfrac{pV}{RT} = \dfrac{110\,000 \times 0.00125}{8.31 \times 375}$

$n = \dfrac{137.5}{3116.25} = 0.04412\,mol$ (to 4 significant figures)

5 mol of gas in the equation ($2BCl_3 + 3CO$) formed from 1 mol of B_2O_3

0.04412 mol of gas is formed from $\dfrac{0.04412}{5} = 0.008824\,mol\ B_2O_3$

M_r of $B_2O_3 = 2(10.8) + 3(16.0) = 69.6$

Mass of $B_2O_3 = n \times M_r = 0.008824 \times 69.6 = 0.6142\,g$ of B_2O_3

Answer is 0.614 g of B_2O_3 to 3 significant figures.

TIP

1250 cm³ in m³ can be put into your calculator as 1250 × 10⁻⁶, usually 1250 EXP (or EE) −6.

TEST YOURSELF 9

1 Calculate the volume in dm^3 that 0.726 mol of nitrogen occupies at a temperature of 200 K and a pressure of 150 kPa. (The gas constant, $R = 8.31\,J\,K^{-1}\,mol^{-1}$). Give your answer to 3 significant figures.

2 Calculate the mass of ammonia in g which has a volume of 20.0 dm^3 at a pressure of 100 kPa and a temperature of 298 K. (The gas constant, $R = 8.31\,J\,K^{-1}\,mol^{-1}$). Give your answer to 3 significant figures.

3 Thionyl chloride ($SOCl_2$) reacts with water to form hydrogen chloride and sulfur dioxide. (The gas constant, $R = 8.31\,J\,K^{-1}\,mol^{-1}$).

$SOCl_2(l) + H_2O(l) \rightarrow SO_2(g) + 2HCl(g)$

2.00 g of $SOCl_2$ reacts completely with water.
a) Calculate the amount, in moles of $SOCl_2$ used.
b) Calculate the total amount, in moles, of gases produced.
c) Calculate the volume of gas present in dm^3 at 300 K and 120 kPa pressure. Give your answer to 3 significant figures.

Determining empirical and molecular formulae

- The formula which is determined from experimental mass (or percentage) data is called the empirical formula.
- The empirical formula is the simplest whole number ratio of the atoms of each element in a compound.
- The molecular formula is the actual number of the atoms of each element in a compound.
- The molecular formula is a simple multiple of the empirical formula.

TIP

Remember to use the M_r to determine how many times the empirical formula you need for the molecular formula. Also remember to cancel down the number of each type of atom to its lowest number to determine the empirical formula.

EXAMPLE 35

The empirical formula of a compound is CH_2O but its M_r is 180.0. Determine the molecular formula of the compound.

Answer

The M_r of CH_2O is 30.0.

So $6 \times CH_2O$ ($6 \times 30.0 = 180.0$) must be present in the compound so the molecular formula is $C_6H_{12}O_6$.

EXAMPLE 36

The molecular formula of a compound is $Na_2S_4O_6$. Determine the empirical formula of the compound.

Answer

The simplest whole number ratio of the atoms is found by dividing the number of each type of atom by 2.

The empirical formula is NaS_2O_3.

Determining formulae of simple compounds

Simple compounds are formed from two elements, for example, sodium chloride and magnesium oxide. However you must also be able to use percentage information by mass to determine the empirical formula of a simple compound.

You can calculate the amount, in moles, of the atoms of each element by dividing the percentage or mass of the element by its A_r (always use the A_r for these type of calculations with elements).

- The amounts in moles are converted to a simple ratio – this is best achieved by making the lowest mole value $= 1$ and then dividing though the other mole values by the lowest mole value.
- In some examples, you may be given the mass of the elements which combine and in other examples you may be given the mass of the compound formed (a simple subtraction will calculate the mass of the second element).
- You also need to be able to plan practically how to carry out these experiments to determine the formula of a simple compound. Most of the experiments involve heating to constant mass but full practical details and apparatus required may be expected.

TIP

Take care with diatomic elements like chlorine and oxygen. Use 35.5 as the A_r of chlorine atoms and 16.0 as the A_r of oxygen atoms. Mistakes are made most often using 71.0 and 32.0 to calculate the amount, in moles, of the elements. It is the amount, in moles, of the atoms that you are trying to find not the amount of the diatomic molecules.

EXAMPLE 37

1.06 g of magnesium combines with oxygen to give 1.76 g of magnesium oxide, calculate the formula of the oxide of magnesium.

Answer

1 Find the mass of the empty crucible: 16.18 g **(1)**
2 Find the mass of the crucible and some magnesium: 17.24 g **(2)**
3 Mass of magnesium = **(2)** – **(1)** = 17.24 – 16.18 = 1.06 g **(3)**
4 Find the mass of the crucible after heating to a constant mass: 17.94 g **(4)**
5 Mass of oxygen combined = **(4)** – **(2)** = 17.94 – 17.24 = 0.70 g

We can now calculate the formula of the oxide of magnesium.

Element	Magnesium	Oxygen
Mass (g)	1.06	0.70
A_r	24.3	16.0
Moles	$\frac{1.06}{24.3} = 0.04362$	$\frac{0.70}{16.0} = 0.04375$
Ratio (÷ 0.04362)	1	1.003 (=1)
Empirical Formula	MgO	

TIP

The masses are measured to 2 decimal places which may allow for some error. This is why the ratio is not exactly 1:1 so do allow some room for error but not too much. For example 1.329 is most likely 1.333 so the moles should be multiplied by 3 to achieve a whole number ratio.

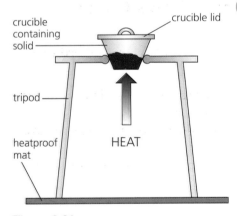

crucible containing solid
crucible lid
tripod
heatproof mat
HEAT

Figure 2.21

EXAMPLE 38

A sample of solid phosphorus was burned in excess oxygen. 0.775 g of phosphorus reacted with 1.00 g of oxygen.

1 Calculate the empirical formula of the oxide of phosphorus formed.
2 Given that the M_r of the oxide of phosphorus is 284, calculate the molecular formula of the oxide.

Answers

1 In this example the masses of the two elements are given directly.

Element	Phosphorus	Oxygen
Mass (g)	0.775	1.00
A_r	31.0	16.0
Moles	$\frac{0.775}{31.0} = 0.0250$	$\frac{1.00}{16.0} = 0.0625$
Ratio (÷ 0.025)	1	2.5
	The ratio works out at 1:2.5 but both are multiplied by 2 to give whole numbers	
	2	5
Empirical Formula	P_2O_5	

2 The M_r of P_2O_5 is 142 and the M_r of the oxide is 284.
So 2 × P_2O_5 must be present in the compound so the molecular formula is P_4O_{10}.

EXAMPLE 39

A compound containing chlorine and oxygen contains 61.2% by mass of oxygen. Calculate the empirical formula of the compound.

Answer

In a percentage by mass question, we assume 100 g of the compound are present so in this example oxygen makes up 61.2 g of the 100 g and the rest of the mass (100 − 61.2 = 38.8 g) is due to chlorine.

Element	Chlorine	Oxygen
Mass (g)	38.8	61.2
A_r	35.5	16.0
Moles	$\dfrac{38.8}{35.5} = 1.093$	$\dfrac{61.2}{16.0} = 3.825$
Ratio (÷ 1.093)	1	3.4995 (= 3.5)
	The ratio works out at 1:3.5 but both are multiplied by 2 to give whole numbers	
	2	7
Empirical Formula	Cl_2O_7	

Determining degree of hydration by heating to constant mass

The method of determining simple formula can also be applied to hydrated compounds.

- If hydrated compounds are heated they lose water of crystallisation so their mass decreases and the anhydrous compound is formed.
- By using the mass of the anhydrous compound and the mass of water lost the degree of hydration can be determined.
- The assembled apparatus used to heat hydrated compounds is shown in Figure 2.22.

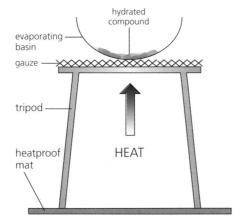

Figure 2.22

EXAMPLE 40

Given that 3.12 g of hydrated copper(II) sulfate, $CuSO_4 . nH_2O$, produces 2.00 g of the anhydrous copper(II) sulfate $CuSO_4$ on heating to constant mass, find the value of n in the formula of the hydrated salt.

Answer

Mass of anhydrous salt $(CuSO_4)$ = 2.00 g

Mass of water lost (H_2O) = 3.12 – 2.00 = 1.12 g

Compound	Copper(II) sulfate	Water
Formula	$CuSO_4$	H_2O
Mass (g)	2.00 g	1.12 g
M_r	159.6	18.0
Moles	$\dfrac{2.00}{159.6} = 0.01253$	$\dfrac{1.12}{18.0} = 0.06222$
Ratio (÷ 0.01253)	1	4.96 (=5)
Empirical Formula	$CuSO_4.5H_2O$	

You can see from the empirical formula that the value of n = 5.

TEST YOURSELF 10

1 An oxide of sulfur contains 40% sulfur by mass. Calculate the empirical formula of the oxide.
2 6.57 g of hydrated nickel(II) sulfate, $NiSO_4.xH_2O$, were heated to constant mass. 3.87 g of solid remained at the end. Calculate the value of x in $NiSO_4.xH_2O$.
3 Calculate the empirical formula of an oxide of lead which contains 90.67% of lead by mass.
4 A compound of nitrogen and hydrogen contains 87.5% nitrogen by mass. The M_r of the compound is 32.0.
   a) Calculate the empirical formula of the compound.
   b) Calculate the molecular formula of the compound.

Atom economies

Atom economy is a measure of how efficiently the atoms in the reactants are used in a chemical reaction.

It can be calculated as a percentage using the following expression:

$$\% \text{ atom economy} = \frac{\text{molecular mass of desired product}}{\text{sum of molecular masses of all reactants}} \times 100$$

EXAMPLE 41

Calculate the percentage atom economy in the addition of bromine to cyclohexene.

Answer

$$C_6H_{10} \quad + \quad Br_2 \quad \rightarrow \quad C_6H_{10}Br_2$$

cyclohexene   bromine   1,2-dibromocyclohexane

This is an addition reaction and by their very nature, they are atom economical because there is only one product.

M_r of reactants		M_r of desired product	
C_6H_{10}	82.0	$C_6H_{10}Br_2$	241.8
Br_2	159.8	-	-
Total	241.8	Total	241.8

% atom economy $= \dfrac{241.8}{241.8} \times 100 = 100\%$

EXAMPLE 42

In the blast furnace, carbon monoxide is used to reduce iron oxide to iron.

Answer

$$Fe_2O_3 \quad + \quad 3CO \quad \rightarrow \quad 2Fe \quad + \quad 3CO_2$$

iron oxide   carbon monoxide   iron   carbon dioxide

Calculate the percentage atom economy of this reaction if iron is the desired product. Give your answer to 3 significant figures.

M_r of reactants		M_r of desired product	
Fe_2O_3	159.6	2Fe	111.6
3CO	84.0	–	–
Total	243.6	Total	111.6

% atom economy $= \dfrac{111.6}{243.6} \times 100 = 45.8\%$

TIP

Chemists now look at having a high percentage yield but also a high atom economy to reduce waste. This is the main thrust of Green Chemistry.

Understanding atom economy and percentage yield

Chemists often use percentage yield to determine the efficiency of a chemical synthesis process. A high percentage yield would indicate that the reaction process is efficient in converting reactants into products. This is important for profit but percentage yield does not take into account any waste products.

Chemistry, like all other industries, is concerned about its effect on the environment, particularly when it comes to waste. A reaction may have a high percentage yield but have a low atom economy. This will mean that other products in the reaction would be waste and with a high percentage yield there are just more of them.

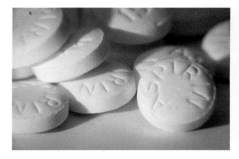

TIP

The mass of salicylic acid can be determined in a tablet by a simple titration with an alkali such as sodium hydroxide solution and using phenolphthalein indicator. Aspirin is a weak monoprotic acid.

Figure 2.23 Aspirin tablets contain the drug salicylic acid. The manufacture of salicylic acid has a 76% atom economy.

ACTIVITY

Finding the formula of titanium oxide

To determine the empirical formula of an oxide of titanium, some titanium metal was heated in a stream of oxygen as shown in Figure 2.24.

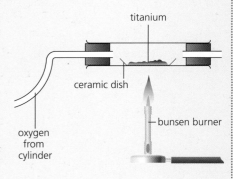

Figure 2.24 Formation of titanium oxide by heating titanium in oxygen.

1 Describe a test which could be carried out, before the cylinder was used, to prove that the gas in it was oxygen.

2 What weighings would be made, before heating to determine the mass of titanium used?

3 In this reaction the titanium could be a solid lump or powdered. State and explain if there is any advantage in using titanium powder.

4 The ceramic container and its contents are repeatedly weighed, heated, reweighed and heated. Suggest and explain the trend in expected results.

5 What safety precautions should be taken in this experiment?

6 How would the reliability of the results be improved?

7 Water vapour reacts with hot titanium to produce titanium oxide and hydrogen. Suggest how you could modify the apparatus to remove any traces of water vapour from the oxygen supply.

8 In this experiment 4.8 g of titanium was oxidised to form 8.0 g of titanium oxide. Deduce the empirical formula of the titanium oxide and suggest its systematic name.

9 a) Titanium oxide is found in the naturally occurring ore rutile. Titanium metal can be extracted from this ore by two different methods. One uses a more reactive metal to displace the titanium:

$$TiO_2 + 2Mg \rightarrow Ti + 2MgO$$

The second method is electrolysis of the ore. The overall reaction for this method is:

$$TiO_2 \rightarrow Ti + O_2$$

Calculate the atom economy for each reaction.

b) Discuss which method is 'greener'? What else might you want to know before making a final decision?

c) Oxygen is a useful product and can be sold. What is the atom economy of the electrolysis if the oxygen is collected and sold?

d) State and explain which method might be more expensive to run in industry.

e) Titanium is used for replacement hip joints. What properties must titanium have to be used in this way?

Figure 2.25 The mineral rutile (titanium oxide), a major ore of the metal titanium. This specimen is from Graves Mountain, Lincoln County, Georgia, USA.

TEST YOURSELF 11

1 Ammonia can be produced by the reaction of water with magnesium nitride.

$$Mg_3N_2(s) + 6H_2O(l) \rightarrow 3Mg(OH)_2(s) + 2NH_3(g)$$

Calculate the percentage atom economy in this reaction if ammonia is the useful product. Give your answer to 3 significant figures.

2 Calculate the percentage atom economy of the following reaction in which hydrogen fluoride is the useful product. Give your answer to 3 significant figures.

$$CaF_2(s) + H_2SO_4(l) \rightarrow CaSO_4(s) + 2HF(g)$$

3 Which of the following reactions gives the highest atom economy for the preparation of magnesium sulfate?

a) $Mg + H_2SO_4 \rightarrow MgSO_4 + H_2$

b) $MgO + H_2SO_4 \rightarrow MgSO_4 + H_2O$

c) $MgCO_3 + H_2SO_4 \rightarrow MgSO_4 + CO_2 + H_2O$

Combined calculations

Some calculations may involve aspects of many different types of calculations where mass, gas volumes, solution volumes and concentrations, pV = nRT are combined.

Percentage yield and reacting mass

9.33 g of zinc carbonate reacts with excess hydrochloric acid.

$$ZnCO_3(s) + 2HCl(aq) \rightarrow ZnCl_2(aq) + CO_2(g) + H_2O(l)$$

1 Calculate the amount, in moles, of $ZnCO_3$ in 9.33 g of zinc carbonate.

M_r of $ZnCO_3 = 65.4 + 12.0 + 3(16.0) = 125.4$

$$\text{moles of } ZnCO_3 = \frac{m}{M_r} = \frac{9.33}{125.4} = 0.07440 \text{ mol}$$

2 Calculate the mass of $ZnCl_2$ produced in this reaction assuming an 85% yield.

ratio of $ZnCO_3:ZnCl_2 = 1:1$ so moles of $ZnCl_2 = 0.07440\,mol$

M_r of $ZnCl_2 = 65.4 + 2(35.5) = 136.4$

Theoretical yield of $ZnCl_2 = 0.07440 \times 136.4 = 10.148\,g$ (to 3 decimal places)

$$\text{percentage yield} = \frac{\text{actual yield}}{\text{theoretical yield}} \times 100$$

$$\text{actual yield} = \frac{\text{percentage yield} \times \text{theoretical yield}}{100}$$

$$= \frac{85 \times 10.148}{100} = 8.626\,g \text{ (to 3 decimal places)}$$

Solution volume and $pV = nRT$

Solid lead reacts with warm nitric acid according to equation:

$$3Pb(s) + 8HNO_3(aq) \rightarrow 3Pb(NO_3)_2(aq) + 2NO(g) + 4H_2O(l)$$

20.0 cm^3 of 2.40 mol dm^{-3} nitric acid reacted completely with solid lead. Some lead was left over.

1 Calculate the amount, in moles, of HNO_3 used.

$$\text{moles of } HNO_3 = \frac{v \times c}{1000} = \frac{20.0 \times 2.40}{100} = 0.0480\,mol$$

2 Calculate the amount, in moles of NO formed.

ratio of $HNO_3:NO = 8:2 = 4:1$

$$\text{moles of NO formed} = \frac{0.0480}{4} = 0.0120\,mol$$

3 Calculate the volume of NO in cm^3 formed at 500 K and 120 kPa (The gas constant, R = 8.31 J K^{-1} mol^{-1}). Give your answer to 3 significant figures.

$pV = nRT$

$p = 120\,kPa = 120000\,Pa$

$n = 0.0120\,mol$

$R = 8.31\,J\,K^{-1}\,mol^{-1}$

$T = 500\,K$

$120000 \times V = 0.0120 \times 8.31 \times 500$

$120000 \times V = 49.86$

$V = \dfrac{49.86}{120000} = 0.0004155\,m^3$

V in cm^3 = V in m^3 $\times 10^6 = 0.0004155 \times 10^6 = 416\,cm^3$

Practice questions

1 Which of the following is the empirical formula of an oxide of manganese that contains 36.8% oxygen by mass.

   A MnO **B** MnO_2

   C Mn_2O_3 **D** Mn_3O_4 *(1)*

2 The following table of results was obtained from titration experiments.

Titration number	Initial volume /cm³	Final volume /cm³
1	0.40	21.40
2	21.40	43.10
3	0.00	21.30
4	21.30	42.70

Which titrations should be used to calculate the average titre?

   A all the titrations   **B** 1 and 2

   C 2 and 4   **D** 3 and 4   *(1)*

3 Which of the following is the mass of H_2SO_4 formed from 2.5 kg of sulfur, assuming a 64% yield and all other reactants are in excess?

$$2S(s) + 3O_2(g) + 2H_2O(l) \rightarrow 2H_2SO_4(aq)$$

   A 3.82 kg   **B** 4.89 kg

   C 7.64 kg   **D** 9.78 kg   *(1)*

4 Which of the following pieces of apparatus would have the highest percentage error in the measurements shown?

   A A volume of 25 cm³ measured using a measuring cylinder which has an error of ± 0.5 cm³

   B A temperature of 45.5 °C measured using a thermometer which has an error of 0.5 °C

   C A mass of 1.20 g measured using a balance with an error of ± 0.001 g

   D A volume of 25 cm³ measured using a pipette which has a percentage error of ± 0.3 cm³.   *(1)*

5 1.24 g of phosphorus were burned completely in oxygen to give 2.84 g of phosphorus oxide.

   a) Calculate the empirical formula of the oxide.   *(3)*

   b) Calculate the molecular formula of the oxide given that 1 mole of the oxide weighs 284.0 g.   *(1)*

6 Write a balanced equation for the thermal decomposition of strontium nitrate into strontium oxide, nitrogen(IV) oxide and oxygen.   *(1)*

7 Calculate the number of molecules of water present in 0.1 g. (The Avogadro constant, $L = 6.02 \times 10^{23}$).   *(2)*

8 Calcium sulfate reacts with carbon to form calcium sulfide according to the equation:

$$CaSO_4(s) + 4C(s) \rightarrow CaS(s) + 4CO(g)$$

   a) Calculate the mass of carbon required to react with 250 kg of calcium sulfate.   *(3)*

   b) Calculate the percentage atom economy of this reaction if calcium sulfide is the desirable product.   *(2)*

9 Ammonia reacts with chlorine according to the equation:

$$8NH_3(aq) + 3Cl_2(g) \rightarrow 6NH_4Cl(s) + N_2(g)$$

25.0 cm³ of 7.80 mol dm⁻³ ammonia solution react completely with chlorine. Give all answers to 3 significant figures.

   a) Calculate the amount, in moles, of NH_3 in 25.0 cm³ of 7.80 mol dm⁻³ solution.   *(1)*

   b) Calculate the amount, in moles, of N_2 gas formed.   *(1)*

   c) Calculate the volume of N_2 gas, in cm³, at 298 K and 100 kPa pressure. (The gas constant, $R = 8.31\,J\,K^{-1}\,mol^{-1}$).   *(3)*

10 A sample of 1.50 g of phosphorus was heated in chlorine in the form of phosphorus(V) chloride, PCl_5. The phosphorus reacts according to the equation

$$P_4(s) + 10Cl_2(g) \rightarrow 4PCl_5(s)$$

8.34 g of PCl_5 were obtained. Calculate the percentage yield. Give your answer to 3 significant figures.   *(4)*

11 A solution of ethanoic acid (CH_3COOH) is diluted by placing $10.0\,cm^3$ of the solution into a $250\,cm^3$ volumetric flask and making the volume up using deionised water. $25.0\,cm^3$ of this diluted solution were placed in a conical flask and titrated against $0.200\,mol\,dm^{-3}$ potassium hydroxide (KOH) solution. $15.7\,cm^3$ of the potassium hydroxide solution were required for neutralisation.

$$CH_3COOH(aq) + KOH(aq) \rightarrow$$
$$CH_3COOK(aq) + H_2O(l)$$

a) Calculate the amount, in moles, of KOH used in this titration. *(1)*

b) Calculate the amount, in moles, of CH_3COOH that reacted with KOH. *(1)*

c) Calculate the concentration, in $mol\,dm^{-3}$, of the diluted CH_3COOH solution. *(1)*

d) Calculate the concentration, in $mol\,dm^{-3}$, of the undiluted CH_3COOH solution. Give your answer to 3 significant figures. *(1)*

12 $1.00\,g$ of impure calcium carbonate was reacted with $50.0\,cm^3$ of $1.00\,mol\,dm^{-3}$ hydrochloric acid. Once the reaction is finished the solution is placed in a $250\,cm^3$ volumetric flask and the volume made up to $250\,cm^3$ using deionised water. A $25.0\,cm^3$ sample of this solution is pipetted into a conical flask and titrated against $0.100\,mol\,dm^{-3}$ sodium hydroxide solution. The average titre was determined to be $34.0\,cm^3$. Calculate the percentage purity of the calcium carbonate to 3 significant figures. *(7)*

13 A sample of hydrated sodium carbonate, $Na_2CO_3.xH_2O$, was heated to constant mass in an evaporating basin. The measurements below are taken at 5 minute intervals.

Mass of evaporating basin $= 122.400\,g$

Mass of evaporating basin and hydrated sample $= 122.900\,g$

Mass of evaporating basin and sample after 5 minutes heating $= 122.714\,g$

Mass of evaporating basin and sample after 10 minutes heating $= 122.612\,g$

Mass of evaporating basin and sample after 15 minutes heating $= 122.612\,g$

a) Calculate the mass of anhydrous sodium carbonate present at the end of the experiment. *(1)*

b) Calculate the amount, in moles, of anhydrous sodium carbonate present at the end of the experiment. *(1)*

c) Calculate the mass of water lost by heating. *(1)*

d) Calculate the amount, in moles, of water lost by heating. *(1)*

e) Determine the value of x in $Na_2CO_3.xH_2O$. *(1)*

14 Lead reacts with nitric acid to form lead(II) nitrate according to the equation.

$$3Pb(s) + 8HNO_3(aq) \rightarrow 3Pb(NO_3)_2(aq) + 2NO(s) + 4H_2O(l)$$

a) Calculate the percentage atom economy if lead(II) nitrate is the desirable product. *(2)*

b) The amount of lead(II) nitrate produced was $0.522\,mol$.

i) Calculate the mass of lead required to produce 0.522 moles of lead(II) nitrate if the nitric acid was in excess. *(2)*

ii) Calculate the volume, in dm^3, of $0.522\,mol$ of NO at $62\,°C$ and $125\,kPa$ pressure. (The gas constant, $R = 8.31\,J\,K^{-1}\,mol^{-1}$.) Give your answer to 3 significant figures. *(3)*

15 $0.0500\,g$ of magnesium ribbon reacts completely with $0.100\,mol\,dm^{-3}$ hydrochloric acid according to the equation:

$$Mg(s) + 2HCl(aq) \rightarrow MgCl_2(aq) + H_2(g)$$

a) Calculate the volume of $0.200\,mol\,dm^{-3}$ hydrochloric acid required to react completely with the magnesium. Give your answer to 3 significant figures. *(3)*

b) Calculate the volume of hydrogen gas, in cm^3, produced in this reaction at $20.0\,°C$ and $1.10 \times 10^5\,Pa$. (The gas constant, $R = 8.31\,J\,K^{-1}\,mol^{-1}$). Give your answer to 3 significant figures. *(4)*

3

Bonding

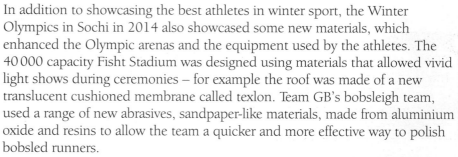

TEST YOURSELF ON PRIOR KNOWLEDGE 1

1 State the name of the following ions:
  a) N^{3-} **b)** Ca^{2+} **c)** S^{2-} **d)** Br^- **e)** Rb^+ **f)** H^-

2 What is meant by a covalent bond?

3 Name the compounds formed between the following elements:
  a) sodium and oxygen   **b)** calcium and fluorine
  c) magnesium and iodine   **d)** potassium and chlorine

4 State how many covalent bonds the following atoms can form:
  a) fluorine **b)** carbon **c)** oxygen

Nature of ionic, covalent and metallic bonds

In addition to showcasing the best athletes in winter sport, the Winter Olympics in Sochi in 2014 also showcased some new materials, which enhanced the Olympic arenas and the equipment used by the athletes. The 40 000 capacity Fisht Stadium was designed using materials that allowed vivid light shows during ceremonies – for example the roof was made of a new translucent cushioned membrane called texlon. Team GB's bobsleigh team, used a range of new abrasives, sandpaper-like materials, made from aluminium oxide and resins to allow the team a quicker and more effective way to polish bobsled runners.

To design new materials chemists need information about the structure – how the atoms or ions in the substance are arranged; and the bonding – how they are held together.

Many materials are pure substances. All pure substances may be classified as elements or compounds. The diagram below shows the main subdivisions of all pure substances. The type of bonding and structure shown by each type of substance with some common examples are also given.

Figure 3.1 Fisht Stadium in Sochi, Russia where the 2014 Winter Olympics were held. The roof is made of a new material called texlon, of which each layer has been engineered to transmit, reflect or scatter the image, enabling the roof to double as a visual display as shown above.

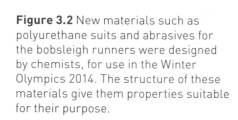

Figure 3.2 New materials such as polyurethane suits and abrasives for the bobsleigh runners were designed by chemists, for use in the Winter Olympics 2014. The structure of these materials give them properties suitable for their purpose.

The main type of structures (metals, ionic compounds, molecular covalent and macromolecular elements and compounds) will be examined. Semi-metals can also be called metalloids and they have properties of both metals and non-metals. Silicon has a macromolecular structure.

Ionic bonding

Ionic compounds are formed because of a transfer of electrons usually from metal atoms (which lose electrons from their outer energy levels) to non-metal atoms (which gain electrons to their outer energy levels).

- Metal atoms lose electrons and non-metal atoms gain electrons. This gives them a noble gas electron configuration.
- The atoms of d block elements do not always achieve a noble gas electron configuration.
- When atoms lose and gain electrons, they are no longer electrically neutral and so have acquired an overall charge. They are now called ions.
- Metal atoms lose their outer electrons and so become positively charged.
- Non-metal atoms gain electrons to fill their outer energy level and so become negatively charged.
- Positive ions are called **cations**. Negative ions are called **anions**.
- The formation of the ions can be shown using electron configurations.
- The attraction between the positive and negative ions is the ionic bond.

Formation of ions

Ionic compounds form when atoms react together and transfer electrons. It is important to understand this as it explains the charge on the ions formed.

Throughout this section you should understand that the **ionic bond** is the electrostatic attraction between oppositely charged ions.

> **TIP**
> The ionic bond is **not** the transfer of electrons. The transfer of electrons forms the ions. This is a common mistake.

EXAMPLE 1

Sodium reacts vigorously with chlorine. Explain how sodium atoms and chlorine atoms react to form ions.

Answer

sodium atom		**sodium ion**
Na	$\rightarrow$	Na$^+$
$1s^2\ 2s^2\ 2p^6\ 3s^1$		$1s^2\ 2s^2\ 2p^6$

chlorine atom		**chloride ion**
Cl	$\rightarrow$	Cl$^-$
$1s^2\ 2s^2\ 2p^6\ 3s^2\ 3p^5$		$1s^2\ 2s^2\ 2p^6\ 3s^2\ 3p^6$

Figure 3.3

- The sodium atom loses its $3s^1$ electron to achieve a noble gas electron configuration. The sodium ion is Na$^+$.

- Chlorine atoms have a $3p^5$ electron configuration and each atom requires one electron to achieve a noble gas electron configuration. The chloride ion is Cl$^-$

- Both ions have a full outer energy level of electrons.

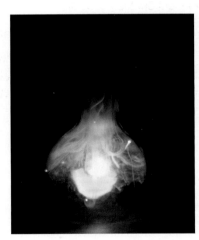

Figure 3.4 Sodium reacts vigorously with chlorine as shown in this photograph. Why?

EXAMPLE 2

Explain how atoms of magnesium react with atoms of oxygen.

Answer

magnesium atom
Mg
$1s^2\ 2s^2\ 2p^6\ 3s^2$ $\rightarrow$

magnesium ion
Mg^{2+}
$1s^2\ 2s^2\ 2p^6$

oxygen atom
O
$1s^2\ 2s^2\ 2p^4$ $\rightarrow$

oxide ion
O^{2-}
$1s^2\ 2s^2\ 2p^6$

Figure 3.5

● The magnesium atom loses its $3s^2$ electrons to form a noble gas electron configuration. The magnesium ion is Mg^{2+}.

● Oxygen atoms have a $3p^4$ electron configuration and each atom requires two electrons to achieve a noble gas electron configuration. The oxide ion is O^{2-}.

● Both ions have a full outer energy level of electrons.

EXAMPLE 3

Explain how atoms of calcium react with atoms of fluorine.

Answer

● In this example two fluorine atoms are required for each calcium atom as each calcium atom has 2 electrons to lose from $4s^2$ to obtain a noble gas configuration.

● Each fluorine atom requires one electron to achieve a noble gas configuration.

● The calcium ion is Ca^{2+} and the fluoride ion is F^-.

● Again the ions formed have a full outer energy level of electrons.

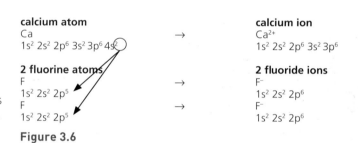

calcium atom
Ca
$1s^2\ 2s^2\ 2p^6\ 3s^2\ 3p^6\ 4s^2$ $\rightarrow$

calcium ion
Ca^{2+}
$1s^2\ 2s^2\ 2p^6\ 3s^2\ 3p^6$

2 fluorine atoms
F
$1s^2\ 2s^2\ 2p^5$ $\rightarrow$
F
$1s^2\ 2s^2\ 2p^5$ $\rightarrow$

2 fluoride ions
F^-
$1s^2\ 2s^2\ 2p^6$
F^-
$1s^2\ 2s^2\ 2p^6$

Figure 3.6

Figure 3.7 Fluorite is mineral composed of calcium fluoride and is the source of most of the world's fluorine. It is colourful both in visible and ultraviolet light and is often used ornamentally.

Unknown elements forming ionic compounds

A compound may be formed from unknown elements named X and Y and you may be told the electron configuration or the number of electrons in their outer shell.

EXAMPLE 4

Element X has 2 electrons in its outermost shell and element Y has 7 electrons in its outermost shell.

Several questions can result from this:

- What is the formula of an ion of X?

  As X has 2 electrons in its outer shell, it will lose these 2 electrons when it forms an ion so the charge on the ion is X^{2+}.

- What is the formula of an ion of Y?

  As Y has 7 electrons in its outer shell, it will gain 1 electron when it forms an ion so the charge on the ion is Y^-.

- What is the formula of the compound formed between X and Y?

  Use the charges of the ions worked out from above: X is 2+ and Y is – so $2Y^-$ are needed for each X^{2+}. The formula of the compound is XY_2.

Ionic crystals

- The ionic bond is the electrostatic attraction between the oppositely charged ions.
- The ionic solid formed has the ions held in a three-dimensional framework called an **ionic lattice**. A **lattice** is a regular repeated three-dimensional arrangement of atoms, ions, or molecules in a metal or other crystalline solid.
- The lattice for NaCl has each Na^+ ions surrounded by 6 Cl^- ions and each Cl^- ions surrounded by 6 Na^+ ions. This type of lattice is said to have a 6:6 configuration. Its lattice is described as a cubic arrangement as it is based on a cube.
- The ionic solid formed has many **strong electrostatic attractions** between the oppositely charged ions.
- The regular pattern of the ions within the structure causes the crystalline nature of ionic compounds. When sodium chloride is heated, it makes a cracking sound which is caused by the ionic crystalline structure breaking up. This is called decrepitation.

You may be asked to draw a sodium chloride lattice with a specified number of ions. Start with a Na^+ ion and surround it with 6 Cl^- ions as shown in Figure 3.9. The lines can be used to give a three-dimensional shape and show the attraction between the ions. The diagram may be extended for more ions.

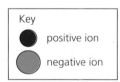

Key

● positive ion

● negative ion

↑ An ionic lattice structure

Figure 3.8 As can be seen from the diagram, each Na^+ ion is surrounded by six Cl^- ions and each Cl^- ion is surrounded by six Na^+ ions. The regular structure continues like this to form the crystal.

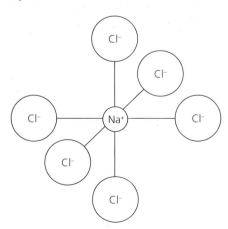

Figure 3.9 Part of an ionic lattice.

The strength of the ionic bond accounts for many of the properties of ionic substances.

1 Ionic compounds have a high melting point or boiling point, or are solid at room temperature

The energy required to melt an ionic solid is large due to the large number of strong electrostatic attractions between the positive and negative ions, which is know as ionic bonding.

The smaller the ions and the higher the charge on the ions, the stronger the ionic bond. Magnesium oxide has a melting point of 2852 °C whereas sodium chloride has a melting point of 797 °C. The attraction between the Mg^{2+} and O^{2-} ions is stronger than the attraction between the Na^+ and Cl^-. Some atomic and ionic radii measured in picometres (pm) are given below.

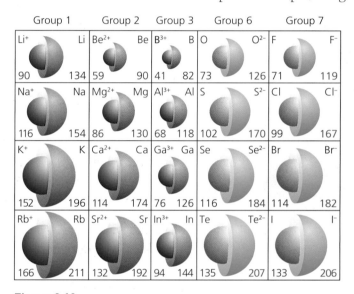

Figure 3.10

As can be seen the ionic radius of Mg^{2+} is 86 pm whereas the ionic radius of Na^+ is 116 pm. The O^{2-} ion has an ionic radius of 126 pm whereas the Cl^- ion is 167 pm. The smaller and higher charged ions in magnesium oxide result in a stronger ionic bond and a higher melting point and boiling point.

> **TIP**
> 1 pm = 10^{-12} m. There are 1 000 000 000 000 nm in 1 m. Sometimes nanometres (nm) are used. 1 nm = 10^{-9} m. There are 1000 pm in 1 nm. It is important to be able to recognise units of this scale when dealing with the size of particles.

Positive ions are generally smaller than the atoms from which they are formed. Negative ions are larger than the parent atom. This results from the metal atoms losing electrons from the outer energy level so the ion has an electron configuration with one less energy level occupied. The effective nuclear charge (ratio of protons to electrons) increases so the electrons are pulled closer to the nucleus.

For negative ions, the ion is larger than the atom as the repulsion between the electrons moves them further apart from each other. Also the effective nuclear charge decreases as there are more electrons with the same number of protons.

2 Ionic compounds are usually soluble in water

Water likes charged substances and likes to surround the ions that have broken out of the lattice. When the moving water molecules hit the ionic lattice they can knock ions off and then water molecules surround the ions.

For some ionic substances like aluminium oxide the electrostatic attraction between the positive and negative ions (the ionic bond) is so strong that water cannot break up the lattice so the compound is insoluble in water.

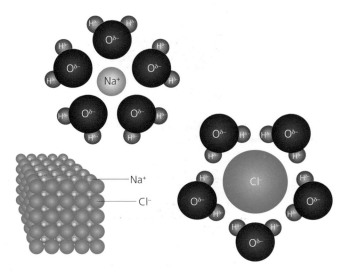

Figure 3.11 Sodium chloride dissolving in water. The sodium ions and the chloride ions become surrounded by water molecules.

TIP
Remember that ionic substances do not like to dissolve in non-polar solvents like hexane. (Polarity and non-polarity will be discussed later.)

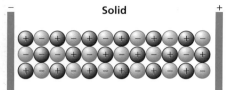

Solid

Ions fixed in lattice and cannot move

Figure 3.12

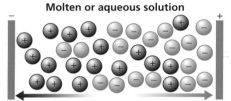

Molten or aqueous solution

Ions can now move and conduct electricity

+ ions move to negative terminal   − ions move to positive terminal

Figure 3.13

TIP

The list of ions will be added to as you progress through the course but you should learn these ions carefully. Remember you need to be able to work from a formula to a name or from a name to a formula. For example, be able to identify $(NH_4)_2SO_4$ as ammonium sulfate, but also be able to write the formula of sodium carbonate as Na_2CO_3.

3 Ionic compounds conduct electricity when molten or when in aqueous solution

In the solid state, ionic compounds cannot conduct electricity. However, in the liquid and aqueous state, the ions are free to move and carry charge. Electrical conduction is caused by the movement of charged particles and for molten ionic compounds and ionic compounds dissolved in water, it is the ions which are free to move and carry a charge.

Figures 3.12 and 3.13 show how, when molten, or dissolved, the ions in an ionic compound can move and carry charge, yet when solid they cannot.

Other ionic compounds

Some ionic compounds contain molecular ions. Molecular ions include the following ions:

Ion	Formula
Sulfate	SO_4^{2-}
Nitrate	NO_3^-
Hydroxide	OH^-

Ion	Formula
carbonate	CO_3^{2-}
hydrogen carbonate	HCO_3^-
ammonium	NH_4^+

Compounds containing molecular ions are ionic and exhibit the properties of ionic compounds. The molecular ions may contain covalent bonds within the ions but their compounds are ionic.

TEST YOURSELF 2

1 Potassium chloride is formed when potassium reacts with chlorine.
  a) Explain, using electron configurations, how atoms of potassium and atoms of chlorine form potassium ions and chloride ions.
  b) Explain why potassium chloride is soluble in water.
  c) State two properties of potassium chloride, apart from solubility, which are typical of ionic compounds.
2 Sodium hydroxide is an ionic compound.
  a) State the charge on a sodium ion.
  b) Write the electron configuration of a sodium ion.
3 Complete the table below giving either the name or formula of the ionic compound or the formulae of the ions present in the compound including charges.

Name of compound	Formula	Formula of positive ion	Formula of negative ion
		Mg^{2+}	O^{2-}
silver(I) fluoride			
		Li^+	CO_3^{2-}
		Zn^{2+}	Br^-

Nature of covalent and dative covalent bonds

A covalent bond consists of one or more shared pairs of electrons between two atoms.

Covalent bonds are found in:

a) molecular elements and compounds

   for example: Cl_2, P_4, S_8, CO_2, H_2O, CH_4, CCl_4, C_2H_6

b) macromolecular (giant) covalent elements and compounds

   for example: C (graphite and diamond), SiO_2

c) molecular ions

   for example: NH_4^+, NO_3^-, SO_4^{2-}, CO_3^{2-}, HCO_3^-, H_3O^+

A single covalent bond is a shared pair of electrons. Normally each atom provides one electron, which will have existed as an unpaired electron in an orbital. The number of unpaired electrons is mostly equal to the number of covalent bonds which the atom can form but beware of the promotion of electrons as shown in the examples which follow.

A single covalent bond is represented as a line between two atoms, for example, H—Cl.

A double covalent bond is two pairs of shared electrons. A double covalent bond is represented as a double line between two atoms, for example, O=C=O.

A triple covalent bond is three pairs of shared electrons. A triple covalent bond is represented as a triple line between two atoms, for example N≡N.

Covalent bonds exist between non-metal atoms (some exceptions do occur where metal atoms can form covalent bonds, i.e. beryllium in beryllium chloride, aluminium in aluminium chloride).

The human body contains carbohydrate, protein, fat and about 70% water. All of these molecules contain covalent bonds.

Dot and cross diagrams

Dot and cross diagrams are used to show the arrangement of electrons in covalently bonded molecules.

A shared pair of electrons may be represented as ×● to show that the two electrons in the bond are from different atoms.

Dot and cross diagrams for covalent molecules, and ions containing covalent bonds, help to determine the shape of the molecule or ion.

TIP

Dot and cross diagrams with up to 12 electrons surrounding a central atom in a molecule or ion can help to explain shape. For atoms of elements in Period 2, the maximum is eight electrons but in Period 3 and beyond the maximum can be 2 × the group number.

How atoms form covalent bonds

A dot and cross diagram shows the covalent bonding in a molecule or ion.

- Atoms use unpaired electrons in orbitals to form covalent bonds.
- The unpaired electron in an orbital of one atom can be shared with an unpaired electron in an orbital of another atom.
- Atoms can promote electrons into unoccupied orbitals in the **same energy level** to form more covalent bonds.
- Atoms may not promote electrons, so often a variety of compounds can be formed, for example phosphorus can form PCl_3 and PCl_5.

✗ Boron trifluoride

Boron trifluoride (BF_3) is shown as an example of how atoms form covalent bonds and how a dot and cross diagram and bonding diagram should be drawn.

Figure 3.14

Elements in Group 3 have a s^2p^1 arrangement of electrons in their outer shell. One of the s electrons in the outer shell can be promoted to the p sub-shell to give the element three unpaired electrons.

Boron in the ground state is shown on the left but when forming covalent bonds, boron promotes one of the paired 2s electrons to the 2p to give three unpaired electrons.

> **TIP**
> This process is called hybridisation and while it is not on the course for A level, it does explain why boron can form three covalent bonds but from its electron configuration it only has one unpaired electron.

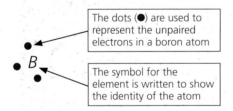

The dots (●) are used to represent the unpaired electrons in a boron atom

The symbol for the element is written to show the identity of the atom

Figure 3.15

The three unpaired electrons in boron are usually represented as shown on the left.

Other atoms with unpaired electrons can form covalent bonds with boron by sharing the unpaired electrons. Fluorine atoms have an electron configuration of $1s^2\ 2s^2\ 2p^5$. Fluorine atoms have one unpaired electron which can form a covalent bond with boron atoms forming BF_3.

> **TIP**
> The boron atom in BF_3 has an empty orbital (one of the 2p orbitals). This will be important later when coordinate bonding is discussed.

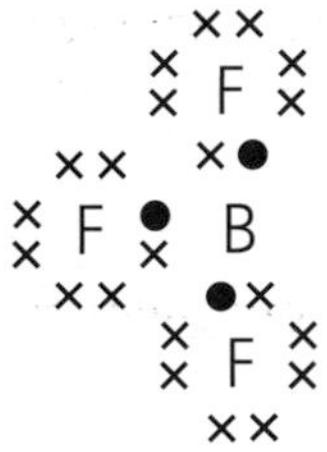

Figure 3.16

The ×● presents the covalent bond. Boron forms three single covalent bonds, one to each fluorine atom. The bonds can be shown in a bonding diagram where a single line is used to represent the covalent bonds.

This is not the shape of the BF_3 molecule but the single lines between the atoms show the single covalent bonds.

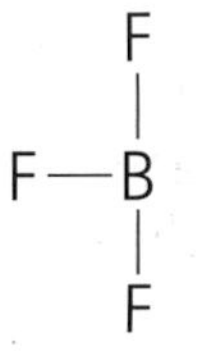

> **TIP**
> × or ● can be used for either atom in a molecule as it is simply to show the pairing of the atoms. Throughout the diagrams which follow ● has been used for the central atom.

```
      F
      |
  F — B
      |
      F
```

Figure 3.17

Beryllium chloride (BeCl₂)

Beryllium chloride is covalent. This is unusual chemistry for a Group 2 element. Beryllium atoms have an electron configuration of $1s^2 2s^2$. Beryllium atoms promote one electron from the 2s to the 2p sub-shell to give the beryllium atoms two unpaired electrons. The dot and cross diagram for beryllium chloride is shown on the left with a diagram showing the covalent bonds in a beryllium chloride molecule.

Dot and cross diagram   Bonding diagram

Figure 3.18

The beryllium atom in beryllium chloride and the boron atom in boron trifluoride do not have a complete outer shell when they form covalent bonds.

Methane (CH₄)

Carbon is in Group 4. Carbon atoms have an electron configuration of $1s^2 2s^2 2p^2$. This is shown below on the left. When forming covalent bonds, carbon atoms can promote one electron from the 2s to the 2p to give four unpaired electrons. Carbon atoms can form four covalent bonds.

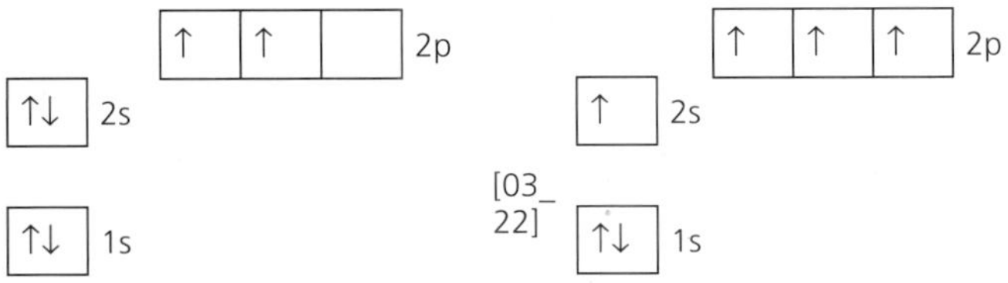

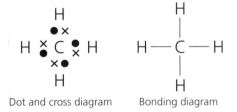

[03_22]

Figure 3.19

Hydrogen atoms have an electron configuration of $1s^1$ so can form one covalent bond by sharing the one electron.

Figure 3.21

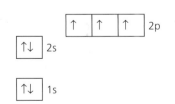

Dot and cross diagram   Bonding diagram

Figure 3.20

The photograph shows a worker at a landfill site in France. This site collects the methane gas released from household waste as it decomposes, and burns it to produce electricity.

Ammonia (NH₃)

Nitrogen is in Group 5 and atoms of nitrogen have an electron configuration of $1s^2\ 2s^2\ 2p^3$.

Nitrogen atoms have three unpaired electrons so can form three covalent bonds. Nitrogen atoms cannot promote one of the 2s electrons as there are no more sub-shells available in the second energy level. However in ammonia, the nitrogen atom will have a pair of electrons in the 2s sub-shell which are not involved in bonding. This is called a lone pair of electrons and should **always** be shown on a diagram of bonding or shape. They are mostly shown as ×× or •• on the atom.

Figure 3.22

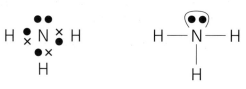

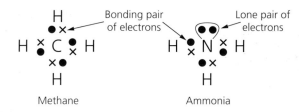

Figure 3.23 Bonding pairs of electrons and lone pairs of electrons.

Methane   Ammonia

Figure 3.24

A bonding pair of electrons is a pair of electrons shared between two atoms.

A lone pair of electrons is an unshared (non-bonding) pair of electrons.

For example in a molecule like methane there are four bonding pairs of electrons whereas an ammonia molecule has 3 bonding pairs of electrons and one lone pair of electrons.

Water (H_2O)

Oxygen is in Group 6 and has an electron configuration of $1s^2 2s^2 2p^4$.

Oxygen atoms have two unpaired electrons so can form two covalent bonds. Again oxygen atoms cannot promote one of the 2s electrons as there are no more sub-shells available in the second energy level. However in water, the oxygen atom will have two pairs of electrons in the 2s and 2p sub-shells which are not involved in bonding. So the oxygen atom in water has two lone pairs of electrons and again these should be shown as $\times\times$ or $\bullet\bullet$ on the atom.

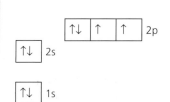

Figure 3.25

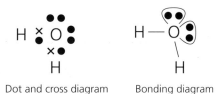

Dot and cross diagram   Bonding diagram

Figure 3.26

In all of the molecules examined so far, the central atom does not have more than eight electrons. For elements in Period 3, the outer electrons are in the third energy level and this allows the 3d sub-shell to be used for hybridisation. This means that for atoms of elements in Groups 5, 6 and 7 from Periods 3 onwards, the atoms can form a maximum of 5, 6 and 7 covalent bonds respectively.

Phosphorus pentafluoride (PF_5)

The electron configuration of atoms of phosphorus is $1s^2 2s^2 2p^6 3s^2 3p^3$.

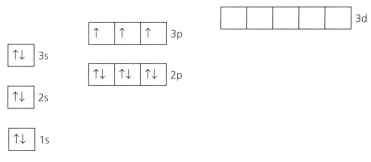

Figure 3.27

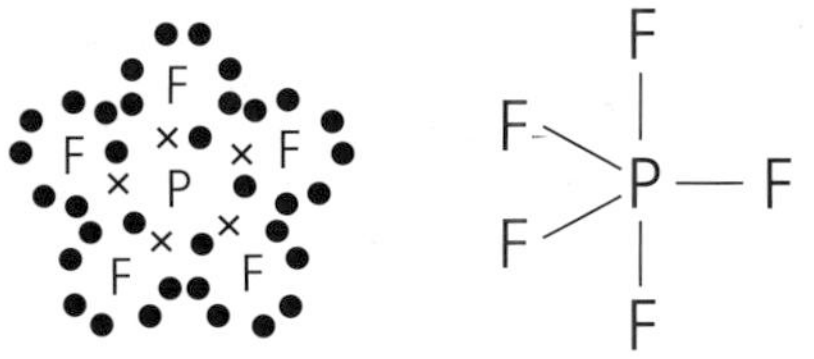

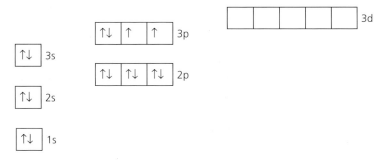

Dot and cross diagram   **Bonding diagram**

Figure 3.28

Atoms of phosphorus can promote one of the 3s electrons to the 3d, which will allow atoms of phosphorus to form five covalent bonds.

There are five bonding pairs of electrons around the central phosphorus atom and no lone pairs.

Sulfur hexafluoride (SF_6)

The electron configuration of atoms of sulfur is $1s^2 2s^2 2p^6 3s^2 3p^4$.

Figure 3.29

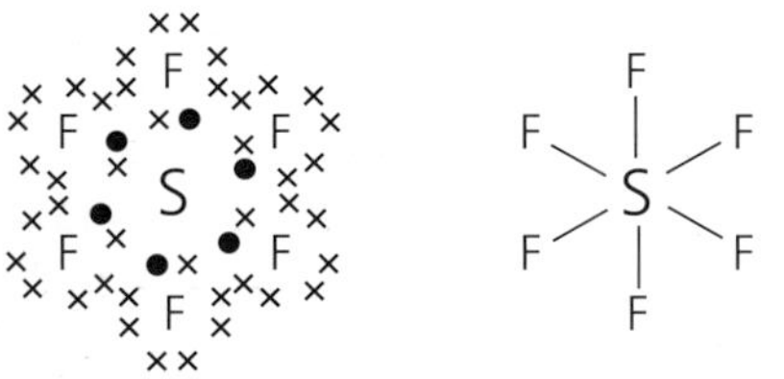

Atoms of sulfur can promote two electrons; one from the 3s electrons to the 3d and one from the 3p to the 3d which will allow atoms of sulfur to form six covalent bonds.

There are six bonding pairs of electrons around the central sulfur atom and no lone pairs.

Dot and cross diagram   **Bonding diagram**

Figure 3.30

> **TIP**
> The examples given showing bonding may be applied to similar molecules with different elements from the same group. For example, silane (SiH_4) has similar bonding to methane (CH_4); phosphine (PH_3) has similar bonding to ammonia (NH_3); hydrogen sulfide (H_2S) has similar bonding to water (H_2O). The dot and cross diagrams for these molecules are the same with only the central atom changed.

Multiple covalent bonds

Some molecules and ions contain multiple covalent bonds. Some examples are given.

Oxygen, O_2

In diatomic oxygen, the two oxygen atoms have two unpaired electrons as discussed for water. These electrons are in the 2p sub-level and oxygen atoms can share electrons in both of the 2p orbitals.

It is not completely necessary to show the lone pairs of electrons on the oxygen atoms. O_2 is often shown as O=O. The double lines between the oxygen atoms represent a double covalent bond.

Dot and cross diagram   **Bonding diagram**

Figure 3.31

Nitrogen, N_2

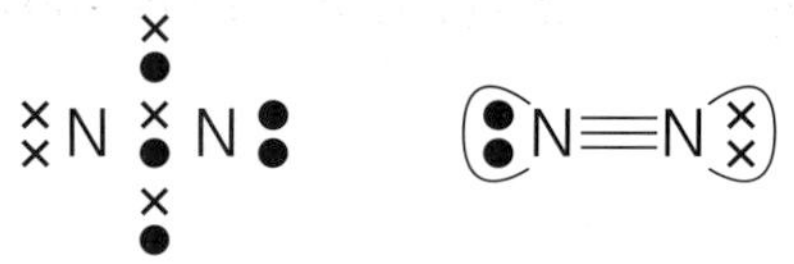

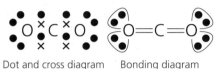

Dot and cross diagram   Bonding diagram

Figure 3.32

In diatomic nitrogen, the two nitrogen atoms have three unpaired electrons as discussed for ammonia. These electrons are in the 2p sub-shell and nitrogen atoms can share electrons between all three of the 2p orbitals.

It is not completely necessary to show the lone pairs of electrons on the nitrogen atoms. N_2 is often shown as N≡N. The triple lines between the nitrogen atoms represent a triple covalent bond.

Carbon dioxide, CO_2

Carbon can form four covalent bonds by promoting one electron from the 2s to the 2p giving it four unpaired electrons. Oxygen can form two covalent bonds (due to two unpaired electrons).

Dot and cross diagram   Bonding diagram

Figure 3.33

Again is it not completely necessary to show the lone pairs of electrons on the oxygen atoms. CO_2 is often shown as O=C=O. CO_2 contains two double covalent bonds.

Coordinate bond (dative covalent bond)

An atom which has a lone pair of electrons can form a coordinate bond with another atom which has an empty orbital. The lone pair of electrons is donated into the empty orbital on another atom to form a coordinate bond. The coordinate bond, once formed, is the same as a normal covalent bond. A **coordinate bond** contains a shared pair of electrons with both electrons supplied by one atom.

An example of this is the ammonium ion, NH_4^+.

When an ammonia (NH_3) molecule reacts with an H^+ ion, a coordinate bond forms between the lone pair of electrons (pair of electrons not involved in bonding) on the N atom and the empty 1s sub-shell in the H^+ ion.

The ammonia molecule has a lone pair of electrons which it can donate and the hydrogen ion has no electrons, so it has an empty orbital available for sharing a pair of electrons. Both electrons in the coordinate bond come from one atom.

When a coordinate bond is formed, it is indistinguishable from a normal covalent bond. When it is necessary to distinguish between a 'covalent bond' and a 'dative covalent bond' (coordinate bond), use → to represent the latter, instead of — .

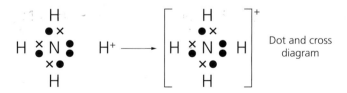

Dot and cross diagram

Figure 3.34

Figure 3.36

A neutral NH_3 molecule and an H^+ ion result in an overall charge of $+$ on the ammonium ion formed. If there is a charge on the final ion, it must be shown.

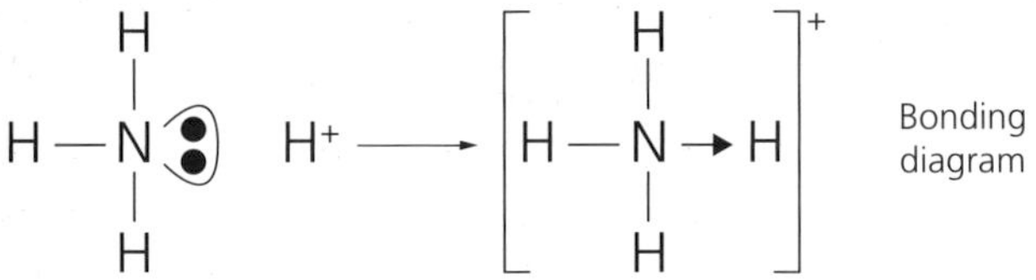

Figure 3.35

The photograph shows white fumes of ammonium chloride particles forming when gases from concentrated ammonia solution (left) and concentrated hydrochloric acid (right) mingle and react. There is a coordinate bond in the ammonium ion. The ammonium and chloride ions are held by ionic bonds

Other examples include:

1 Hydronium ion, H_3O^+, formed from water and H^+.

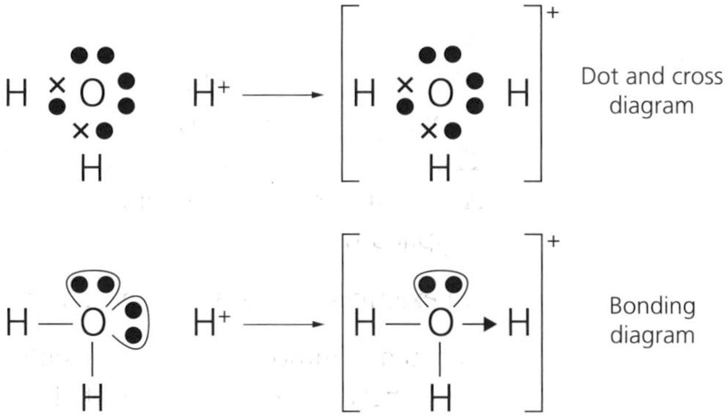

Figure 3.37

2 NH_3 reacting with BF_3

Earlier it was stated that the empty orbital in BF_3 would be important later. The lone pair of electrons on the ammonia molecule is able to form a coordinate bond with the empty orbital in BF_3. This forms a molecule with the formula NH_3BF_3.

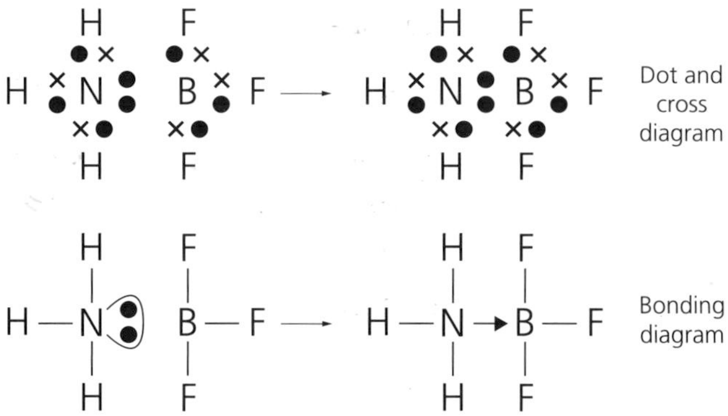

Figure 3.38

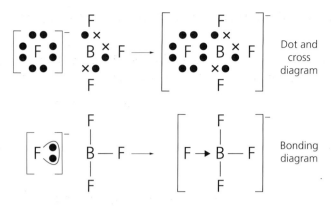

Figure 3.39

Dot and cross diagram

Bonding diagram

3 BF$_4^-$ formed from BF$_3$ and F$^-$

Fluoride ions have four lone pairs of electrons. They can donate one of the pairs of electrons to BF$_3$ to form BF$_4^-$. The ion formed has an overall negative charge due to the charge on the fluoride ion.

The fluoride ions form a coordinate bond with BF$_3$.

Once a coordinate bond is formed it has the same properties as a covalent bond and although both electrons in the bond come from the same atoms, these electrons are now treated as a bonding pair of electrons.

TEST YOURSELF 3

1 What is a covalent bond?
2 From the list below:
   CO$_2$ H$_2$O   N$_2$ H$_3$O$^+$ O$_2$ NH$_3$ CH$_4$ BF$_3$ NH$_4$Cl
   a) Which contain multiple covalent bonds?
   b) Which contain coordinate bonds?
3 Draw a dot and cross diagram of a molecule of ammonia and label a bonding pair of electrons and a lone pair of electrons.
4 Explain how a coordinate bond forms.

Metallic bonding

As well as looking at the bonding in metallic compounds and non-metallic elements and compounds, we must also look at the bonding in metals.

- Metals are generally solids and have their particles packed close together.
- The atoms are packed in layers and the outer shell electrons are not bound to an individual atom. In fact these outer shell electrons can move about between the layers.
- These electrons are referred to as **delocalised electrons** as they are not confined to any one atom.
- As these delocalised electrons can move, this explains why metals can conduct electricity and heat.
- The atoms in the layers are now without their outer shell electrons and so they are ions. There is confusion created when discussing the particles in metals (whether to refer to them as atoms or ions). In general when discussing bonding, the metal particles are referred to as positive ions. However, usually when discussing structure and reactivity, the convention is to discuss metal atoms.
- The metallic bond is the **electrostatic attraction** between the **delocalised electrons** and the **positive metals ions in the lattice**.

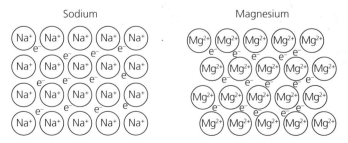

Figure 3.40

This metallic structure is often described as a lattice of positive metal ions in a sea of delocalised electrons.

The arrangement of the ions in sodium and magnesium is different as they have a different metallic lattice structure but this is beyond the A-level course.

The structure and bonding in a metal can be used to explain the physical properties of metals.

Properties of metals

1 Metals conduct electricity

There are delocalised electrons in the metal structure and these electrons can move. An electric current flows because of the movement of electrons or charged particles. The delocalised electrons can carry the charge.

2 Metals conduct heat

Heat is conducted when particles can move and are close enough together to pass on the heat energy from one to another. The delocalised electrons in the metal structure enable heat energy to be passed through the metal.

Silver is an excellent conductor of heat, so one of its uses is in the rear-window defrosters of cars. The tiny silver/ceramic lines conduct heat onto the glass, clearing frost, ice and condensation.

3 Metals are ductile and malleable

Metals can be drawn out into wires or hammered into shape. This is due to the layered structure of the lattice as the layers can slide over each other without disrupting the bonding.

Figure 3.41

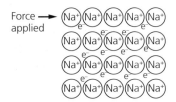

Figure 3.42

The top layer moves with the force applied to it. It is also important to note that the bonding is not disrupted by the movement of the layer as it just shifts over one and is still held together by the delocalised electrons. The strong attraction between the positive ions and the delocalised electrons hold the structure together.

4 Metals have high densities

In metals the positive ions are packed tightly together and so the density is high. Tungsten wire is used as winding wire for musical instruments such as the cello and viola. The high density of tungsten allows the strings to be thinner and yet withstand the frequency of vibration.

5 Most metals have high melting points

Any discussion of melting points is related to the strength of bonding within the structure being examined. Metals have a large regular structure with strong forces of attraction between the positive ions and the delocalised electrons. It is these attractive forces which must be overcome in order for the metal to melt. This requires a large amount of energy in the form of heat.

Figure 3.43

The melting point of the first three elements in Period 3 (Na, Mg and Al) are 98°C, 649°C and 660°C. The increase in melting point of the metals is due to an increase in the strength of the metallic bond. Sodium forms Na^+ ions in the metallic lattice with only one delocalised electron per sodium. The metallic bond in aluminium is stronger as the aluminium ion is Al^{3+} and there are three electrons which may be delocalised per aluminium; also the aluminium ion is smaller than the Mg^{2+} which is smaller than the Na^+ ion. The attraction between the smaller Al^{3+} ions in the metallic lattice and the delocalised electrons is stronger than the attraction between the larger Na^+ ions and the delocalised electrons.

Transition metals have much higher melting points than the main group metals. This is due again to the large number of d sub-shell electrons which may be delocalised creating an even stronger metallic bond. For example iron melts at 1535°C and tungsten melts at 3410°C (the highest melting point of any metal).

TEST YOURSELF 4

1 Draw the structure of the metal magnesium indicating how the metal is bonded.
2 Explain the difference between the bonding in a metal and the structure of a metal.
3 Explain why iron, like most metals, is a good conductor of electricity.
4 Explain the meaning of the following terms:
   a) malleable
   b) ductile
5 Sodium has the typical properties of a metal.
   a) Explain why sodium is malleable.
   b) Explain why sodium conducts electricity.

Bonding and physical properties

There are two main types of covalent substances.

1 Molecular (sometimes called molecular covalent or simple covalent)

2 Macromolecular (sometimes called giant covalent)

Both these types of substances contain covalent bonds but they differ in their structure.

Elements and compounds can be described as molecular if they exist as simple discrete molecules. For example, chlorine molecules are Cl_2; water molecules are H_2O; sulfur hexafluoride molecules are SF_6, ethane molecules are C_2H_6. The formula of these elements and compounds are molecular formulae and they show exactly how many atoms of each element are present in one molecule of the compound, for example the molecular formula of ethene is C_2H_4 which means that each molecule of ethene contains two carbon atoms and four hydrogen atoms.

Molecular covalent crystalline substances

Molecular covalent substances exist as single molecules, i.e. I_2, S_8, CCl_4, H_2O, CH_4, O_2, Cl_2.

These substances exist as gases (CH_4, O_2 and Cl_2) or liquids (H_2O and CCl_4) or low melting point solids (I_2 and S_8) at room temperature and pressure.

Many solid molecular covalent substances form crystalline structures which are called **molecular covalent crystals**.

I_2 and H_2O (ice) are covalent crystalline substances due to the attractions between the molecules.

Iodine

The large iodine molecules pack together into a regular arrangement causing the crystalline form of iodine.

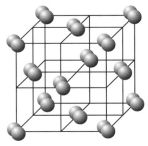

Figure 3.44 Iodine crystals are shown on the left. The structure of iodine on the right shows iodine molecules as small units which pack together in a regular lattice giving rise to the crystalline nature of solid iodine.

Ice

The structure of ice is shown below. The molecules of water are arranged in a regular arrangement forming a crystalline structure.

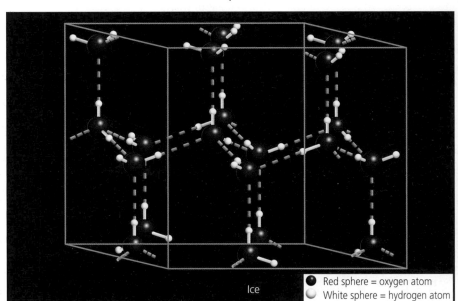

Ice

● Red sphere = oxygen atom
○ White sphere = hydrogen atom

Figure 3.45

Molecular covalent substances like iodine and ice which exist in a crystalline form are referred to as **molecular covalent crystals**.

Properties of molecular covalent crystals

- Molecular covalent crystalline substances have low melting points (I_2 114°C and ice 0°C).
- They are also brittle as they do not have the strong bonds holding them together like other crystalline substances such as ionic crystals and diamond.
- They do not conduct electricity as there are no charged particles to carry charge.

Macromolecular (giant covalent) structures

Some non-metallic elements and compounds can form a giant structure of covalent bonds. These structures are called macromolecular (or giant covalent). They include diamond and graphite (forms of the element carbon). The regular arrangement of atoms causes the crystalline form.

Carbon

There are two forms of the element carbon. These forms have identical atoms but it is the way in which the atoms are bonded together that makes the forms different. Different forms of the same element in the same physical state are called **allotropes**. Two allotropes of carbon are **diamond** and **graphite**.

Figure 3.46 Diamond and graphite, two very different substances, but both are the same element.

The physical properties of macromolecular (giant covalent) substances mostly depend on the many strong covalent bonds within the structure.

Diamond

- Diamond is the hardest naturally occurring substance due to the many strong covalent bonds and the rigid three-dimensional structure holding surface atoms in place.
- Diamond has a very high melting point (3550°C) as it has many strong covalent bonds which require a lot of energy to break.
- Diamond tipped tools are used for cutting glass/drilling/engraving.
- Each carbon atom is strongly bonded to four others in a tetrahedral arrangement and the bond angle is 109.5°.
- Diamond does not conduct heat or electricity as there are no charged particles which can move.

The diagram on the left shows only a small part of the structure of diamond.

> **TIP**
> The structure of macromolecular (giant covalent) substances is described as giant or macromolecular but the bonding is covalent.

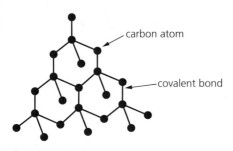

carbon atom

covalent bond

Figure 3.47

Figure 3.48 A dental drill is a small, high-speed drill used to remove decay and shape the tooth structure prior to the insertion of a filling or crown. The drill bit is known as a burr and often has a diamond coating.

Graphite

- Graphite conducts electricity due to delocalised electrons between the layers which can move and carry charge.
- Graphite has a very high melting point (approximately 3600 °C) as it has many strong covalent bonds which require a lot of energy to break.
- Graphite has a layered structure with weak forces of attraction between the layers. This means that the layers can slide over each other. This accounts for the flakiness of graphite and its use in pencil lead and as a lubricant.
- Each carbon atom in graphite is bonded strongly to three others in a hexagonal arrangement and the bond angle is 120°.
  Carbon atoms have four unpaired electrons and so can form four covalent bonds. In diamond all the four unpaired electrons are used in bonding as each carbon atom is bonded to four others. In graphite only three of these electrons are used in covalent bonding and the fourth electron becomes delocalised between the layers, providing the weak forces of attraction between the layers. The presence of the delocalised electrons explains why graphite can conduct electricity.
- The forces of attraction between the layers are weak and so can be broken easily hence allowing the layers to slide over each other. However the strong covalent bonds in the layers give graphite its high melting point as they require a large amount of energy to break them.

The structure of graphite is shown in Figure 3.50 where the layered structure can be clearly seen.

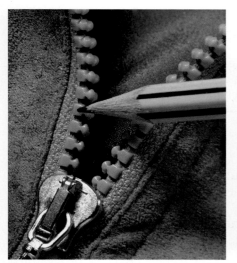

Figure 3.49 The graphite in a pencil being used to lubricate a zip fastener as the layers can slide over each other.

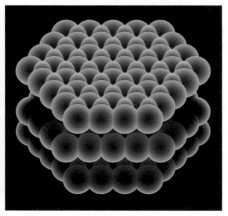

Figure 3.50 Layered structure of graphite.

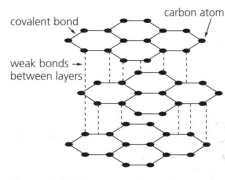

Figure 3.51 Structure of graphite.

Types of crystalline substances

Crystalline substances can be divided into the following categories:

1 Metals

2 Ionic compounds

3 Molecular (simple) covalent substances

4 Macromolecular (giant) covalent substances

A discussion on the types of crystalline substance and their properties may be found in the individual sections on metals, ionic crystals, molecular covalent crystals and giant covalent crystals.

Table 3.1 summarises the four types of crystalline substances.

Table 3.1 Crystalline substances.

Type of crystalline substance	Metals	Ionic compounds	Molecular (simple) covalent	Macromolecular (giant) covalent
Common examples	Magnesium	Sodium chloride	Ice, iodine	Diamond, graphite
Bonding	Metallic	Ionic	Covalent within the molecules and intermolecular forces between the molecules	Covalent (graphite has weak bonds between the layers in its structure)
Electrical conductivity	Conduct electricity when solid and molten	Does not conduct electricity when solid. Conducts electricity when molten or when dissolved in water	Does not conduct electricity	Does not conduct electricity (graphite does conduct electricity as a solid)
Melting point	Generally high melting points	High melting points	Low melting points	High melting points
Solubility in water	Insoluble in water (some metals will react with water)	Generally soluble in water	Mostly insoluble in water (some polar substances dissolve in water and some non-polar substances react with water)	Insoluble in water

> **TIP**
> You should be able to draw simple examples for each type of crystalline substance and explain the properties. Intermolecular forces will be examined in more detail later in the chapter.

States of matter

Matter exists in three states: solid, liquid and gas. As a substance changes from one state to another there is an energy change.

Melting

Melting is the change of state from solid to liquid. The temperature at which melting occurs is called the melting point. The melting point can be measured in degrees Celsius (°C) or kelvin (K).

Energy is taken in when a substance melts to overcome the forces or bonds. The stronger the forces or bonds in a substance, the greater the energy required to melt the substance.

The melting point is often used as a comparison between the strength of the forces or bonds in crystalline substances such as metals, ionic compounds, molecular (simple) covalent crystals and macromolecular (giant) covalent substances.

Freezing

Freezing is the change of state from liquid to solid. The temperature at which a substance freezes is the same as its melting point.

Energy is released when a substance freezes as forces or bonds are formed. The stronger the forces or bonds formed on freezing, the more energy is released.

Boiling

Boiling is the change of state from liquid to gas. The temperature at which a substance boils is called the boiling point. Again the boiling point can be measured in degrees Celsius (°C) or kelvin (K).

Energy is taken in when a substance boils. The stronger the bonds in the liquid substance, the greater the energy required to boil the substance.

Boiling overcomes all attractions between the particles or molecules in a substance so it is often used as a good measure of the strength of the bonding or forces in a structure particularly for molecular covalent substances where the intermolecular forces are broken on boiling.

Condensing

Condensing is the change of state from gas to liquid. The temperature at which a substance condenses is the same as its boiling point.

Energy is released when a substance condenses. The stronger the bonds or forces formed on condensing, the more energy is released.

Subliming

Subliming is the change of state from solid to gas on heating or from gas to solid on cooling. Substances which sublime are solid iodine and solid carbon dioxide, which is called dry ice.

Figure 3.52 Iodine sublimes when heated.

TEST YOURSELF 5

1 Name two molecular covalent crystalline substances.
2 State the type of bonds present in an iodine crystal.
3 From the following ions:
   Cl⁻ F⁻ Na⁺ O²⁻ Mg²⁺
   a) Which one is the smallest?
   b) Which two would form an ionic compound with the highest melting point?
4 Name the type of crystals shown by the following substances.
   a) graphite
   b) potassium iodide
   c) sulfur
   d) magnesium
5 Explain why a large amount of energy is needed to melt diamond.

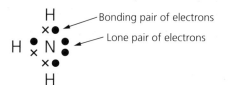

Bonding pair of electrons

Lone pair of electrons

Figure 3.53 Example of bonding pair and lone pair.

Shapes of simple molecules and ions

The shape of a covalent molecule or an ion depends on the repulsion of the electrons around a central atom. The electron pairs are charge clouds around an atom and they repel each other as far as possible. There are two types of electron pair, a bonding pair of electrons and a lone (non-bonding) pair of electrons.

The shape of the molecule or ion is determined from:

- the total number of electron pairs around a central atom
- the number of bonding pairs of electrons
- the number of lone pairs of electrons.

You must be able to identify the lone pairs and bonding pairs of electrons in any molecule or ion. Lone pairs are held closer to the central atom so they have a greater repulsive effect on the other pairs of electrons.

Figure 3.54 The order of strength of the repulsions experienced by the electron pairs.

This means that lone pairs of electrons repel lone pairs of electrons more than they repel bonding pairs of electrons. The lowest level of repulsion is between bonding pairs of electrons. The molecule or ion will take up a shape which minimises these repulsions. The shape depends on the arrangement of atoms around a central atom.

In questions relating to shapes of molecules and ions you may be asked for any combination of the following:

- a sketch of the shape
- the name of the shape
- the bond angle
- an explanation of the shape.

For the examples that follow, all of the above will be given for each molecule or ion.

Examples with only bonding pairs of electrons

sketch of the shape

Figure 3.55

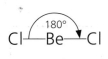

Figure 3.56

Bond angle = 180°
Shape = Linear
Explanation: two bonding pairs of electrons repel each other equally and the molecule takes up this shape to minimise repulsions

Beryllium chloride (BeCl$_2$)

Around the beryllium atom there are two bonding pairs of electrons only. These repel each other equally so the molecule takes up a **linear** shape to minimise the effect of the repulsions.

The bond angle is the angle between the two covalent bonds, which in beryllium chloride is 180°.

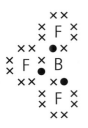

Sketch of the shape

120°

Bond angle = 120°
Shape = Trigonal planar
Explanation: three bonding pairs of electrons repel each other equally and the molecule takes up this shape to minimise repulsions

Figure 3.57

Boron trifluoride (BF₃)

There are three bonding pairs of electrons around the boron atom in BF_3. These repel each other equally and so the molecule takes up a **trigonal planar** shape with a bond angle of 120°.

The two shapes encountered so far can be drawn easily as they are two dimensional. When there are four or more pairs of electrons, the arrangement becomes three-dimensional, which requires a little more skill in drawing.

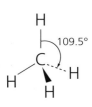

Sketch of the shape

109.5°

Bond angle = 109.5°
Shape = Tetrahedral
Explanation: four bonding pairs of electrons repel each other equally and the molecule takes up this shape to minimise repulsions

Figure 3.58

Methane (CH₄)

There are four bonding pairs of electrons around the carbon atom in methane. These repel each other equally and the molecule takes up a **tetrahedral** shape to minimise repulsions. The bond angle is 109.5°.

The term tetrahedral stems from the fact that the solid shape formed when all the hydrogen atoms are connected would be a triangular-based pyramid with four sides, which is called a tetrahedron. The shaded area is the triangular base of the pyramid.

tetrahedron

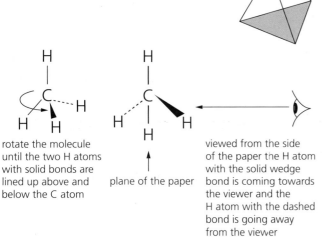

rotate the molecule until the two H atoms with solid bonds are lined up above and below the C atom

plane of the paper

viewed from the side of the paper the H atom with the solid wedge bond is coming towards the viewer and the H atom with the dashed bond is going away from the viewer

Figure 3.59

TIP
When drawing a three-dimensional shape, such as the tetrahedral shape of methane, three types of lines are drawn to show the three-dimensional arrangement of the atoms. Bonds in the plane of the page are shown as normal lines (—). Bonds coming towards the viewer out of the plane of the page are drawn using a solid wedge getting thicker as it comes out towards the atom at the end of the bond (◀). Bonds going backwards from the plane of the paper are shown using a dashed line (- - -).

If a molecule of methane were viewed from the side and rotated with two of the bonds appearing to be vertical in one plane with the carbon atom, the other bonds would be with one going into the plane of the paper and the other coming out.

99

TIP
Try this with a molymod kit to make sure you can see the tetrahedral shape. It should look like Figure 3.60. Again it is clear that with two H atoms in the one plane, there is an H atom in front of that plane and one behind.

Figure 3.60

Carbon dioxide (CO_2)

The carbon in carbon dioxide has two sets of bonding pairs of electrons. A double bonding pair of electrons repels in the same way as a single bonding pair. The two sets of bonding pairs of electrons repel each other equally so CO_2 takes up a linear shape to minimise repulsions.

TIP
The identical and symmetrical polar bonds explain the lack of polarity in the carbon dioxide molecule, even though it contains polar bonds. This will be explained soon.

hydrogen cyanide

Bond angle = 180°
Shape = Linear
Explanation: two (sets of) bonding pairs of electrons repel each other equally and the molecule takes up this shape to minimise repulsions

Figure 3.61

A double bonding pair of electrons or a triple bonding pair of electrons repel in the same way as a single bonding pair. This can be seen from the bonding diagram for hydrogen cyanide (HCN) and ethene (C_2H_4).

Hydrogen cyanide is **linear** due to equal repulsions of the triple bonding pair of electrons and the single bonding pair of electrons. Around the carbon atoms in ethene the shape is **trigonal planar** due to the equal repulsion of the three sets of bonding pairs of electrons (even though one is a double set).

ethene

Figure 3.62

It is thought that receptors in the nasal cavity identify the shape of a molecule and use this to sense their odour. However as shown in the photograph benzaldehyde and hydrogen cyanide have very different shapes, yet they both smell of bitter almonds. Recent research has shown that molecular vibrations rather than shape may be used by receptors to identify smell.

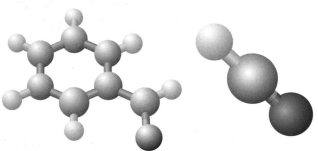

Figure 3.63 Benzaldehyde (left) and hydrogen cyanide (right).

Phosphorus pentafluoride (PF_5)

In PF_5 there are five bonding pairs of electrons around the central phosphorus atom. These bonding pairs of electrons repel each other equally and the molecule takes up a **trigonal bipyramidal** shape. There are two bond angles in a trigonal bipyramid, 90° and 120°.

Again if the points where the fluorine atoms are placed are connected the shape formed is a triangle with a pyramid above and below. This is called a trigonal bipyramid.

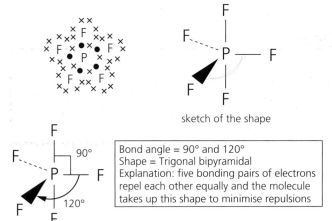

sketch of the shape

Bond angle = 90° and 120°
Shape = Trigonal bipyramidal
Explanation: five bonding pairs of electrons repel each other equally and the molecule takes up this shape to minimise repulsions

Figure 3.64

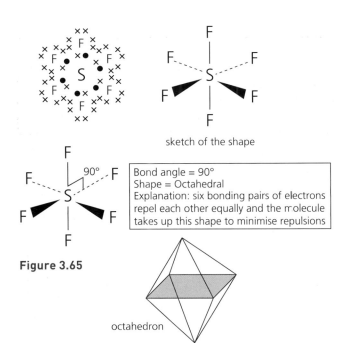

sketch of the shape

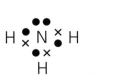

Bond angle = 90°
Shape = Octahedral
Explanation: six bonding pairs of electrons repel each other equally and the molecule takes up this shape to minimise repulsions

Figure 3.65

octahedron

Bond angle = 107°
Shape = Pyramidal
Explanation: three bonding pairs of electrons and one lone pair of electrons; the lone pair of electrons has a greater repulsion than the bonding pairs of electrons and the molecule takes up this shape to minimise repulsions

Figure 3.66

sketch of the shape

Bond angle = 104.5°
Shape = Bent
Explanation: two bonding pairs of electrons and two lone pairs of electrons; the lone pairs of electrons have a greater repulsion than the bonding pair of electrons; the molecule takes up this shape to minimise repulsions

Figure 3.67

Sulfur hexafluoride (SF₆)

In SF_6 there are six bonding pairs of electrons around the central sulfur atom. These repel each other equally and the molecule takes up an octahedral shape to minimise the repulsions.

The octahedral shape is also called square bipyramidal. The term octahedral comes from the fact that the solid shape formed from connecting all the fluorine atoms forms an eight sided figure called an octahedron. The shaded area is the central square.

Examples with bonding pairs of electrons and lone pairs of electrons

The following examples all have four pairs of electrons around the central atom. These pairs of electrons take up a tetrahedral shape like CH_4. However out of the four pairs of electrons, some are bonding pairs of electrons and some are lone pairs of electrons. Remember a lone pair of electrons has a greater repulsion than a bonding pair of electrons.

Ammonia (NH₃)

There are three bonding pairs of electrons and one lone pair of electrons around the central nitrogen atom in NH_3. The basic arrangement of the electron pairs is tetrahedral around the nitrogen but as there is no atom attached to the lone pair all you see is the bottom of the tetrahedron which looks like a pyramid. The extra repulsion from the lone pair squeezes the bonding pairs of electrons closer together decreasing the bond angle to 107°.

Water (H₂O)

The basic arrangement of the electron pairs is tetrahedral around the oxygen atom but as there is no atom attached to the lone pairs all you see is two bonds of the tetrahedron, which makes it appear bent. The extra repulsion from the lone pairs squeezes the bonds closer giving a bent (or V) shape and decreasing the bond angle to 104.5°.

TIP
Remember that with any shape with lone pairs of electrons, you should show the lone pairs on the central atom. This should be done using ⌣ or ●●. You can also do this using ●● or ××. See Figure 3.68.

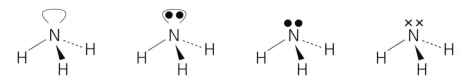

Figure 3.68 Different ways to show a lone pair of electrons.

Examples involving coordinate bonds

When a coordinate bond forms it converts a lone pair of electrons into a bonding pair of electrons.

Ammonia reacts with hydrogen ions to form the ammonium ion, NH_4^+.

Ammonia	Ammonium ion
Bond angle = 107° Shape = Pyramidal Explanation: three bonding pairs of electrons and one lone pair of electrons; lone pair of electrons has a greater repulsion than the bonding pair of electrons; the molecule takes up this shape to minimise repulsions	Bond angle = 109.5° Shape = Tetrahedral Explanation: four bonding pairs of electrons repel each other equally and the ion takes up this shape to minimise repulsions

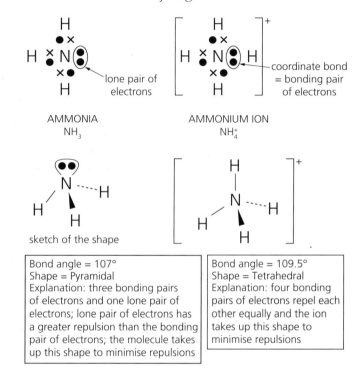

Figure 3.69

The formation of the coordinate bond causes a change in the shape. Remember ammonia (NH_3) is pyramidal (three bonding pairs of electrons and one lone pair of electrons) but the ammonium ion is tetrahedral (four bonding pairs of electrons).

The H_3O^+ ion is formed when H_2O reacts with H^+. The H_3O^+ ion has three bonding pairs of electrons and one lone pair of electrons around the nitrogen atom so it takes up a pyramidal shape (bond angle 107°) to minimise repulsions.

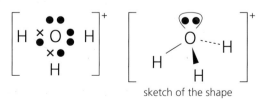

Bond angle = 107°
Shape = Pyramidal
Explanation: three bonding pairs of electrons and one lone pair of electrons; lone pair of electrons has greater repulsion than the bonding pair of electrons; the ion takes up this shape to minimise repulsions

Figure 3.70

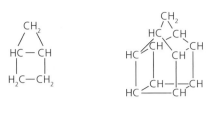

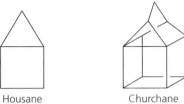

Housane   Churchane

Figure 3.72 Housane, C_5H_8 and churchane, $C_{11}H_{12}$.

The BF_4^- ion is formed when BF_3 reacts with F^-. The BF_4^- ion has four bonding pairs of electrons around the central boron atom so it takes up a tetrahedral shape (bond angle 109.5°) to minimise repulsions.

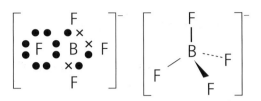

Bond angle = 109.5°
Shape = Pyramidal
Explanation: four bonding pairs of electrons repel each other equally and the ion takes up this shape to minimise repulsions

Figure 3.71

Some unusual molecules get their name from their 3D shapes. Housane, C_5H_8 and churchane, $C_{11}H_{12}$ are shown in Figure 3.72.

More unusual examples

With some more complex molecules and ions it is important to be able to visualise the total number of electrons around the central atom. This helps with the basic shape. The number of lone pairs of electrons and bonding pairs of electrons dictate the shape and the bond angle but it is important to remember in a shape like a trigonal bipyramid that the lone pairs will take up positions as far away from each other as possible. They will also push the bonding pairs of electrons closer together. Each lone pair typically reduces the bond angle by around 2 to 2.5°.

Bromine trifluoride (BrF_3)

- Bromine (outer electron configuration, $4s^2\ 4p^5$) has one unpaired electron so to form the three covalent bonds required for BrF_3, one electron in the bromine atom is promoted to a higher sub-level.
- This gives bromine three unpaired electrons in this compound, which can form three bonding pairs of electrons and leaves two lone pairs of electrons.
- Five pairs of electrons would suggest a trigonal bipyramidal general shape with two of the pairs being lone pairs of electrons. The dots in the dot and cross diagram represent the bromine electrons.
- The shape is described as T-shaped as the lone pairs of electrons take up positions 120° from each other. The three fluorine atoms take up the three other positions in the trigonal bipyramid
- The repulsion from the lone pairs of electrons is greater than the repulsion from the bonding pairs of electrons so the 90° angle is reduced to 86°.

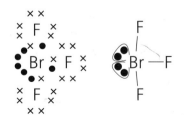

Bond angle = 86°
Shape = T shaped
Explanation: three bonding pairs of electrons and two lone pairs of electrons; basic shape is trigonal bipyrimidal but lone pairs of electrons have greater repulsion than the bonding pair of electrons; the molecule takes up this shape to minimise repulsions

Figure 3.73 Bromine trifluoride.

The shape could be described as trigonal planar if the three fluorine atoms took up the positions in Figure 3.74. However due to the greater repulsion of the lone pairs of electrons the T-shape is more correct.

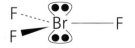

trigonal planar shape sketch

Figure 3.74 Trigonal planar arrangement.

BrF_3 reacts to form BrF_2^+ and BrF_4^-. Determine the shape of both BrF_2^+ and BrF_4^-.

Bond angle = 104.5°
Shape = bent
Explanation: two bonding pairs of electrons and two lone pairs of electrons; the lone pairs of electrons have a greater repulsion than the bonding pair of electrons; the molecule takes up this shape to minimise repulsions

Figure 3.75 BrF_2^+ ion.

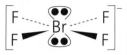

Bond angle = 90°
Shape = square planar
Explanation: four bonding pairs of electrons and two lone pairs of electrons; the lone pairs of electrons have a greater repulsion than the bonding pair of electrons; the molecule takes up this shape to minimise repulsions

Figure 3.76 BrF_4^- ion.

Bond angle = 180°
Shape = Linear
Explanation: two bonding pairs of electrons and three lone pairs of electrons; basic shape is trigonal bipyramidal but lone pairs of electrons have a greater repulsion than the bonding pairs of electrons; the molecule takes up this shape to minimise repulsions

Figure 3.77 XeF_2.

In BrF_2^+, BrF_3 has lost an F^-. BrF_3 contains two lone pairs of electrons and three bonding pairs of electrons.

BrF_2^+ will contain two lone pairs of electrons and two bonding pairs of electrons. With four pairs of electrons the basic shape is tetrahedral but like water the shape will be bent with a bond angle of 104.5°.

In BrF_4^-, BrF_3 has gained a F^- so BrF_4^- has two lone pairs of electrons but four bonding pairs of electrons. This gives six electron pairs and a basic octahedral shape where the atoms and lone pairs will be at 90° to each other.

To get as far away from each other as possible, the lone pairs of electrons take up positions above and below the Br atom. This leaves the four F atoms in the ion arranged in a square planar arrangement with a bond angle of 90°.

Xenon difluoride (XeF_2)

Determine the shape of XeF_2.

- Xe promotes one electron to allow it to have two unpaired electrons and to form two covalent bonds.
- Xe will have three lone pairs of electrons and two bonding pairs of electrons.
- The basic shape is trigonal bipyramidal.
- The three lone pairs take up position as far away from each other as possible (120° from each other). The bonding pairs take up position above and below the Xe atom so the molecule is linear with a bond angle of 180°

Table 3.2 Summary table for shapes of molecules.

Total number of electron pairs around central atom	Number of bonding pairs of electrons	Number of lone pairs of electrons	Shape	Bond angle	Examples
2	2	0	linear	180°	$BeCl_2$, CO_2
3	3	0	trigonal planar	120°	BF_3
4	4	0	tetrahedral	109.5°	CH_4, NH_4^+
4	3	1	pyramidal	107°	NH_3, H_3O^+
4	2	2	bent	104.5°	H_2O, BrF_2^+
5	5	0	trigonal bipyramid	90° and 120°	PF_5
5	3	2	T-shaped	86°	BrF_3
5	2	3	linear	180°	XeF_2
6	6	0	octahedral	90°	SF_6
6	4	2	square planar	90°	BrF_4^-

1 What is the shape and bond angle for the following molecules?

a) H_2O **b)** SiH_4 **c)** PH_3 **d)** BCl_3

2 What is the shape and bond angle for the following ions?

a) H_3O^+ **b)** PH_4^+ **c)** BF_4^- **d)** BF_2^+

Bond polarity

The **electronegativity** of an element is the power of an atom to attract the pair of electrons in a covalent bond.

Every covalent bond contains electrons and the electrons in the bond move in the bonded orbitals. However the electron distribution in the bond may not be even and the electrons may be found more often closer to one atom than the other. This atom is more electronegative.

The Pauling Electronegativity scale is commonly used with F (4.0) as the most electronegative element and Cs (0.7) as the least electronegative (or most electropositive). There are no units of electronegativity.

The Periodic Table in Figure 3.78 gives values of electronegativity of some elements.

1	2											3	4	5	6	7
							H 2.1									
Li 1.0	Be 1.5											B 2.0	C 2.5	**N** 3.0	**O** 3.5	**F** 4.0
Na 0.9	Mg 1.2											Al 1.5	Si 1.8	P 2.1	S 2.5	**Cl** 3.0
K 0.8	Ca 1.0	Sc 1.3	Ti 1.5	V 1.6	Cr 1.6	Mn 1.5	Fe 1.8	Co 1.8	Ni 1.8	Cu 1.9	Zn 1.6	Ga 1.6	Ge 1.8	As 2.0	Se 2.4	Br 2.8
Rb 0.8	Sr 1.0											In 1.7	Sn 1.8	Sb 1.9	Te 2.1	I 2.5
Cs 0.7	Ba 0.9											Tl 1.8	Pb 1.8	Bi 1.9	Po 2.0	At 2.2

Figure 3.78 Electronegativities.

Electronegativity is dependent upon three factors:

1 The distance of the bonding electrons from the attractive power of the nucleus. *This can be taken to be the same as the atomic radius.*

2 The size of the nuclear charge. *This can be measured as the atomic number.*

The atomic radius decreases across a period so the electrons in the outer energy levels, which would be forming covalent bonds, are closer to the nucleus in oxygen compared to nitrogen. Oxygen also has a greater nuclear charge than nitrogen.

3 The attractive power of the nucleus being shielded by inner electrons. Going down a group there are more inner energy levels of electrons shielding the attractive power of the nucleus, which, combined with the increase in atomic radius, leads to a decrease in electronegativity despite the increase in nuclear charge.

Trends in electronegativity

In the Periodic Table, the following trends in electronegativity values are observed:

1 Electronegativity increases across a period.

- This is because atomic radius decreases across a period, giving a progressively stronger attraction between the positive nucleus and the two electrons in the covalent bond.
- Also the nuclear charge increases across the period and this will cause a greater attraction for the electrons in the covalent bond.

2 Electronegativity decreases down a group.

- This is because the atomic radius increases down a group, giving a progressively weaker attraction between the positive nucleus and the electrons in the covalent bond.
- The shielding of the nuclear charge increases down a group as there are more electrons in inner energy levels.

The variation in electronegativity is most noticeable across the period. The decreasing trend is observed going down a group but it is not as noticeable due to the increasing nuclear charge.

Electronegativity as a guide to the polarity of bonds

In a covalent bond the two atoms at either end of the bond have an electronegativity value. For example in H—Cl, the electronegativity values are: H = 2.1; Cl = 3.0

This means that the electrons in this bond are drawn closer to Cl than to H as Cl has a higher electronegativity value. This is represented by the use of $\delta+$ (delta plus) and $\delta-$ (delta minus) above the atoms in the bond. The $\delta-$ is placed above the atom which has the higher electronegativity value and the $\delta+$ is placed above the atom with the lower electronegativity value. The more electronegative atom has the greater pull on the two electrons hence it is $\delta-$; the other atom is $\delta+$.

Ionic and covalent character

For compounds composed of two different elements, the difference in electronegativity of the different atoms of the elements dictates the type of compound formed (i.e. ionic or covalent) and, if the compound is covalent, also determines the polarity of the molecule.

1 No difference or a very small difference in electronegativity results in a non-polar molecule. For example: Br_2; I_2; Cl_2; O_2; CH_4.

2 A small difference in electronegativity **usually** results in a polar molecule. For example: HF; HCl; H_2O; NH_3.

3 A large difference in electronegativity results in an ionic compound. For example: NaCl; MgO; CaF_2.

Table 3.3 Properties and examples of covalent and ionic bonds.

Non-polar covalent bond	Polar covalent bond	Ionic bond
A—A	$\overset{\delta+ \;\; \delta-}{A-B}$	A⁺B⁻
The electronegativities of the two atoms in the covalent bond are the same. This means that the bonding electrons are shared equally, i.e. electron distribution is symmetrical.	B is more electronegative than A. The electrons in the covalent bond are closer to B, i.e. electron distribution is not symmetrical. So A will be slightly positive (δ+) and B slightly negative (δ−). δ+ and δ− are used to show the polarity of the bonds and are written above the atoms in the bond.	B is much more electronegative than A. The electron cloud of B⁻ is not distorted at all by A⁺. The bond consists of the attraction between the oppositely charged ions A⁺ and B⁻.
Examples: Cl—Cl H—H O=O N≡N	Examples $\overset{\delta+ \;\; \delta-}{H-Cl}$ $\overset{\delta+ \;\; \delta-}{C-Cl}$ $\overset{\delta+ \;\; \delta-}{Be-Cl}$ $\overset{\delta+ \;\; \delta-}{C=O}$	Examples Li⁺F⁻ K⁺Cl⁻ Cs⁺Cl⁻ Na⁺F⁻

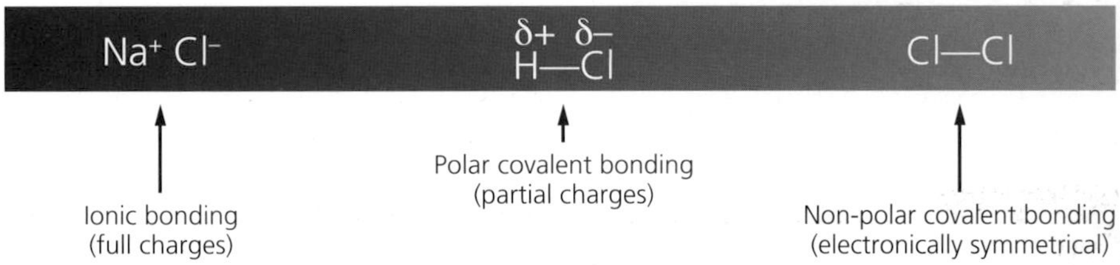

Na⁺ Cl⁻ → Ionic bonding (full charges)

$\overset{\delta+ \;\; \delta-}{H-Cl}$ → Polar covalent bonding (partial charges)

Cl—Cl → Non-polar covalent bonding (electronically symmetrical)

Figure 3.79

TIP
The overall polarity of a molecule is often called a dipole. Carbon dioxide is non-polar or can be said not to possess a dipole.

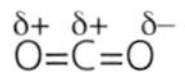

$\overset{\delta+ \;\;\; \delta+ \;\;\; \delta-}{O=C=O}$

Figure 3.80 Non-polar carbon dioxide.

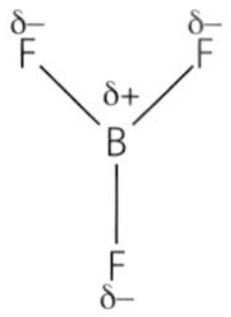

Figure 3.81 Non-polar boron trifluoride.

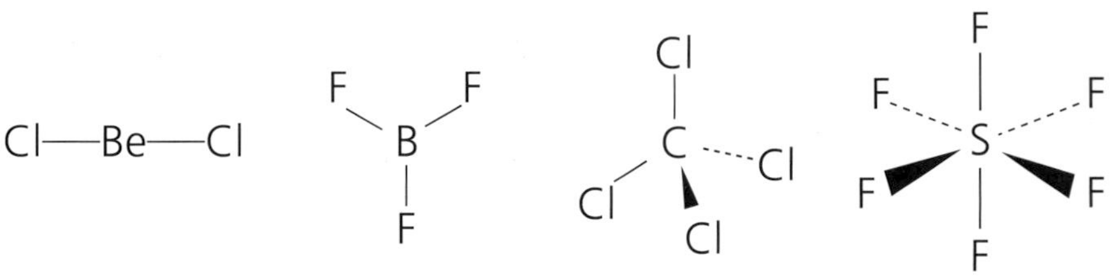

A molecule may be described as polar or non-polar. A molecule such as water, which contains two polar O—H bonds, is a polar molecule. However some substances may contain polar bonds but the overall molecule is non-polar.

Carbon dioxide contains polar C=O bonds but carbon dioxide molecules are non-polar. This is because the polarities of the bonds are equal, both are C=O bonds and the molecule has the two C=O bonds arranged symmetrically so the polarities of the bonds cancel each other out.

BF_3 contains polar B—F bonds but BF_3 is non-polar. This is again due to the equally polar bonds being arranged symmetrically around the boron atom so the polarities of the bonds cancel each other out.

Other non-polar molecules due to the cancelling out of the effect of equally polar bonds are: $BeCl_2$, CCl_4 and SF_6. Generally, equal polar bonds arranged in a linear, trigonal planar, tetrahedral or octahedral arrangements create non-polar molecules.

Figure 3.82 Examples of non-polar molecules containing polar bonds.

ACTIVITY

Testing a liquid for polarity

The polar nature of a liquid can be demonstrated by using the following experimental method.

- Fill a burette with the liquid under investigation and place a beaker under the jet.
- Rub an acetate rod with a duster to create static charge.
- Open the burette tap and bring the charged rod close to the stream of water.
- Is the stream deflected from its vertical path?

1 Water was attracted to the charged rod because it is polar and contains polar bonds.

  a) What is meant by the phrase 'polar bond'?

  b) Explain why the O—H bond in water is polar.

2 Draw diagrams to show a water molecule attracted to a positively-charged rod and to a negatively-charged rod. Show the polarity of the water molecule.

3 The results table for the experiment is shown below.

  a) For each of the liquids ethanol, hexane and trichloromethane, state and explain if it is a polar molecule.

  b) Complete the results table.

Liquid	Is the stream deflected?
Water	yes
Ethanol, CH_3CH_2OH	
Hexane, C_6H_{14}	
Trichloromethane, $CHCl_3$	

4 Tetrachloromethane, CCl_4, is not safe to use in a school laboratory.

  a) Tetrachloromethane contains polar bonds. Identify the polar bond in tetrachloromethane.

  b) Predict the effect of a charged rod on a stream of tetrachloromethane and explain your answer.

Forces between molecules

Intermolecular forces are the attractive forces between covalent molecules. (Note that **intramolecular forces** are the bonds that exist within the molecule itself, i.e. covalent or coordinate.)

There are three types of intermolecular force:

- induced dipole-dipole forces (van der Waals' forces)
- permanent dipole-dipole forces
- hydrogen bonds.

Induced dipole-dipole forces (van der Waals' forces)

Induced dipole-dipole forces or van der Waals' forces are attractive forces between all molecules and atoms. They are caused by induced dipoles, which are temporary and constantly shifting. These attractive forces are also called London forces or dispersion forces.

All molecules are composed of atoms which have electrons orbiting the nucleus. These electrons are in rapid motion and at any one time they may be distributed more on one side of the molecule than the other. Another molecule approaching this side of the molecule will have its electrons repelled. This creates a temporary or induced dipole.

These induced dipoles act one way and then another, continually forming and disappearing as a result of electron movement. The forces are always attractive forces. Even though the average dipole on every molecule is zero, the forces between the molecules at any instant are not zero.

Figure 3.83 shows the distribution of electrons in two neighbouring atoms. The electrons keep moving and the dots show the position of the electrons.

In this figure, the electrons are evenly distributed and the atoms have no induced dipole.

In Figure 3.84, the electrons on the atom of the left are closer to the right hand side. The electrons in the right hand atom are repelled to the right and both atoms possess a temporary and instantaneous dipole.

The electrons are in continuous movement and so the dipoles keep shifting. This creates an attraction between all atoms in close proximity. The larger a molecule is, the higher the M_r and the greater the number of electrons. The more electrons there are, the greater the induced dipoles, so the greater the van der Waals' forces.

Figure 3.83 Even distribution of electrons.

Figure 3.84 A temporary dipole is created.

> **TIP**
> Remember that a larger molecule (greater M_r) has more electrons. More electrons mean greater van der Waals' forces between molecules.

Figure 3.85

Why can geckos walk upside down on a ceiling? Geckos have tiny hair-like structures on their feet. When they touch a surface, van der Waals' forces between these hairs and the surface draw the materials together and cause adhesion. These forces are relatively weak compared to normal chemical bonds, but they provide the requisite adhesive strength that allows the gecko to support its body upside down or to scurry along walls and ceilings.

Scientists have tried to develop various materials that mimic the gecko's remarkable stickiness. Researchers at the University of Kiel in Germany have created silicone tape with tiny hairs which not only boasts impressive bonding strength, but can also be attached and detached thousands of times without losing its adhesive properties. A $20\,cm^2$ piece of the new GSA (gecko-inspired synthetic adhesive) tape was able to support the weight of one team member dangling from the ceiling.

Evidence for van der Waals' forces

Halogens are non-polar diatomic molecules. As shown in Table 3.4, there is a definite trend in their boiling points, melting points and their states as the size of the molecules increases.

Table 3.4

Halogen	Formula	M_r	State	Melting point (°C)	Boiling point (°C)
fluorine	F_2	38	gas	-220	-188
chlorine	Cl_2	71	gas	-101	-34
bromine	Br_2	160	liquid	-7	58
iodine	I_2	254	solid	114	183

This is again explained by the increase in the M_r which causes an increase in the number of electrons. This leads to an increased induced dipole-dipole force thus leading to increased van der Waals' forces between the molecules.

- For small non-polar molecules such as CCl_4, Cl_2, CH_4 and CO_2, the van der Waals' forces are very small so these tend to be gases or volatile liquids.
- For large non-polar molecules such as polymers (e.g. polythene) there are many contacts between molecules which means that the van der Waals' forces are reasonably large.

Noble gases are monatomic. They are non-polar as they have no permanent dipole and do not form any bonds, yet they will condense if the temperature is low enough. Table 3.5 gives the boiling points of the Noble gases.

Table 3.5 Boiling points of Noble gases.

	Helium	Neon	Argon	Krypton	Xenon	Radon
Boiling point (°C)	−268.8	−245.9	−185.8	−151.7	−106.6	−61.7

The increase in boiling point suggests that forces do exist between these atoms in the liquid state. The increase in boiling point may be explained by:

- larger atoms
- more electrons
- greater induced dipole-dipole forces
- greater van der Waals' forces between the atoms.

Van der Waals' forces are much weaker than covalent bonds, permanent dipole-dipole forces or hydrogen bonds.

Permanent dipole-dipole forces

Molecules which have a permanent dipole within the molecule have an additional type of intermolecular force between their molecules. These permanent dipoles cause the molecules to be attracted to each other. The δ+ end of the dipole on one molecule is attracted to the δ- end of the dipole on another molecule. The attraction is called a permanent dipole-dipole force.

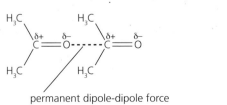

Trichloromethane

permanent dipole-dipole force

Propanone

permanent dipole-dipole force

Figure 3.86 Permanent dipole-dipole force.

The permanent dipole-dipole forces are responsible for holding together the molecules in polar substances but it should be remembered that there are also van der Waals' forces between these polar molecules as well.

Hydrogen bonds

Hydrogen bonds occur between a $\delta+$ H atom (which is covalently bonded to O, N or F) of one molecule and the lone pair of electrons of an O, N or F atom of another molecule. The strength of a hydrogen bond is about $\frac{1}{10}$ the strength of a covalent bond. O—H, N—H and F—H bonds are the most polar bonds but only O—H and N—H are found in covalent molecules as H—F bonds are only found in hydrogen fluoride.

Hydrogen bonds explain many properties of simple covalent molecules.

1 Water

Figure 3.87 shows the hydrogen bond between two water molecules.

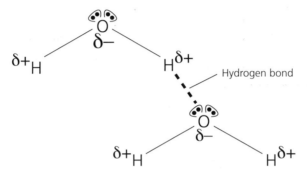

Figure 3.87 Hydrogen bonding in water

The important features of drawing a diagram like this are:

● show $\delta+$ and $\delta-$ on all atoms
● show the lone pair of electrons forming the hydrogen bond
● draw a dashed line to show the hydrogen bond between the lone pair and the $\delta+$ H atom.

The hydrogen bonds between water molecules explain many of its physical properties.

a) Water is fluid as the hydrogen bonds can break and reform allowing water molecules to move around each other.

b) Water has a higher boiling point than would be expected for a Group 6 hydride. The graph in Figure 3.88 on page 112 shows the boiling points of all the Group 6 hydrides.

As can be seen clearly, water has a much higher boiling point than all the other Group 6 hydrides. It would be expected based on the other hydrides that water should have a boiling point of around $-90\,°C$. This would be the case if the only intermolecular forces between water molecules were van der Waals' forces of attraction and permanent dipole-dipole forces. However the hydrogen bonds between water molecules mean that it requires substantially more energy to separate the water molecules into the gaseous state.

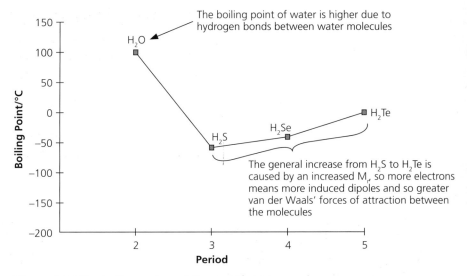

Figure 3.88 The boiling points of Group 6 hydrides.

The increase seen from H_2S to H_2Se and to H_2Te is caused by an increase in the van der Waals' forces of attraction. As you descend Group 6, each atom has more electrons so the induced dipoles increase and hence the van der Waals' forces between molecules increase as well.

This pattern repeats for the Group 5 and Group 7 hydrides but not for the Group 4 hydrides as CH_4 is non-polar.

Figure 3.89 shows the boiling points of all the hydrides of elements in Groups 4 to 7 in Periods 2 to 5.

- H_2O has a higher boiling point than H_2S.
- HF has a higher boiling point than HCl.
- NH_3 has a higher boiling point than PH_3.
- With CH_4, however, its boiling point is as expected based solely on van der Waals' forces of attraction; there are no hydrogen bonds between CH_4 molecules.

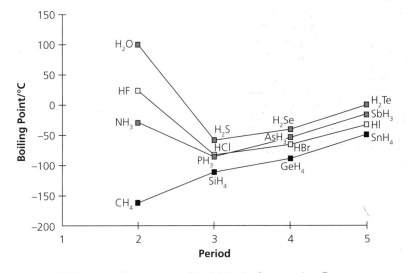

Figure 3.89 The boiling points of hydrides in Groups 4 to 7.

c) Ice has a lower density than water.

When water freezes the hydrogen bonds hold the water molecules in a more open 3D crystalline structure. The water molecules are further apart from each other than they were in liquid water and so ice has a lower density.

As can be seen from the figures, the molecules in ice are much further apart than the molecules in water. This is caused by the hydrogen bonds holding the water molecules in this more open tetrahedral arrangement. This more open arrangement means that ice has a lower density than water (fewer particles in the same space) and so ice floats on water.

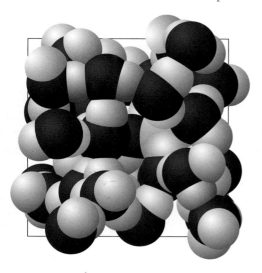

water

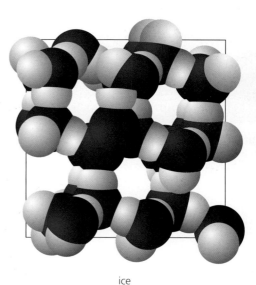

ice

Figure 3.90 Water and ice.

Figure 3.91 Iceberg.

Many solids are denser than water and sink in it. The iceberg in Figure 3.91 is off the coast of Newfoundland in Canada. Why does ice float on water?

2 Ethanol (CH_3CH_2OH)

a) Ethanol is soluble (or miscible) in water. The reason being that water can form hydrogen bonds with ethanol molecules as shown in Figure 3.92.

Again, when asked to draw this diagram follow the important features noted for drawing hydrogen bonds between water molecules on page 111.

b) Ethanol has a higher boiling point than expected for a molecule of its M_r. Ethanol has a boiling point of 79 °C whereas propane (C_3H_8) has a similar M_r but has a boiling point of −42 °C. This can be explained by the presence of hydrogen bonds between ethanol molecules. Whereas the only forces of attraction between propane molecules are van der Waals' forces.

113

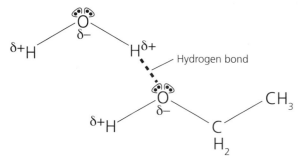

Figure 3.92 Hydrogen bonding between ethanol and water.

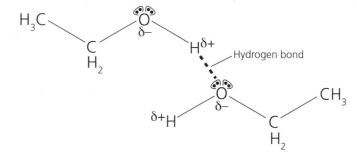

Figure 3.93 Hydrogen bonding in ethanol.

Figure 3.94 A glass of gin and tonic water. Why are they miscible?

TIP
The order of strength of the intermolecular forces is very important.

3 Propanone (CH_3COCH_3)

a) Propanone has a higher boiling point than would be expected for a molecule with its M_r. However its boiling point would be lower than an alcohol of similar M_r.

Molecule	M_r	Boiling point (°C)
Butane (C_4H_{10})	58	0.5
Propanone (CH_3COCH_3)	58	57
Propan-1-ol ($CH_3CH_2CH_2OH$)	60	97.1

The permanent dipole-dipole forces between propanone molecules cause its boiling point to be higher than an alkane with a similar M_r. However its boiling point is lower than an alcohol with a similar M_r.

The van der Waals' forces between the molecules in butane are weaker than the van der Waals' forces and permanent dipole-dipole forces between the molecules in propanone. These are weaker than the van der Waals' forces and hydrogen bonds between the molecules in propan-1-ol.

b) Solubility of propanone in water (or miscibility of propanone and water)

Permanent dipole-dipole interactions and van der Waals' forces exist between molecules of propanone as the C=O bond is polar (Figure 3.95). However when propanone mixes with water, hydrogen bonds are formed (Figure 3.96). This matches the definition of a hydrogen bond as the $\delta+$ H atom in one molecule (bonded to oxygen, nitrogen or fluorine) bonds with the lone pair of an oxygen, nitrogen or fluorine atom in another molecule. The ability of propanone to form hydrogen bonds with water molecules explains its solubility in water.

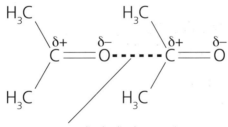

Figure 3.95 Intermolecular forces between propanone molecules.

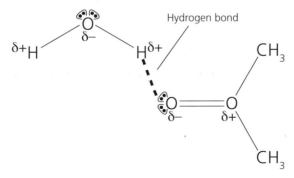

Figure 3.96 Hydrogen bonding between propanone and water.

ACTIVITY

Comparing the viscosity of liquids

Crude oil is very viscous, it does not flow easily. The viscosity of liquids can be compared by measuring their flow rate, in $cm^3 s^{-1}$. This may be carried out by measuring the volume of liquid flowing from a burette in a certain time interval.

The alcohols studied were:

CH_3CH_2OH CH_2OHCH_2OH $CH_2OHCHOHCH_2OH$
ethanol   ethane-1,2-diol   propane-1,2,3-triol

1 Draw a labelled diagram showing the assembled apparatus used to carry out this experiment.
2 Explain how you would carry out this experiment. State the measurements you would take, and explain how you would use these to determine the flow rate of each alcohol.
3 In the experiment the flow rate of ethanol was found to be $2.0\,cm^3 s^{-1}$. Calculate a value for the molar flow rate of ethanol in $mol\,min^{-1}$. Ethanol has density of $0.79\,g\,cm^{-3}$.
4 Hydrogen bonding has an effect on flow rate.
   a) Explain what is meant by the term hydrogen bonding.
   b) Show using a diagram, the hydrogen bonding between two ethanol molecules.
5 The results of this experiment show that propane-1,2,3-triol has a greater viscosity than ethane-1,2-diol which in turn has a greater viscosity that ethanol. Explain this trend.
6 It was suggested that the flow rate depended on the relative molecular mass (M_r) of the molecule, rather than on hydrogen bonding. Explain how you would carry out an experiment to determine that M_r did not affect flow rate, but hydrogen bonding did.

TEST YOURSELF 7

1 What causes van der Waals' forces of attraction?
2 What is needed in a molecule to allow it to form hydrogen bonds?
3 What is the strongest type of intermolecular force which exists between the molecules in the liquid state of the following substances?
   a) CH_4;   **b)** H_2O;
   c) C_2H_5OH;   **d)** H_2S;
   e) CCl_4;   **f)** NH_3
4 Explain fully why ice floats on water.
5 Explain why the halogens change from gases (fluorine and chlorine) to liquid (bromine) to solid (iodine).

Practice questions

1 Which of the following substances only shows van der Waals' forces between its molecules?

   A HF   **B** CH_4

   C NH_3 **D** H_2O *(1)*

2 What is the shape of the ammonium ion?

   A bent   **B** pyramidal

   C trigonal planar   **D** tetrahedral   *(1)*

3 What type of bond is formed when ammonia reacts with hydrogen chloride?

   A coordinate   **B** covalent

   C ionic   **D** metallic   *(1)*

4 Which atom has the lowest electronegativity value?

   A F   **B** H

   C N   **D** O   *(1)*

5 The graph below shows the boiling points of the Group 5 hydrides.

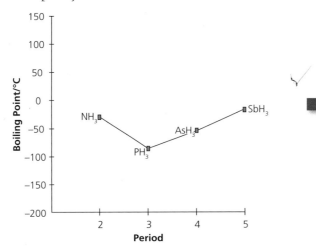

Figure 3.97

a) Explain why the boiling point of ammonia (NH_3) is much higher than that of phosphine (PH_3). *(3)*

b) Explain why there is a general increase in boiling point from phosphine (PH_3) to stibine (SbH_3). *(3)*

6 The density of water is $1 \, g \, cm^{-3}$ whereas the density of ice is $0.92 \, g \, cm^{-3}$. Explain why the density of ice is less than the density of water. *(2)*

7 Sodium chloride has a melting point of 797 °C and will not conduct electricity in the solid form but will when molten.

a) Explain why sodium chloride solution conducts electricity but solid sodium chloride does not. *(2)*

b) State the type of structure shown by sodium chloride. *(1)*

c) Explain why sodium chloride has a high melting point. *(2)*

8 Propanone (CH_3COCH_3) and butane ($CH_3CH_2CH_2CH_3$) both have a relative molecular mass of 58.

The boiling point of propanone is 56 °C whereas the boiling point of butane is 0 °C.

a) Name all the intermolecular forces which exist between molecules of butane in the liquid state. *(1)*

b) Name all the intermolecular forces which exist between molecules of propanone in the liquid state. *(1)*

c) Explain why the boiling point of butane is much lower than the boiling point of propanone. *(3)*

9 The diagram below shows the bonding and shape of a sulfate ion.

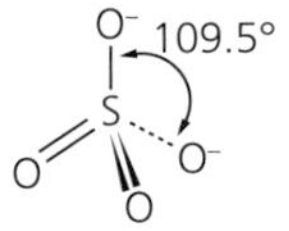

Figure 3.98

a) What name is given to the shape shown for the sulfate ion? *(1)*

b) Identify one neutral molecule which has the same shape as the sulfate ion. *(1)*

c) Identify one positive ion which has the same shape as the sulfate ion. *(1)*

d) Explain the shape of the sulfate ion based on the bonding diagram shown above using your understanding of electron pair repulsion theory. *(4)*

10 The melting points of magnesium and some of its compounds are given below.

	Melting point (°C)
Magnesium	650
Magnesium oxide	2852
Magnesium chloride	714
Magnesium iodide	637

a) Explain why magnesium has a higher melting point than sodium (melting point 98 °C). *(2)*

b) State the type of crystal structure shown by magnesium. *(1)*

c) Why is a very large amount of energy required to melt magnesium oxide. *(3)*

d) Explain why magnesium iodide has a lower melting point than magnesium chloride. *(3)*

11 a) Sulfur exists as S_8 molecules. It melts at 115 °C.

   i) Explain why the melting point of sulfur is low. *(3)*

   ii) State the type of bonding and structure present in sulfur. *(2)*

b) Sulfur difluoride, SF_2, reacts with chlorine in the presence of sodium fluoride. The reaction forms sulfur(IV) fluoride and sodium chloride.

   i) Write an equation for this reaction. *(1)*

   ii) Using your understanding of electron pair repulsion theory, describe the shape of the SF_2 and predict its bond angle. *(3)*

   iii) State the type of structure shown by sulfur dichloride and sodium fluoride. *(2)*

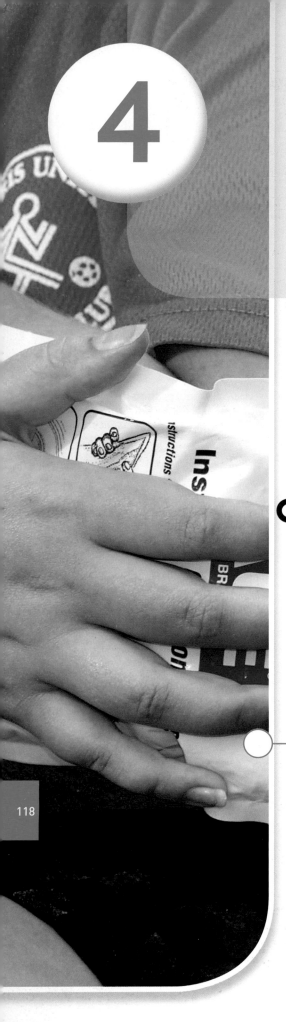

Energetics

Chemical reactions may be described as exothermic or endothermic depending on the heat exchange with their surroundings.

PRIOR KNOWLEDGE

- If a chemical reaction releases heat to the surroundings (the reaction vessel feels warmer) it is described as exothermic. If a chemical reaction absorbs heat from the surroundings, and so feels colder, it is described as endothermic.
- All combustion reactions are exothermic.
- Many oxidation and neutralisation reactions are exothermic.
- Exothermic reactions are used in self-heating food cans and in some hand warmers.
- If a reaction is reversible and is endothermic in the forward direction, it will be exothermic in the reverse direction. So if a reversible reaction is exothermic in one direction, it will be endothermic in the opposite direction.
- Thermal decomposition reactions are endothermic.
- Some sports injury packs which cool quickly use an endothermic reaction.

TEST YOURSELF ON PRIOR KNOWLEDGE 1

1 Which of the following types of reactions are exothermic and which are endothermic?
   a) thermal decomposition
   b) neutralisation
   c) combustion
2 Classify the following reactions as exothermic or endothermic.
   a) $CH_4(g) + 2O_2(g) \rightarrow CO_2(g) + 2H_2O(l)$
   b) $2NaOH(aq) + H_2SO_4(aq) \rightarrow Na_2SO_4(aq) + 2H_2O(l)$
   c) $2Mg(s) + O_2(g) \rightarrow 2MgO(s)$
   d) $MgCO_3(s) \rightarrow MgO(s) + CO_2(g)$

Enthalpy

Enthalpy is defined as a thermodynamic property of a system linked to internal energy. It is represented by the capital letter H. The enthalpy relates to the energy of the bonds broken and made during a chemical reaction. Some of the important points about enthalpy are listed below.

- Reactants – the enthalpy of the reactants in a chemical reaction is given as H_1. This relates to the energy of the bonds in the reactants.
- Products – the enthalpy of the products in a chemical reaction is given as H_2. This relates to the energy of the bonds in the products.
- ΔH is the change in enthalpy.
- ΔH = enthalpy of the products – enthalpy of the reactants.

Figure 4.1 Sports coaches often use an instant cold pack to reduce the pain of a sporting injury. The pack consists of two bags; one containing water inside a bag containing solid ammonium chloride. When the inner bag of water is broken by squeezing the package, the water dissolves the ammonium chloride in an endothermic change. This reaction absorbs heat from the surroundings and the pack becomes cold.

....

Enthalpy The enthalpy change is the heat energy change at constant pressure.

- For an exothermic reaction ΔH is negative; for an endothermic reaction ΔH is positive.
- Enthalpy changes can be measured and calculated but enthalpy cannot.

ΔH is measured under stated conditions. For example ΔH_{298} is the enthalpy change at a temperature of 298 K and a pressure of 100 kPa.

Some chemical reactions are reversible. For a reversible reaction the ΔH value for the reverse reaction has the same numerical value as the forward reaction but the sign is changed. This means that for an exothermic reaction in the forward direction ΔH is negative, but in the reverse direction the ΔH value has the same numerical value but it would be positive.

For example, the enthalpy changes for the reactions below show that the sign changes if the direction of the reaction is reversed.

$$N_2(g) + 3H_2(g) \rightarrow 2NH_3(g) \quad \Delta H = -92 \, kJ \, mol^{-1}$$

$$2NH_3(g) \rightarrow N_2(g) + 3H_2(g) \quad \Delta H = +92 \, kJ \, mol^{-1}$$

In many of the examples of calculations seen in the following sections, you will reverse a reaction theoretically to determine another enthalpy change.

Reaction profiles

A **reaction profile** is a diagram of the enthalpy levels of the reactants and products in a chemical reaction. The vertical (y) axis is enthalpy but *not* ΔH.

The horizontal (x) axis is progress of reaction, reaction coordinate or extent of reaction. Two horizontal lines are drawn and labelled with the names or formulae of reactants and products. These represent the enthalpy of the reactants (on the left) and the enthalpy of the products (on the right).

In an endothermic reaction the product line is at a higher enthalpy value than the reactant line as the reaction has absorbed energy. Conversely, in an exothermic reaction the product line is at a lower enthalpy value than the reactant line as the reaction has released energy. The difference between the lines is labelled ΔH (change in enthalpy). In an endothermic reaction this has a positive value. In an exothermic reaction this has a negative value. All values are measured in $kJ \, mol^{-1}$. If the actual reactants and products are known, the lines should be labelled with their names or formulae. Otherwise the labels 'reactants' and 'products' are sufficient. Figure 4.2 shows enthalpy level diagrams for both an endothermic reaction (Figure 4.2a) and an exothermic reaction (Figure 4.2b).

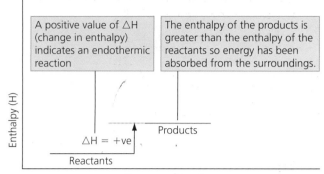

A positive value of △H (change in enthalpy) indicates an endothermic reaction

The enthalpy of the products is greater than the enthalpy of the reactants so energy has been absorbed from the surroundings.

Enthalpy (H)

△H = +ve

Products

Reactants

Progress of reaction

The value of the enthalpy change (△H) in this endothermic reaction is positive. This is because there has been an increase in enthalpy from reactants to products. This is a standard feature of **endothermic** reactions and it must be remembered that **△H is positive.**

Figure 4.2a An enthalpy level diagram for an endothermic reaction.

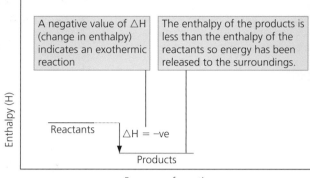

A negative value of △H (change in enthalpy) indicates an exothermic reaction

The enthalpy of the products is less than the enthalpy of the reactants so energy has been released to the surroundings.

Enthalpy (H)

Reactants

△H = −ve

Products

Progress of reaction

The value of the enthalpy change (△H) in this exothermic reaction is negative. This is because there has been a decrease in enthalpy from reactants to products. This is a standard feature of **exothermic** reactions and it must be remembered that **△H is negative.**

Figure 4.2b An enthalpy level diagram for an exothermic reaction.

Figure 4.3 A self-warming coffee can uses an inner chamber to hold the coffee and an outer chamber to hold calcium oxide and water, separated by a barrier. When the ring can is pulled, the chemicals come into contact and, due to an exothermic reaction, release heat to warm the coffee.

Reaction pathways

The **reaction pathway** is shown as a line from reactants to products on an enthalpy level diagram. It represents the route in terms of enthalpy from reactants to products. Reaction pathways require an input of energy to break bonds in the reactants before new bonds can form in the products. This amount of energy is the maximum height of the pathway above the enthalpy level of the reactants. This is called the **activation energy**.

For an exothermic reaction, a typical labelled enthalpy level diagram showing the reaction pathway would appear as shown in Figure 4.4.

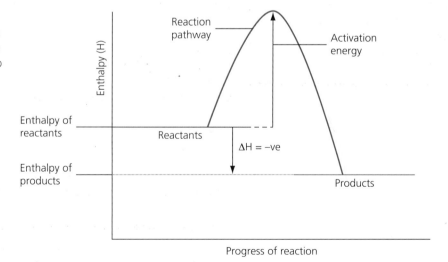

This shows that the activation energy is the minimum amount of energy which the reactants must have in order to react. Some reactions have low activation energy and can obtain enough energy at room temperature to raise the reactants to the required enthalpy value to allow the reaction to proceed.

Enthalpy (H)

Reaction pathway

Activation energy

Enthalpy of reactants

Reactants

△H = −ve

Enthalpy of products

Products

Progress of reaction

Figure 4.4 The reaction pathway for an exothermic reaction.

For an endothermic reaction, a typical labelled enthalpy level diagram showing reaction pathway would appear as shown in Figure 4.5.

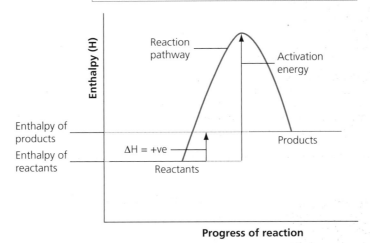

This shows that the activation energy is the amount of energy which the reactants must have in order to react. Many endothermic reactions have a high activation energy and cannot obtain enough energy at room temperature to raise the reactants to the required enthalpy value to allow the reaction to proceed.

Figure 4.5 The reaction pathway for an endothermic reaction.

Standard enthalpy values

Standard enthalpy values are the ΔH values for enthalpy changes of specific reactions measured under standard conditions.

- Standard conditions are represented by the symbol $\ominus$. This symbol is used after ΔH to indicate that an enthalpy changes occurs under standard conditions.
- Standard conditions are 100 kPa pressure and a stated temperature.
- There are three basic enthalpy changes for which the definitions must be learnt and you must be able to write equations to represent the reactions.
- State symbols should always be included in chemical equations to represent enthalpy changes.

Standard enthalpy of reaction ($\Delta_r H^\ominus$)

This is the enthalpy change when substances react under standard conditions in quantities given by the equation for the reaction.

For example:

$$CaO(s) + H_2O(l) \rightarrow Ca(OH)_2(s) \qquad \Delta_r H^\ominus = -63.7 \, kJ \, mol^{-1}$$

This means that when 1 mole of calcium oxide reacts with 1 mole of water to form 1 mole of calcium hydroxide, 63.7 kJ of heat would be released.

Standard enthalpy of formation ($\Delta_f H^\ominus$)

This is the enthalpy change when 1 mole of a compound is formed from its constituent elements with all reactants and products in standard states under standard conditions.

Figure 4.6 The reaction between calcium oxide and water is the reaction used in self-heating coffee cans (Figure 4.3). It is a vigorous reaction and here, steam is seen rising from the hot reaction mixture. The standard enthalpy of reaction for the calcium oxide and water is $\Delta_r H^\ominus = -63.7 \, \text{kJ mol}^{-1}$.

For example:

$$Ca(s) + C(s) + 1\tfrac{1}{2}O_2(g) \rightarrow CaCO_3(s)$$

$\Delta_f H^\ominus = -1128.8 \, \text{kJ mol}^{-1}$. This enthalpy change is per mole of calcium carbonate formed.

or

$$2C(s) + 2H_2(g) \rightarrow C_2H_4(g)$$

$\Delta_f H^\ominus = +52.5 \, \text{kJ mol}^{-1}$. This enthalpy change is per mole of ethene formed.

The enthalpy of formation of an element is zero if the element is in its standard state. The enthalpy of formation of oxygen, $O_2(g)$, is zero.

Standard enthalpy of combustion ($\Delta_c H^\ominus$)

This is the enthalpy change when 1 mole of a substance is burned completely in excess oxygen with all reactants and products in their standard states under standard conditions.

For example:

$$CH_4(g) + 2O_2(g) \rightarrow CO_2(g) + 2H_2O(l)$$

$\Delta_c H^\ominus$ for this reaction (per mole of methane burned) $= -890 \, \text{kJ mol}^{-1}$

The enthalpy change for the reaction:

$$2CH_4(g) + 4O_2(g) \rightarrow 2CO_2(g) + 4H_2O(l)$$

is $\Delta_r H^\ominus = 2 \times \Delta_c H^\ominus = -1780 \, \text{kJ}$

Note that as with all standard combustion reactions and enthalpy values, the combustion must be **per mole of substance burned**.

Figure 4.7 This image shows the most recent launch of the space shuttle which took place in July 2011. A variety of fuels with large enthalpies of combustion are used to get the rocket off the pad.

Some important points to note:

- It is vital that the standard enthalpy values of formation and combustion are kept to *per mole* of what they refer.
- If two moles of a fuel are combusted then the standard enthalpy of reaction is the standard enthalpy of combustion multiplied by 2.
- If four moles of a compound are formed from its elements in their standard states, then the standard enthalpy of formation value must be multiplied by 4 to get the standard enthalpy value for this reaction.
- The standard enthalpy of combustion of carbon ($C(s) + O_2(g) \rightarrow CO_2(g)$) has the same value as the standard enthalpy of formation of carbon dioxide. It is the same reaction and all the reactants and products are in standard states in both equations. This also applies to the standard enthalpy of formation of water and the standard enthalpy of combustion of hydrogen ($H_2(g) + \tfrac{1}{2}O_2(g) \rightarrow H_2O(l)$).

TIP

When asked to write an equation for a standard enthalpy change of combustion or formation, remember that for formation it is per mole of the compound formed and for combustion it is per mole of the substance burned.

TEST YOURSELF 2

1 State the conditions required for enthalpies of formation to be quoted as standard values.
2 Explain why the value of the standard enthalpy of formation of water is the same as the standard enthalpy of combustion of hydrogen.
3 Write an equation to represent the standard enthalpy of formation of ammonia.
4 Write equations, including state symbols, to represent the standard enthalpies of formation given below:
   a) formation of calcium oxide, $CaO(s)$
   b) formation of sodium chloride, $NaCl(s)$
   c) formation of water, $H_2O(l)$
   d) formation of carbon monoxide, $CO(g)$
   e) formation of ethanol, $C_2H_5OH(l)$
   f) formation of butane, $C_4H_{10}(g)$.
5 Write equations, including state symbols, to represent the standard enthalpies of combustion given below:
   a) combustion of carbon monoxide, $CO(g)$
   b) combustion of ethene, $C_2H_4(g)$
   c) combustion of methane, $CH_4(g)$
   d) combustion of potassium, $K(s)$
   e) combustion of hydrogen, $H_2(g)$
   f) combustion of ethanol, $C_2H_5OH(l)$.

Experimental determination of enthalpy changes

During a chemical reaction the enthalpy change in the reaction causes a change in the temperature of the surroundings. For many reactions, this change in temperature can be measured using a thermometer or a temperature probe. The energy released or absorbed from the reaction can be used to increase or decrease the temperature of a sample of water or the solution in which the reaction occurs.

Temperature change (ΔT) may be converted to energy change (q) using the expression:

$$q = mc\Delta T$$

where

q = change in heat energy in joules

m = mass in grams of the substance to which the temperature change occurs (usually water (for combustion) or a solution)

c = specific heat capacity (energy required to raise the temperature of $1\,g$ of a substance by $1\,°C$)

ΔT = temperature change in $°C$ or kelvin (K).

TIP

The specific heat capacity of water is $4.18\,J\,K^{-1}\,g^{-1}$. For many solutions in neutralisation determinations and other reactions the value is assumed to be the same. Do not be concerned if the value is quoted as $4.18\,J\,°C^{-1}\,g^{-1}$ as the temperature change of $1\,K$ (kelvin) is the same as a temperature change of $1\,°C$.

Figure 4.8 Substances with high specific heat capacities take a lot of heat energy, and therefore a long time to heat up and also a long time to cool down. One interesting effect is the way in which the land heats up quicker than the sea. The specific heat capacity of sea water is greater than that of the land and so more heat energy is needed to heat up the sea by the same amount as the land and so it takes longer. It also takes longer to cool down.

Calculating enthalpy change in a solution

The formula $q = mc\Delta T$ can be used to calculate the enthalpy change per mole of a substance which dissolves in water to form a solution. When an acid reacts with an alkali, a neutralisation reaction occurs. The enthalpy change of the neutralisation reaction can be calculated per mole of water formed in the reaction. These two types of calculations will be discussed in greater depth below.

Calculating enthalpy of solution
Some substances dissolve in water exothermically while others dissolve endothermically. The enthalpy of solution of sulfuric acid is very exothermic and so dilution should always be performed by adding the acid to the water, rather than the water to the acid. The following example shows how $q = mc\Delta T$ can be used to calculate the enthalpy change when one mole of a substance dissolves in water.

EXAMPLE 1

0.0770 moles of solid potassium iodide were dissolved in 25.0 cm³ (25.0 g) of deionised water in an open polystyrene cup. The temperature of the water was observed to decrease by 13.7 °C. Assuming that the temperature drop was due to the dissolution of potassium iodide, calculate a value for the enthalpy change to 3 significant figures in kJ mol⁻¹. (The specific heat capacity of water is 4.18 J K⁻¹ g⁻¹.)

Answer

$\Delta T = 13.7\,°C$

$q = mc\Delta T = 25.0 \times 4.18 \times 13.7 = 1431.65\,J$

number of moles $(n) = 0.0770$

energy change per mol of potassium iodide $= \dfrac{1431.65}{0.0770} = 18\,592.857\,J\,mol^{-1}$

enthalpy change in kJ mol⁻¹ $= 18\,592.857/1000 = +18.6\,kJ\,mol^{-1}$ (to 3 significant figures)

This enthalpy change is endothermic so a '+' is placed in front of the value to indicate an endothermic enthalpy change.

Calculating enthalpy of neutralisation

The enthalpy change that occurs during a neutralisation reaction when an acid reacts with an alkali is a specific type of enthalpy change in solution. An example of such a reaction is given in the following activity.

ACTIVITY

Calculating enthalpy of neutralisation

1 Measure accurately using a pipette 25.0 cm³ of a known concentration (usually 1.00 mol dm⁻³) of an acid. *This could be a strong acid or a weak acid. The enthalpy of neutralisation of a weak acid with a strong base is less than the enthalpy of neutralisation of a strong acid with a strong base as some energy is used in dissociating the acid fully; sodium hydroxide solution and ethanoic acid is –56.1 kJ mol⁻¹ while sodium hydroxide and hydrochloric acid – a strong base and strong acid – is –57.9 kJ mol⁻¹. The enthalpy of neutralisation of a weak base with a strong acid is similar to that of a weak acid with a strong base (ammonia and hydrochloric acid is –53.4 kJ mol⁻¹). The enthalpy of neutralisation of a weak acid with a weak base is less exothermic (ammonia and ethanoic acid is –50.4 kJ mol⁻¹). The energy difference depends on how weak the acid or base actually is.*

2 Place the acid in a polystyrene cup.

3 Calculate the number of moles of acid used:

$$\text{moles of acid} = \frac{\text{solution volume (25.0 cm}^3) \times \text{concentration (1.00 mol dm}^{-3})}{1000}$$

4 Measure accurately using a pipette 25.0 cm³ of a known concentration (usually 1.00 mol dm⁻³) of alkali.

5 Calculate the number of moles of alkali used as for the acid in part 3.

6 Write a balanced symbol equation for the reaction of the acid with the alkali to determine the number of moles of water formed. (If the acid is sulfuric acid, the moles of water formed will be $2H_2O$ whereas other acids usually form 1 mole of H_2O for each mole of the acid and alkali used.)

7 From the balanced symbol equation calculate the number of moles of water formed (n).

8 Place the thermometer in the acid and record its initial temperature (T_1).

9 Add the alkali and record the highest temperature reached (T_2).

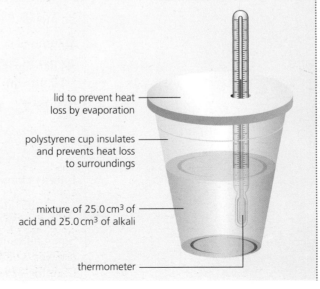

lid to prevent heat loss by evaporation

polystyrene cup insulates and prevents heat loss to surroundings

mixture of 25.0 cm³ of acid and 25.0 cm³ of alkali

thermometer

Figure 4.9

Method of calculation

1 Calculate temperature change $(\Delta T) = T_2 - T_1$. (This may be given in the question.)
2 Calculate the heat energy change in joules using $q = mc\Delta T$ (c is the specific heat capacity, i.e. the amount of energy required to raise the temperature of 1 g of a substance (in this case the solution) by 1 K (1 °C). This will be quoted in the question. m is the mass of the solution heated and as the total volume is 50 cm³, you will be given the density or told to assume the density is 1 g cm⁻³. Remember mass = density × volume.)
3 Convert to J mol⁻¹ by dividing q by n.
4 Convert to kJ mol⁻¹ by dividing by 1000. The answer is ΔH_n for the neutralisation. (All enthalpy values for neutralisation are exothermic and should have a negative sign.)

EXAMPLE 2

25.0 cm³ of 1.00 mol dm⁻³ hydrochloric acid were placed in a polystyrene cup and the initial temperature recorded as 22.7 °C. 25.0 cm³ of 1.00 mol dm⁻³ sodium hydroxide solution were added.

The highest temperature recorded was 29.3 °C. Assume the specific heat capacity of the solution is 4.18 J K⁻¹ g⁻¹ and the density of the solution is 1.00 g cm⁻³. Calculate a value for the enthalpy change when one mole of water is formed, to 3 significant figures.

Answer

Experimental data given in the question are in bold in the answer below.

$$\text{Moles of acid} = \frac{25.0 \times 1.00}{1000} = 0.0250 \text{ mol}$$

$$\text{Moles of alkali} = \frac{25.0 \times 1.00}{1000} = 0.0250 \text{ mol}$$

Neither reactant is in excess.

Balanced symbol equation:

$$NaOH(aq) + HCl(aq) \rightarrow NaCl(aq) + H_2O(l)$$

1 mole of H_2O is produced for each mole of NaOH and HCl

so moles of water formed = 0.0250 = $\boldsymbol{n}$

Volume of solution = 50.0 cm³

As density is assumed to be 1.00 g cm⁻³, the mass of solution heated = 50.0 g = $\boldsymbol{m}$

$$\Delta T = \boldsymbol{T_2} - \boldsymbol{T_1} = 29.3 - 22.7 = 6.6 °C$$

$$q = mc\Delta T = 50.0 \times 4.18 \times 6.6 = 1379.4 \text{ J}$$

$$\boldsymbol{n} = 0.0250$$

Energy change per mol of water formed $= \dfrac{1379.4}{0.0250} = 55\,176 \text{ J mol}^{-1}$

$$\Delta_n H^{\ominus} = \frac{55176}{1000} = -55.2 \text{ kJ mol}^{-1} \text{ (to 3 significant figures)}$$

Since all values in this calculation are given to 3 significant figures, the answer would be expected to be to 3 significant figures. You can refer to Chapter 17 for more details on mathematical precision.

EXAMPLE 3

In a neutralisation reaction between hydrochloric acid and sodium hydroxide solution, 0.0125 moles of water were formed and the temperature rose by 5.4 °C. The total mass of solution was 30.0 g and the specific heat capacity of the solution is assumed to be 4.18 J K^{-1} g^{-1}. Calculate a value for the enthalpy change in kJ mol^{-1} when one mole of water is formed in this neutralisation reaction. Give your answer to 2 decimal places.

Answer

ΔT = **5.4** °C

$q = mc\Delta T = 30.0 \times 4.18 \times 5.4 = 677.16$ J

n = 0.0125

Energy change per mol of water formed $= \dfrac{667.16}{0.0125} = 54\,172.8$ J mol^{-1}

Enthalpy change in kJ mol^{-1} $= \dfrac{54172.8}{1000} = -54.173$ kJ mol^{-1}

Enthalpy change to 2 decimal places = -54.17 kJ mol^{-1}

Calculating enthalpy change of combustion

The following procedure explains how to determine and calculate a value for the enthalpy of combustion of a liquid fuel from experimental data. The procedure can be examined in practical-style questions and may include a diagram of the apparatus used and any sources of errors due to heat loss. Values determined during the experiment are given in bold.

ACTIVITY

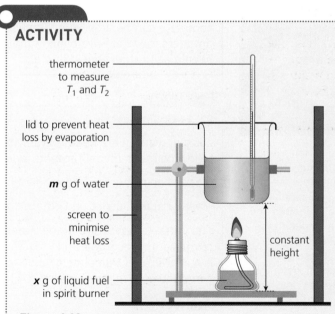

thermometer to measure T_1 and T_2

lid to prevent heat loss by evaporation

m g of water

screen to minimise heat loss

constant height

x g of liquid fuel in spirit burner

Figure 4.10

1 Set up a beaker containing a known volume of deionised water $= m$ cm$^3 = $ **m** g of water. (The mass of water may be given but as the density of water is 1 g cm^{-3}, the mass and volume are the same value, i.e. 100 g of water is 100 cm^3. Other solutions or liquids being heated may have a different density which will be given.)

2 Measure initial temperature of water using a thermometer (T_1).

3 Burn the fuel below the beaker of water. The mass of the burner containing the fuel can be measured before and after combustion to calculate the mass of fuel burned. The container should be insulated from heat loss and evaporation as shown in the diagram. The volume of the liquid fuel may be given along with the density of the fuel: mass (g) = volume (cm^3) × density (g cm^{-3}).

4 Measure the highest temperature reached by the water using a thermometer (T_2).

Method of calculation

1 Calculate temperature change (ΔT) = $T_2 - T_1$. (The temperature change may be given in the question, so use this as ΔT.)

2 Calculate the heat energy change in joules, $q = mc\Delta T$

3 Calculate moles of fuel used by dividing x by M_r of fuel = n moles. (The moles of fuel burned may be given in the question instead of the mass so this is n.)

4 Convert to J mol^{-1} (J per mol) by dividing q by n.

5 Convert to kJ mol^{-1} by dividing by 1000. (The temperature will increase in this example so the final enthalpy change should have a negative sign.)

The answer is a value for the standard enthalpy of combustion of the substance undergoing combustion.

Figure 4.11 Flambéing is often associated with the tableside presentation of certain liqueur-drenched dishes, such as Bananas Foster, when the alcohol is ignited and results in a flare of blue-tinged flame. By rapidly burning off the volatile alcohol, flambéing can infuse a dish with additional aroma and flavour and moderates the harshness of raw, high-proof spirits.

TIP
Remind yourself about significant figures by reading pages 348–350 in Chapter 17.

TIP
When enthalpy values are calculated from experimental data, the values may be less than expected as some heat is lost to the surroundings and also by evaporation. Also the substance may not have undergone complete combustion (due to an inadequate supply of oxygen) or the products may not have been in their standard states (water may have been a gas rather than a liquid).

EXAMPLE 4

200 g of water were heated by burning ethanol in a spirit burner. The following mass measurements were recorded:

Mass of spirit burner and ethanol (before burning) = 58.25 g

Mass of spirit burner and ethanol (after burning) = 57.62 g

The initial temperature of the water was 20.7 °C and the highest temperature recorded was 41.0 °C. The specific heat capacity of water is 4.18 J K^{-1} g^{-1}. Calculate a value for the standard enthalpy change of combustion of ethanol in kJ mol^{-1} to 3 significant figures.

Answer

Experimental data given in the question are in bold in the answer below.

$\Delta T = T_2 - T_1 = 41.0 - 20.7 = 20.3$ °C

m = 200 g of water

x = mass of ethanol burned = 58.25 - 57.62 = 0.63 g

$$n = \frac{\text{Mass}}{M_r} = \frac{0.63}{46.0} = 0.01370 \, \text{mol}$$

$$q = mc\Delta T = 200 \times 4.18 \times 20.3 = 16\,970.8 \, \text{J}$$

Heat energy change per mol of ethanol burned $= \dfrac{16970.8}{0.01370}$

$$= 1\,238\,744.526 \, \text{J mol}^{-1}$$

Standard enthalpy of combustion of ethanol per mole of ethanol =

$$\frac{1\,238\,744.526}{1000} = -1240 \, \text{kJ mol}^{-1} \text{ (to 3 significant figures)}$$

Calculating other quantities from enthalpy changes

The expression $q = mc\Delta T$ can be rearranged to calculate the temperature change given the standard enthalpy change of combustion or neutralisation.

EXAMPLE 5

The standard enthalpy of combustion of propane is $-2202 \, \text{kJ mol}^{-1}$. Given that 0.015 mol of propane are burned completely and the fuel is used to heat 200 g of water (specific heat capacity $4.18 \, \text{J g}^{-1} \text{K}^{-1}$), calculate the theoretical temperature change which would be measured. Give your answer to 3 significant figures.

Answer

For 1 mol of propane, 2 202 000 J of energy would be released on complete combustion.

For 0.015 mol of propane, heat energy released

$$= 0.015 \times 2\,202\,000$$

$$= 33\,030 \, \text{J} \; (= q)$$

$$q = mc\Delta T$$

$$\Delta T = \frac{q}{mc}$$

$$m = 200 \, \text{g}$$

$$c = 4.18 \, \text{J K}^{-1} \text{g}^{-1}$$

$$\Delta T = \frac{33030}{200 \times 4.18} = 39.5 \, \text{K}$$

When 0.015 mol of propane are burned completely in oxygen, the temperature of 200 g of water should rise by 39.5 K or 39.5 °C.

TIP

Remember that this is the theoretical temperature change assuming that complete combustion of the fuel occurred and that there was no heat loss to the surroundings, to the container or by evaporation.

REQUIRED PRACTICAL

Measuring and evaluating the enthalpy change for the neutralisation of sodium hydroxide and hydrochloric acid

Sodium hydroxide can be neutralised by adding hydrochloric acid. This reaction is exothermic. A student carried out an experiment to determine enthalpy of neutralisation for this reaction. The method followed was:

- Transfer 25.0 cm³ of 1.0 mol dm⁻³ hydrochloric acid to a plastic cup.
- Record the temperature of the hydrochloric acid to 1 decimal place.
- Transfer 25.0 cm³ of sodium hydroxide to a second clean, dry plastic cup and place the plastic cup inside a beaker. Stir the sodium hydroxide with a thermometer and record the temperature to 1 decimal place.
- Every minute for a further 3 minutes stir the solution, measure the temperature and record.
- At the fourth minute add the 25.0 cm³ of hydrochloric acid from the plastic cup. Stir the mixture but do not record the temperature.
- Continue to stir the mixture and measure the temperature at the fifth minute, and then every subsequent minute for a further 5 minutes. Record each temperature in a table like the one given below. Temperature of the hydrochloric acid at the start = 21.1 °C.

Table 4.1 Experiment results.

Time/min	Temperature/°C	Time/min	Temperature/°C
0	21.1	6	31.8
1	21.0	7	32.0
2	21.2	8	31.4
3	21.1	9	30.9
4		10	30.6
5	31.2		

1 Plot a graph of temperature (y-axis) against time (x-axis) using the results in Table 4.1. Draw a line of best fit for the points before the fourth minute. Draw a second line of best fit for the points after the fourth minute. Extrapolate the lines to the fourth minute, as shown in Figure 4.12.

2 Use your graph to determine an accurate value for the temperature of the sodium hydroxide at the fourth minute *before* mixing.

3 Use your answer from Question 2 and the temperature of the hydrochloric acid before mixing to calculate the average value for the temperature of the two solutions before mixing (T_1).

4 Use your graph to determine an accurate value for the temperature of the reaction mixture at the fourth minute (T_2).

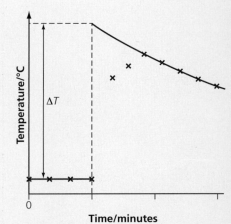

Figure 4.12 Estimating the maximum temperature of a neutralisation reaction.

5 Determine an accurate value for the temperature rise at the fourth minute ($T_2 - T_1$). Record your value to the appropriate precision.

6 Use your answer from Question 5 to calculate the heat given out during this experiment. Assume that the reaction mixture has a density of $1.00\,g\,cm^{-3}$ and a specific heat capacity of $4.18\,J\,K^{-1}\,g^{-1}$.

7 In the experiment $25.0\,cm^3$ of $1.00\,mol\,dm^{-3}$ hydrochloric acid were neutralised with $25.0\,cm^3$ of $1.00\,mol\,dm^{-3}$ sodium hydroxide. Calculate the amount, in moles, of HCl present in $25.0\,cm^3$ and the amount in moles of NaOH present in $25.0\,cm^3$.

8 Use your answers from Questions 6 and 7 to calculate the enthalpy change, in $kJ\,mol^{-1}$, for the reaction between one mole of HCl and the sodium hydroxide.

9 The maximum total error in using the thermometer in this experiment is $\pm0.1\,^{\circ}C$. The error takes into account multiple measurements made using the thermometer. Use your answer from Question 5 to calculate the percentage error in your value for the temperature rise.

10 Consider your graph. State whether your lines of best fit are good enough for you to extrapolate with confidence. Explain your answer.

11 Explain why the experiment should be repeated several times in order to determine an accurate value for the enthalpy change.

12 Suggest one reason why your value for the enthalpy change using the sodium hydroxide might differ from a data book value for the enthalpy change of neutralisation of sodium hydroxide.

13 Suggest why the plastic cup was placed inside a beaker in the practical.

14 Suggest how the $25.0\,cm^3$ of HCl was initially measured out.

TEST YOURSELF 3

1 In an experiment the temperature of $120\,g$ of water rose by $10.1\,^{\circ}C$ when 0.0170 moles of methanol were burned in air and the heat was used to warm the water. Calculate a value for the enthalpy change when one mole of methanol is burned. (The specific heat capacity of water is $4.18\,J\,K^{-1}\,g^{-1}$.) Give your answer to 3 significant figures.

2 $0.600\,g$ of propane (C_3H_8) were completely burned in air. The heat evolved raised the temperature of $100\,g$ of water by $65.0\,^{\circ}C$. (The specific heat capacity of water is $4.18\,J\,K^{-1}\,g^{-1}$.)
   a) Calculate the number of moles of propane burned.
   b) Calculate the heat energy released during the combustion in joules (J).
   c) Calculate a value for the enthalpy of combustion of propane in $kJ\,mol^{-1}$. Give your answer to 3 significant figures.

3 In an experiment, $1.00\,g$ of propanone (CH_3COCH_3) was completely burned in air. The heat evolved raised the temperature of $150\,g$ of water from $18.8\,^{\circ}C$ to $64.3\,^{\circ}C$. Use this data to calculate a value for the enthalpy of combustion of propanone in $kJ\,mol^{-1}$. (The specific heat capacity of water is $4.18\,J\,K^{-1}\,g^{-1}$.) Give your answer to 3 significant figures.

4 In a neutralisation reaction between nitric acid and potassium hydroxide solution, 0.050 moles of water were formed. The temperature rose by 13.0°C. The specific heat capacity of the solution is assumed to be $4.18 \, J \, K^{-1} \, g^{-1}$. Calculate a value for the enthalpy change in $kJ \, mol^{-1}$ when one mole of water is formed if the total mass of the solution is 50.0 g.

5 0.0150 moles of solid sodium hydroxide were dissolved in 25.0 cm³ (25.0 g) of deionised water in an open polystyrene cup. The temperature of the water increased by 6.10°C. The specific heat capacity of the water is $4.18 \, J \, K^{-1} \, g^{-1}$. Assume that all the heat released is used to raise the temperature of the water. Calculate a value for the enthalpy change in $kJ \, mol^{-1}$. Give your answer to an appropriate number of significant figures.

Hess's Law

The **principle of conservation of energy** states that energy cannot be created or destroyed, only changed from one form into another.

After practising medicine for several years in Irkutsk, Russia, Germain Hess became professor of chemistry at the University of St Petersburg in 1830. In 1840 Hess published his law of constant heat summation, which we refer to today as 'Hess's Law'.

Hess's Law states that the enthalpy change for a chemical reaction is independent of the route taken and depends only on the initial and final states.

Hess's Law can be used to calculate enthalpy changes for chemical reactions from the enthalpy changes of other reactions. This is useful as some reactions cannot be carried out in reality, but a theoretical enthalpy change can be determined for these reactions.

EXAMPLE 6

Calculate a value for the standard enthalpy of formation of propanone, $CH_3COCH_3(l)$, given the following standard enthalpy changes of combustion:

	$\Delta_c H^\ominus / kJ \, mol^{-1}$
C(s)	−394
$H_2(g)$	−286
$CH_3COCH_3(l)$	−1821

Answer

There are two ways to approach this question: using a Hess's Law diagram or using the sum of the enthalpy changes.

Solving by Hess's Law diagram
Start by writing the main equation for the enthalpy change you are to calculate and then write equations for the standard enthalpy changes you have been given to answer the question.

Main equation:

$$3C(s) + 3H_2(g) + \tfrac{1}{2}O_2(g) \rightarrow CH_3COCH_3(l)$$

Equations for given enthalpy changes of combustion:

$C(s) + O_2(g)$	$\rightarrow$	$CO_2(g)$	$-394\,kJ\,mol^{-1}$
$H_2(g) + \tfrac{1}{2}O_2(g)$	$\rightarrow$	$H_2O(l)$	$-286\,kJ\,mol^{-1}$
$CH_3COCH_3(l) + 4O_2(g)$	$\rightarrow$	$3CO_2(g) + 3H_2O(l)$	$-1821\,kJ\,mol^{-1}$

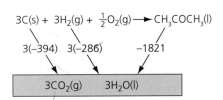

Figure 4.13 Hess's Law diagram.

The substances which do not appear in the main equation but do appear in these equations for the given enthalpy changes are $CO_2(g)$ and $H_2O(l)$. These are these link substances which will allow you to draw a Hess's Law diagram. Below the main equation put $3CO_2(g)$ and $3H_2O(l)$. Draw arrows for the enthalpy changes given to you in the question in the direction of the change. Write values on the arrows and remember to multiply by the number of moles of substance burned.

The final calculation is from the reactants $3C(s)$, $3H_2(g)$ and $\tfrac{1}{2}O_2(g)$ to the products in the box and then from the box to propanone $CH_3COCH_3(l)$. The oxygen will balance as the arrows to the box all represent combustion reactions.

A value for the standard enthalpy of formation of propanone (main equation) can be calculated by going from the reactants to $CO_2(g)$ and $H_2O(l)$ and then reversing the combustion of propanone.

$\Delta_f H^\ominus CH_3COCH_3(l)$

$= +3(-394) + 3(-286) - (-1821)$

$= -1182 - 858 + 1821$

$= -219\,kJ\,mol^{-1}$

Using the sum of the enthalpy changes

$$\Delta_f H = (\text{sum } \Delta_c H \text{ reactants}) - (\text{sum } \Delta_c H \text{ products})$$

This method only works for an example like this in which the enthalpy of formation is calculated from enthalpy of combustion values. The number of moles of the reactants and products must be taken into account.

For this example:

$$\Delta_f H = \Sigma \Delta_c H \text{ (reactants)} - \Sigma \Delta_c H \text{ (products)}$$

where Σ is 'the sum of' which means adding all the values together.

Reactants $= 3C(s) + 3H_2(g)$

$\Sigma \Delta_c H \text{ (reactants)} = 3 \times (-394) + 3(-286)$
$= -2040\,kJ\,mol^{-1}$

Products $= CH_3COCH_3(l)$

$\Sigma \Delta_c H \text{ (products)} = -1821\,kJ\,mol^{-1}$

$\Delta_f H = \Sigma \Delta_c H \text{ (reactants)} - \Sigma \Delta_c H \text{ (products)}$

$\Delta_f H = -2040 - (-1821)$

$= -2040 + 1821$

$= -219\,kJ\,mol^{-1}$

TIP

When using standard enthalpy changes of combustion it is not necessary to balance the $O_2(g)$ as the same number of moles of $O_2(g)$ will be added on each side. In this example, 3 moles of $O_2(g)$ together with the $\tfrac{1}{2}O_2(g)$ (total $3\tfrac{1}{2}O_2(g)$) react on the left-hand side going to the box and $3\tfrac{1}{2}O_2(g)$ react on the right-hand side going to the box so the moles of $O_2(g)$ added cancel out.

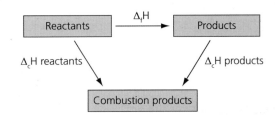

Figure 4.14 Enthalpy cycle using enthalpy of combustion.

EXAMPLE 7

Calculate a value for the standard enthalpy change of combustion of ethane from the following standard enthalpy changes of formation:

$\Delta_f H^{\ominus}/kJ\,mol^{-1}$

$CO_2(g)$	–394
$H_2O(l)$	–286
$C_2H_6(g)$	–85

Answer

Solving by Hess's Law diagram

Main equation:

$$C_2H_6(g) + 3\tfrac{1}{2}O_2(g) \rightarrow 2CO_2(g) + 3H_2O(l)$$

Equations for the given enthalpy changes:

$$C(s) + O_2(g) \quad \rightarrow \quad CO_2(g)$$
$$H_2(g) + \tfrac{1}{2}O_2(g) \quad \rightarrow \quad H_2O(l)$$
$$2C(s) + 3H_2(g) \quad \rightarrow \quad C_2H_6(g)$$

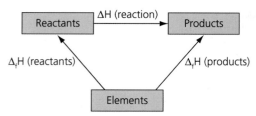

Figure 4.15 Enthalpy cycle using enthalpy of formation.

From the equation for the given enthalpy changes the substances which do not appear in the main equation are $C(s)$ and $H_2(g)$. This cycle can be thought of as ethane being converted to its elements in their standard states and the elements reacting with oxygen to form $CO_2(g)$ and $H_2O(l)$.

$$\Delta_c H^{\ominus}(C_2H_6(g)) = -(-85) + 2(-394) + 3(-286)$$
$$= +85 - 788 - 858$$
$$= -1561\,kJ\,mol^{-1}$$

Using the sum of the enthalpy changes
$$\Delta_c H = (\text{sum }\Delta_f H \text{ products}) - (\text{sum }\Delta_f H \text{ reactants})$$

This method only works in an example in which the enthalpy of combustion or the enthalpy of reaction is calculated from enthalpy of formation values. The number of moles of the reactants and products must be taken into account.

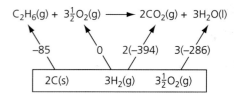

Figure 4.16

For this example:

$$\Delta_c H = \Sigma\Delta_f H \text{ (products)} - \Sigma\Delta_f H \text{ (reactants)}$$

Reactants $= C_2H_6(g) + 3\tfrac{1}{2}O_2(g)$

$\Sigma\Delta_f H$ (reactants) $= -85\,kJ\,mol^{-1}$

Products $= 2CO_2(g) + 3H_2O(l)$

$\Sigma\Delta_f H$ (products) $= 2(-394) + 3(-286)$

$$= -1646\,kJ\,mol^{-1}$$

$$\Delta_c H = \Sigma\Delta_f H \text{ (products)} - \Sigma\Delta_f H \text{ (reactants)}$$

$$\Delta_c H = -1646 - (-85)$$

$$= -1561\,kJ\,mol^{-1}$$

TIP
Oxygen is an element and so its enthalpy of formation value is zero.

TIP
These sum methods work to calculate enthalpy of formation or reaction from enthalpy of combustion values (Example 6) and to calculate enthalpy of combustion or reaction values from enthalpy of formation values.

Figure 4.17 Iron is extracted from its ore in a blast furnace. Heat radiated from the metal is reflected away by the shiny surface of the protective suit, which covers all of the body.

EXAMPLE 8

Iron is extracted from its ore according to the following equation:

$$Fe_2O_3(s) + 3CO(g) \rightarrow 2Fe(s) + 3CO_2(g)$$

Calculate a value for the standard enthalpy change of this reaction from the following standard enthalpy changes of formation:

	$\Delta_fH^{\ominus}$/kJ mol^{-1}
$Fe_2O_3(s)$	−826
$CO(g)$	−111
$Fe(s)$	0
$CO_2(g)$	−394

Answer

Solving by Hess's Law diagram

This cycle can be thought of as Fe_2O_3 and CO being converted to their elements in their standard states and C reacting with O_2 to form CO_2.

$$Fe_2O_3(s) + 3CO(g) \longrightarrow 2Fe(s) + 3CO_2(g)$$

$$-826 \quad 3(-111) \quad 0 \quad 3(-394)$$

$$\boxed{2Fe(s) \qquad 3C(s) \qquad 3O_2(g)}$$

Figure 4.18

$$\Delta_rH^{\ominus} = -(-826) - 3(-111) + 3(-394)$$

$$= +826 + 333 - 1182 = -23 \text{ kJ mol}^{-1}$$

Using the sum of the enthalpy changes

$$\Delta_rH = \Sigma\Delta_fH \text{ (products)} - \Sigma\Delta_fH \text{ (reactants)}$$

Reactants = $Fe_2O_3(s) + 3CO(g)$

$$\Sigma\Delta_fH \text{ (reactants)} = -826 + 3(-111)$$

$$= -1159 \text{ kJ mol}^{-1}$$

Products = $2Fe(s) + 3CO_2(g)$

$$\Sigma\Delta_fH \text{ (products)} = 3(-394)$$

$$= -1182 \text{ kJ mol}^{-1}$$

$$\Delta_cH = \Sigma\Delta_fH \text{ (products)} - \Sigma\Delta_fH \text{ (reactants)}$$

$$\Delta_cH = -1182 - (-1159)$$

$$= -23 \text{ kJ mol}^{-1}$$

TIP

Questions are often asked about why a substance would have zero as the value of its standard enthalpy change of formation. The answer is that this substance is an element.

TIP

Fe(s) is an element in its standard state and so its enthalpy of formation value is zero.

EXAMPLE 9

Vanadium metal can be obtained from ore using calcium metal.
The reaction is represented by the equation:

$$V_2O_5(s) + 5Ca(s) \rightarrow 2V(l) + 5CaO(s)$$

a) Use the standard enthalpies of formation in the table below to calculate a value for the standard enthalpy change of this reaction.

	$V_2O_5(s)$	$Ca(s)$	$V(l)$	$CaO(s)$
$\Delta_fH^\circ/kJ\,mol^{-1}$	−1551	0	+23	−635

b) Explain why the standard enthalpy of formation of Ca(s) is zero but the standard enthalpy of formation of V(l) is **not** zero.

Answers

a)

Solving by Hess's Law diagram

$\Delta_rH^\circ = -(-1551) + 2(+23) + 5(-635)$

$= +1551 + 46 - 3175$

$= -1578\,kJ\,mol^{-1}$

Using the sum of the enthalpy changes

$\Delta_rH = \Sigma\Delta_fH\,(products) - \Sigma\Delta_fH\,(reactants)$

Reactants $= V_2O_5(s) + 5Ca(s)$

$\Sigma\Delta_fH\,(reactants) = -1551\,kJ\,mol^{-1}$

Products $= 2V(l) + 5CaO(s)$

$\Sigma\Delta_fH\,(products) = 2(+23) - 5(-635)$

$= -3129\,kJ\,mol^{-1}$

$\Delta_rH = \Sigma\Delta_fH\,(products) - \Sigma\Delta_fH\,(reactants)$

$\Delta_rH = -3129 - (-1551)$

$= -1578\,kJ\,mol^{-1}$

b) Ca(s) is an element in its standard state whereas V(l) is an element but it is not in its standard state under standard conditions. V(s) is the element in its standard state.

$V_2O_5(s) + 5Ca(s) \longrightarrow 2V(l) + 5CaO(s)$

−1551   0   2(+23)   5(−635)

| $2V(s)$ | $2\frac{1}{2}O_2(g)$ | $5Ca(s)$ |

Figure 4.20

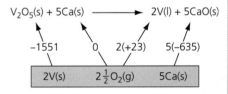

TIP
Ca(s) is an element in its standard state and so its enthalpy of formation value is zero.

TEST YOURSELF 4

1 Using the following values for the standard enthalpy changes, calculate the value for the reactions given below.

$$\Delta_c H^\ominus H_2(g) = -286 \, kJ \, mol^{-1}$$
$$\Delta_f H^\ominus NaCl(s) = -411 \, kJ \, mol^{-1}$$
$$\Delta_f H^\ominus CO_2(g) = -394 \, kJ \, mol^{-1}$$

  a) $2H_2(g) + O_2(g) \rightarrow 2H_2O(l)$
  b) $2Na(s) + Cl_2(g) \rightarrow 2NaCl(s)$
  c) $C(s) + O_2(g) \rightarrow CO_2(g)$
  d) $2C(s) + 2O_2(g) \rightarrow 2CO_2(g)$
  e) $3H_2(g) + 1\frac{1}{2}O_2(g) \rightarrow 3H_2O(l)$

2 Use the standard enthalpy of formation data from the table and the equation for the combustion of methane to calculate a value for the standard enthalpy of combustion of methane.

	$CH_4(g)$	$O_2(g)$	$CO_2(g)$	$H_2O(g)$
$\Delta_f H^\ominus/kJ \, mol^{-1}$	−75	0	−394	−286

3 Use the standard enthalpy of combustion data from the table to calculate a value for the standard enthalpy of formation of sucrose, $C_{12}H_{22}O_{11}(s)$. The equation for the formation of sucrose is given below.

$$12C(s) + 11H_2(g) + 5\frac{1}{2}O_2(g) \rightarrow C_{12}H_{22}O_{11}(s)$$

	$C(s)$	$H_2(g)$	$C_{12}H_{22}O_{11}(s)$
$\Delta_c H^\ominus/kJ \, mol^{-1}$	−394	−286	−5640

Bond enthalpies

The **mean bond enthalpy** is a measure of the energy required to break one mole of a covalent bond measured in the gaseous state in $kJ \, mol^{-1}$ (the 'per mole' is per mole of the covalent bond) averaged across many compounds containing the bond. For example, the mean bond enthalpy of a C—H bond is $412 \, kJ \, mol^{-1}$.

Bond breaking is endothermic and bond making is exothermic.

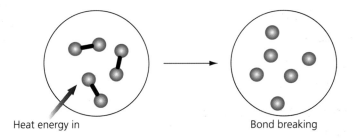

Heat energy in   Bond breaking

Figure 4.21 Breaking bonds takes in heat energy and is endothermic.

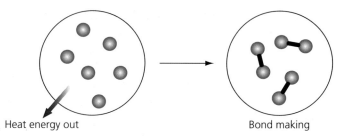

Heat energy out

Bond making

Figure 4.22 Making bonds gives out heat energy and is exothermic.

- Breaking one mole of C—H bonds requires 412 kJ of energy so the value is written as a positive value, $+412\,kJ\,mol^{-1}$. It is positive as it is endothermic.
- Making one mole of C—H bonds releases 412 kJ of energy so the value is written as a negative value, $-412\,kJ\,mol^{-1}$. It is negative as it is exothermic.

General points for bond enthalpies

Mean bond enthalpies relate to the strength of a covalent bond. A higher bond enthalpy value means a stronger covalent bond.

- Generally the shorter the covalent bond, the stronger the bond.
- Triple covalent bonds are generally shorter than double covalent bonds which are shorter than single covalent bonds.
- Triple covalent bonds are generally stronger than double covalent bonds which are stronger than single covalent bonds.

Table 4.2 shows the bond enthalpies and bond lengths of carbon—carbon covalent bonds.

Table 4.2 Bond enthalpies and bond lengths for carbon—carbon bonds.

Covalent bond	Mean bond enthalpy/ kJ mol⁻¹	Bond length/nm
C—C	348	0.154
C=C	611	0.134
C≡C	838	0.120

The C—C is the longest of the three carbon–carbon covalent bonds and is also the weakest. The C≡C is the shortest of the three bonds and the strongest.

The bond length and bond strength (expressed often as the mean bond enthalpy) are important in organic chemistry and the chemistry of the halogens. The bond length and bond enthalpies for the covalent bonds in halogen molecules are given in Table 4.3.

Table 4.3 Bond lengths and bond enthalpies for covalent bonds in halogen molecules.

Covalent bond	Bond enthalpy/kJ mol⁻¹	Bond length/nm
F—F	158	0.142
Cl—Cl	242	0.199
Br—Br	193	0.228
I—I	151	0.267

Figure 4.23 The trend in bond enthalpies for the elements of Group 7.

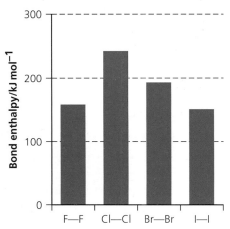

The bond length increases down the group as would be expected as the atomic radius of the atoms involved in the covalent bond increases. The strength of the covalent bond decreases down the group from Cl_2 to I_2. The bond enthalpy of the F—F bond is lower than expected as the bond is very short and there are repulsions between the lone pairs of electrons on the fluorine atoms. This lowers the energy required to break the F—F bond.

The bond length and bond enthalpies for the C—X (carbon–halogen) bonds will be considered in halogenoalkanes.

Calculations involving mean bond enthalpies

These involve calculations of enthalpy changes using mean bond enthalpies. The enthalpy change for a reaction can be calculated by adding together the mean bond enthalpies for all the bonds broken in the reactant molecules and subtracting the total of all the mean bond enthalpies of the bonds made in the products. The reactions are considered in the gas phase so that only the covalent bonds are involved in the calculations. In liquid or solid phase, intermolecular forces would also be involved. Any deviation of the answer to a calculation from the quoted value may be due to the mean bond enthalpy values used or the reactants not being in their gaseous states in the calculation.

ΔH = sum of mean bond enthalpies of bonds broken – sum of mean bond enthalpies of bonds made

EXAMPLE 10

The reaction below represents the formation of hydrogen chloride.

$H_2(g) + Cl_2(g) \rightarrow 2HCl(g)$

Use the following bond enthalpies to calculate a value for the enthalpy change of formation of hydrogen chloride. Explain whether the reaction is exothermic or endothermic.

Bond	Bond enthalpy/kJ mol^{-1}
H—H	436
Cl—Cl	242
H—Cl	432

Figure 4.24 Bonds are broken in hydrogen and chlorine and new bonds are formed in hydrogen chloride.

Answer

Calculate the energy required for moles of bond broken in the reactants:

1 mole of H—H = 436

1 mole of Cl—Cl = 242

Total energy required for bonds broken = 678 kJ mol^{-1}

Calculate the energy released for moles of bonds formed in the products:

2 moles of H—Cl = 2(432) = 864

Total energy released for bonds made = 864 kJ mol^{-1}

ΔH = sum of bond enthalpies of bonds broken – sum of bond enthalpies of bonds made

ΔH = 678 – 864 = –186 kJ mol^{-1}

This value is for the formation of 2 moles of HCl(g). The enthalpy of formation is for 1 mole of HCl so the enthalpy of formation is –93 kJ mol^{-1}.

The enthalpy of formation of hydrogen chloride is exothermic as ΔH is negative.

TIP

The term 'mean bond enthalpy' is not used in this example as the H–H, Cl–Cl and H–Cl bonds only occur in these molecules and so are not averaged across molecules containing the bond.

139

EXAMPLE 11

Calculate a value for the standard enthalpy of combustion of methane from the mean bond enthalpies given.

Bond	Mean bond enthalpy/kJ mol^{-1}
C—H	412
O—H	463
O=O	496
C=O	803

Answer

In the combustion of methane, 1 mole of methane reacts with excess oxygen to form carbon dioxide and water.

$$CH_4(g) + 2O_2(g) \rightarrow CO_2(g) + 2H_2O(g)$$

The following bonds are broken:

4 moles of C—H $\qquad\qquad$ 2 moles of O=O

and the following bonds are made:

2 moles of C=O $\qquad\qquad$ 4 moles of O—H

If you are unsure about the bonds broken or made from the equation for the reaction, draw out the structures of the molecules to make it clearer (Figure 4.25).

The sum of all the mean bond enthalpies of the bonds broken and the sum of all the mean bond enthalpies of the bonds made are calculated from the given mean bond enthalpies.

Sum of mean bond enthalpies of bonds broken

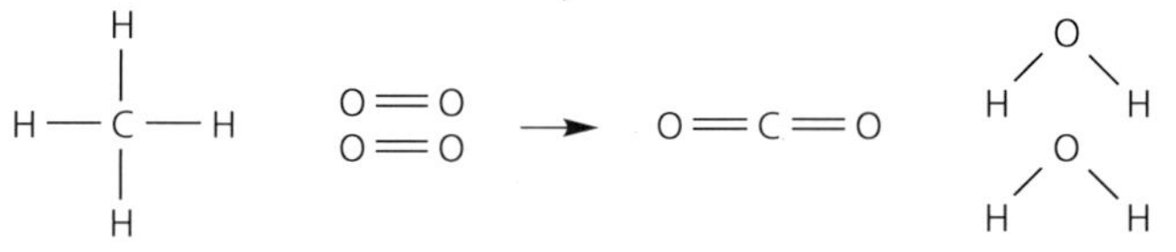

Figure 4.25

$$= 4(412) + 2(496) = 2640\,kJ\,mol^{-1}$$

Sum of mean bond enthalpies of bonds made
$$= 2(803) + 4(463) = 3458\,kJ\,mol^{-1}$$

$$\Delta H = \text{sum of mean bond enthalpies of bonds broken} - \text{sum of mean bond enthalpies of bonds made}$$

Standard enthalpy of combustion of methane
$$= +2640 - 3458 = -818\,kJ\,mol^{-1}$$

The reaction is exothermic as ΔH is negative. The quoted value in data books for the standard enthalpy change of combustion of methane is $-890\,kJ\,mol^{-1}$. Enthalpy changes determined from mean bond enthalpies often differ from the quoted values as the mean bond enthalpies are not specific to the bonds in the molecules in this reaction. They are averaged across many different molecules containing that particular bond. Also, for standard enthalpy changes, water would be a liquid, but when using mean bond enthalpies, the substances are in the gaseous state.

Figure 4.26 Butanol may be used as a fuel in an internal combustion engine. Because its longer hydrocarbon chain causes it to be fairly non-polar, it is more similar to petrol than it is to ethanol. Butanol has been demonstrated to work in vehicles designed for use with gasoline without modification. It can be produced from biomass as 'biobutanol' and is a useful renewable fuel.

EXAMPLE 12

Calculate a value for the standard enthalpy of combustion of butan-1-ol $C_4H_9OH(g)$ using the following mean bond enthalpies.

Bond	Mean bond enthalpy/kJ mol^{-1}
C—C	348
C—H	412
O—H	463
C—O	360
O=O	496
C=O	803

Answer

Equation for combustion of butan-1-ol

$C_4H_9OH(g) + 6O_2(g) \rightarrow 4CO_2(g) + 5H_2O(g)$ (per mole of C_4H_9OH)

Structural equation showing all the moles of covalent bonds:

Figure 4.27

Calculate the energy required for moles of bonds broken in the reactants:

$$3 \text{ moles of C—C} = 3(348)$$
$$9 \text{ moles of C—H} = 9(412) = 3708$$
$$1 \text{ mole of C—O} = 360$$
$$1 \text{ mole of O—H} = 463$$
$$6 \text{ moles of O=O} = 6(496) = 2976$$

Total energy required for bonds broken = 8551 kJ mol^{-1}

Calculate the energy released for moles of bonds formed in the products:

$$8 \text{ C=O} = 8(803) = 6424$$
$$10 \text{ O—H} = 10(463) = 4630$$

Total energy released for bonds made = 11 054 kJ mol^{-1}

ΔH = sum of mean bond enthalpies of bonds broken – sum of mean bond enthalpies of bonds made

$\Delta H = 8551 - 11054 = -2504$ kJ mol^{-1}

TEST YOURSELF 5

1 What is meant by the term 'mean bond enthalpy'?

2 The table gives the bond lengths and mean bond enthalpies of some covalent bonds.

a) What is the general relationship between bond length and mean bond enthalpy?

b) By determining the bonds broken and bonds made during the combustion of propane, determine a value for the standard enthalpy of combustion of propane in $kJ\,mol^{-1}$.

$C_3H_8(g) + 5O_2(g) \rightarrow 3CO_2(g) + 4H_2O(g)$

Bond	Bond length/nm	Mean bond enthalpy/kJ mol^{-1}
H—H	0.074	436
H—O	0.096	463
C—C	0.154	348
C=C	0.134	611
C—H	0.108	412
C—O	0.143	360
C=O	0.122	803
O=O	0.121	496

3 By determining the bonds broken and made during the combustion of gaseous ethanol, $C_2H_5OH(g)$, determine a value for the standard enthalpy of combustion of ethanol.

$C_2H_5OH(g) + 3O_2(g) \rightarrow 2CO_2(g) + 3H_2O(g)$

4 In the following reaction ethane undergoes complete combustion:

$C_2H_6(g) + 3\frac{1}{2}O_2(g) \rightarrow 2CO_2(g) + 3H_2O(l)$

a) Using the mean bond enthalpies given in the table, calculate a value for the standard enthalpy of combustion of ethane in $kJ\,mol^{-1}$.

b) The standard enthalpy of combustion of ethane is quoted as $-1559.8\,kJ\,mol^{-1}$. Explain why there is a difference between this value and the value you obtained in **(a)** above.

Practice questions

1 Which of the following equations represents the standard enthalpy change of formation of hydrogen iodide?

A $H_2(g) + I_2(s) \rightarrow 2HI(g)$

B $H_2(g) + I_2(g) \rightarrow 2HI(g)$

C $\frac{1}{2}H_2(g) + \frac{1}{2}I_2(s) \rightarrow HI(g)$

D $\frac{1}{2}H_2(g) + \frac{1}{2}I_2(g) \rightarrow HI(g)$ *(1)*

2 The standard enthalpy change for the formation of hydrogen fluoride is $-269\,kJ\,mol^{-1}$. What is the enthalpy change for the reaction

$2HF(g) \rightarrow F_2(g) + H_2(g)$?

A -269 **B** $+269$

C -538 **D** $+538$ *(1)*

3 The standard enthalpy of combustion of methane, $CH_4(g)$, is $-690\,kJ\,mol^{-1}$. What is the temperature change when 0.02 mol of methane is burned completely in air and the heat released used to heat 250 g of water? The specific heat capacity of water is $4.18\,J\,K^{-1}\,g^{-1}$.

A 0.66 K   **B** 1.32 K

C 13.2 K   **D** 55.2 K   *(1)*

4 In an experiment, 1.00 g of hexane (C_6H_{14}) was completely burned in air. The heat evolved raised the temperature of 200 g of water from 293.5 K to 345.1 K. Use this data to calculate a value for the enthalpy of combustion of hexane to an appropriate number of significant figures. (The specific heat capacity of water is $4.18\,J\,K^{-1}\,g^{-1}$.)   *(4)*

5 a) **i)** Give the meaning of the term 'standard enthalpy of combustion'.   *(1)*

     ii) Write an equation to represent the standard enthalpy of combustion of butan-1-ol, $CH_3CH_2CH_2CH_2OH(l)$.   *(1)*

b) 0.600 g of butan-1-ol were burned under a beaker containing 250 g of water. The temperature of the water rose by 19.40°C. Using the specific heat capacity of water as $4.18\,J\,K^{-1}\,g^{-1}$, calculate a value for the enthalpy of combustion of butan-1-ol to 2 decimal places.   *(4)*

6 In an experiment, 0.750 g of benzene (C_6H_6) were completely burned in air. The heat raised the temperature of 200 g of water by 43.7°C. Use this data to calculate the enthalpy of combustion of benzene. (The specific heat capacity of water is $4.18\,J\,K^{-1}\,g^{-1}$.) Give your answer to 1 decimal place.   *(4)*

7 Barium carbonate decomposes on heating according to the equation:

$$BaCO_3(s) \rightarrow BaO(s) + CO_2(g)$$

Standard enthalpy of formation data is provided in the table below.

	$BaCO_3(s)$	$BaO(s)$	$CO_2(g)$
$\Delta_f H^\ominus/kJ\,mol^{-1}$	−1216	−554	−394

a) Write equations to represent the reactions for the following standard enthalpies of formation.

     i) formation of barium carbonate   *(1)*

     ii) formation of barium oxide   *(1)*

     iii) formation of carbon dioxide   *(1)*

b) Use the standard enthalpies of formation in the table above to calculate a value for the enthalpy change of the reaction:

$$BaCO_3(s) \rightarrow BaO(s) + CO_2(g) \quad (3)$$

8 a) Using the standard enthalpy of formation data in the table below, calculate a value for the enthalpy for the following reaction:

$$4FeS_2(s) + 11O_2(g) \rightarrow 2Fe_2O_3(s) + 8SO_2(g) \ (3)$$

	$FeS_2(s)$	$O_2(g)$	$Fe_2O_3(s)$	$SO_2(g)$
$kJ\,mol^{-1}$	−178	0	−824	−297

b) Explain why the standard enthalpy of formation of $O_2(g)$ is zero.   *(1)*

9 Calculate a value for the enthalpy change of this reaction given the following enthalpies of formation:

$$2NaNO_3(s) \rightarrow 2NaNO_2(s) + O_2(g)$$

$$\Delta_f H^\ominus NaNO_3(s) = -468\,kJ\,mol^{-1}$$

$$\Delta_f H^\ominus NaNO_2(s) = -359\,kJ\,mol^{-1} \quad (3)$$

10 Propan-2-ol, $C_3H_7OH(g)$, undergoes complete combustion in the following reaction:

$$C_3H_7OH(g) + 4\tfrac{1}{2}O_2(g) \rightarrow 3CO_2(g) + 4H_2O(g)$$

Calculate a value for the standard enthalpy of combustion of propan-2-ol using the mean bond enthalpies below:

C—H	412	O—H	463
C—O	360	C—C	348
O=O	496	C=O	803

(3)

5

Kinetics

TEST YOURSELF ON PRIOR KNOWLEDGE 1

1 State three factors which would increase the rate of a chemical reaction.
2 Magnesium reacts with hydrochloric acid
   a) How would the rate of reaction be different when using powdered magnesium compared to magnesium ribbon when all other factors are kept the same.
   b) Explain your answer to a) in terms of particles.
3 The gases nitrogen and hydrogen react together to form ammonia.

   $N_2(g) + 3H_2(g) \rightleftharpoons 2NH_3(g)$

   a) How would increasing the pressure affect the rate of this reaction?
   b) How would increasing the temperature affect the rate of this reaction?

Rates of reaction vary enormously and depend on different factors.

For example:

● Cookies bake faster at higher temperatures. Bread dough rises more quickly in a warm place than in a cool one (temperature).
● The human body cannot adequately digest lactose without the catalytic assistance of the lactase enzyme (catalyst).
● There is a risk of explosion in factories when handing fine combustible powders such as custard powder (custard powder).
● By adjusting the concentrations of the two chemicals inside, manufacturers can produce glow sticks that either glow brightly for a short amount of time or more dimly for an extended length of time (concentration of reactants).

Temperature, catalysts, surface area and concentration all affect the rate of these different chemical reactions.

Figure 5.1 Cooking is a chemical reaction in which it is important to control the rate.

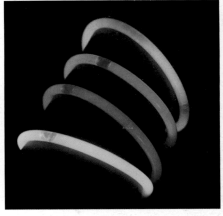

Figure 5.2 Dropping a glow stick into hot water makes it glow more intensely, showing that the reaction runs faster at higher temperature.

Collision theory

Most collisions are unsuccessful. Only a very small proportion of collisions is successful and causes a reaction. Collision theory is used to explain the main factors which affect the rate of a chemical reaction.

1 Concentration:

- If concentration of a reactant is increased
- More particles of that reactant are present
- Which leads to more successful collisions between reactant particles in a given period of time
- This causes the rate of reaction to increase.

2 Pressure:

- If pressure on a gaseous reaction system is increased
- The particles are forced closer together
- Which leads to more successful collisions between reactant particles in a given period of time
- This causes the rate of reaction to increase.

3 Temperature:

- If temperature is increased, the particles gain energy and move faster
- Which leads to more frequent and more successful collisions (collisions with energy greater than the activation energy) between reactant particles in a given period of time
- This causes the rate of reaction to increase.

Figure 5.3 Collision theory states that molecules must collide with each other in order to react. This is similar to judo and wrestling wherein the two competing players must contact one another to score a point.

The **rate of reaction** is the change in concentration of a reactant or product in a given period of time.

The **activation energy** is the minimum amount of energy which the reacting particles require for a successful collision.

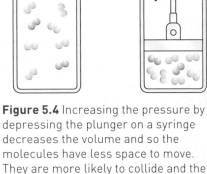

Figure 5.4 Increasing the pressure by depressing the plunger on a syringe decreases the volume and so the molecules have less space to move. They are more likely to collide and the reaction rate increases.

4 Surface area of solid reactants:

Increasing the surface area of solid reactants (by grinding them up) increases the exposed surface of the reactants. This increases the number of successful collisions in a given period of time which increases the rate of the reaction.

Hydrogen ions can hit the outer layer of atoms...

With the same number of atoms now split into lots of smaller bits, there are hardly any magnesium atoms which are inaccessible to the hydrogen ions.

... but not these in the centre of the lump

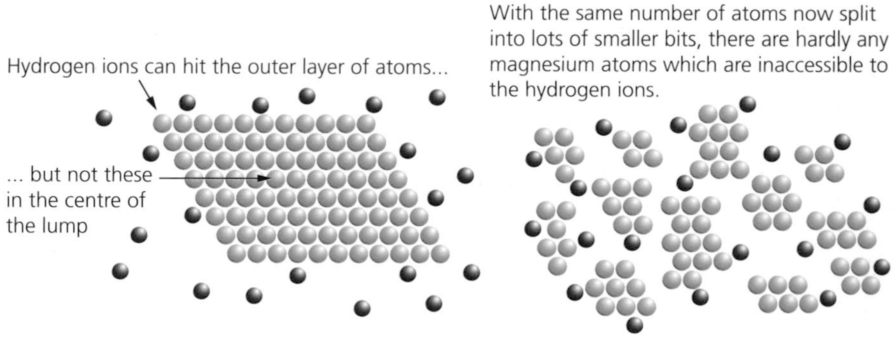

Figure 5.5 Rate of reaction of magnesium with hydrochloric acid for a larger solid piece of magnesium on the left and smaller solid pieces of magnesium on the right.

5 Presence of a catalyst:

- A **catalyst** increases the rate of reaction without being used up.
- A catalyst works by providing an alternative reaction pathway of lower activation energy. By lowering the activation energy, more collisions are successful in a given period of time and so the rate of the reaction increases.

Catalysts are important in many industrial processes by speeding up the process and so reducing costs.

Enthalpy profile diagrams

Enthalpy or reaction profile diagrams were examined for an exothermic and an endothermic reaction. These diagrams also help to explain the definition of a catalyst.

Figure 5.6 A successful jump can be compared to a successful reaction occurring when the activation energy barrier (bar) is overcome.

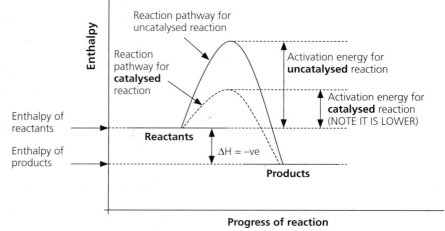

Reaction pathway for uncatalysed reaction

Reaction pathway for **catalysed** reaction

Enthalpy

Activation energy for **uncatalysed** reaction

Activation energy for **catalysed** reaction (NOTE IT IS LOWER)

Enthalpy of reactants

Enthalpy of products

Reactants

$\Delta H = -ve$

Products

Progress of reaction

Figure 5.7 Exothermic reaction enthalpy profile.

The same is true for an endothermic reaction. The catalyst simply provides an alternative reaction pathway of lower activation energy.

You may have to interpret the differences between the four levels on this diagram.

EXAMPLE 1

Using the letters A, B, C and D, identify the following on the diagram below.

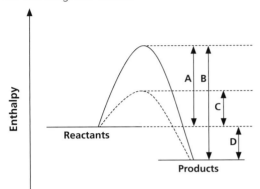

Figure 5.8

- Activation energy of the catalysed reaction
- Enthalpy change
- Activation energy of the reverse uncatalysed reaction
- Activation energy of the uncatalysed reaction

Answers

It is important to be able to recognise or label these changes on an enthalpy level diagram.

- The distance from the peak of the reaction pathway to the enthalpy level of the reactants is the activation energy of the forward reaction.
- The distance from the peak of the reaction pathway to the enthalpy level of the products is the activation energy of the reverse reaction.
- The peak may be for the uncatalysed reaction pathway (higher) or the catalysed reaction pathway (lower).
- The distance between the enthalpy levels of the reactants and products if the enthalpy change.

A is the activation energy of the uncatalysed reaction

B is the activation energy of the reverse uncatalysed reaction

C is the activation energy of the catalysed reaction

D is the enthalpy change

TEST YOURSELF 2

1 What is meant by the term *activation energy*?
2 State the effect of a catalyst on the rate of reaction.
3 Explain, in terms of collisions, how increasing the temperature increases the rate of reaction.
4 Explain what must happen for a reaction to occur in a mixture of molecules of two different gases.
5 Explain, in terms of collisions, how increasing the concentration of a reactant increases the rate of reaction.

Distribution of molecular energies

A Maxwell–Boltzmann distribution is a plot of the number of gaseous molecules against the energy they have at a fixed temperature. It is also called a molecular energy distribution graph or a distribution of molecular energies.

A single plot on the graph shows the distribution of molecular energies at a **constant temperature**.

It should appear as a roughly normal distribution which is asymptotic to the horizontal axis (gets closer and closer but never touches the axis). The main points from a Maxwell–Boltzmann distribution at constant temperature are shown below.

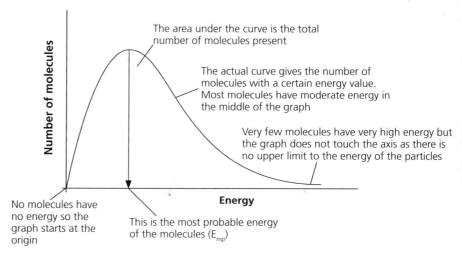

Figure 5.9

The **activation energy**, often represented as (E_a), is the minimum amount of energy which the reacting particles require for a successful collision.

Activation energy on a Maxwell–Boltzmann distribution

As the Maxwell–Boltzmann Distribution represents the energy of the reactant molecules there will be an energy value on the *x*-axis which is the activation energy.

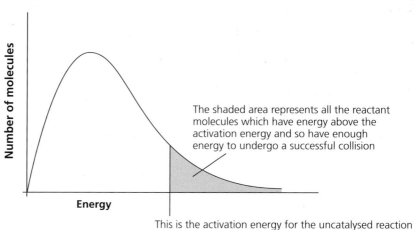

Figure 5.10

A catalyst increases the rate of the reaction by providing an alternative reaction route or pathway with a lower activation energy. Adding a catalyst to the above reaction lowers the value of the activation energy as can be seen from the position of the activation on both distributions. The area above the activation energy on the second distribution is greater. This indicates that there are more molecules with sufficient energy to undergo a successful collision. This increases the rate of the reaction.

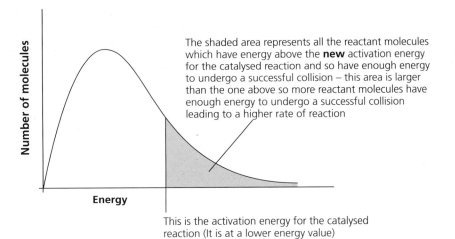

The shaded area represents all the reactant molecules which have energy above the **new** activation energy for the catalysed reaction and so have enough energy to undergo a successful collision – this area is larger than the one above so more reactant molecules have enough energy to undergo a successful collision leading to a higher rate of reaction

This is the activation energy for the catalysed reaction (It is at a lower energy value)

Figure 5.11

A catalyst does not affect the shape of the distribution as long as the temperature and total number of molecules are not changed.

Maxwell–Boltzmann distribution at different temperatures

When the temperature of the reactant molecules is increased, this will increase the energy of the gaseous reactant molecules. This will change the shape of the Maxwell–Boltzmann distribution.

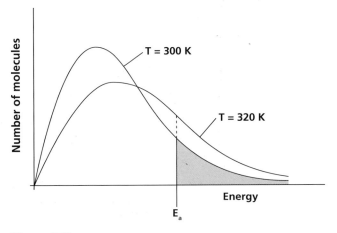

Figure 5.12

TIP

An asymptote is a line of a graph which gets closer and closer to an axis but does not touch the axis. In reality it would touch the axis at infinity which is obviously not within the scale of the graphs.

You will often be asked to sketch a distribution for the same sample of gas at a lower or higher temperature. Note the following:

- Lower temperature distributions are moved to the left and the peak is higher.
- Higher temperature distributions are moved to the right and the peak is lower. The graphs should only cross once.
- The curves should always start at the origin and should end up being asymptotic to the energy axis at higher energy values.

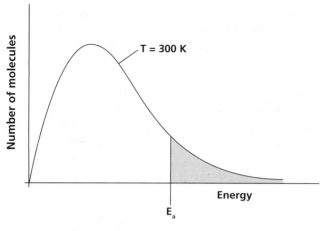

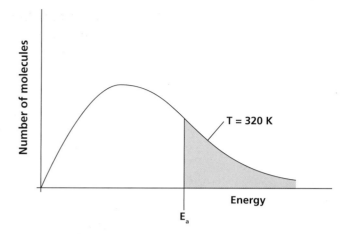

Figure 5.13

If we compare the shaded areas above the activation energy under the distribution curves, it can be seen that at the higher temperature there are many more reactant molecules with enough energy to undergo a successful collision. This explains why there is a higher rate of reaction at a higher temperature.

Taking the shaded areas out of the graph it shows that the shaded area is significantly larger at 320 K than at 300 K.

  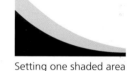

Shaded area from distribution at 300K

Shaded area from distribution at 320K

Setting one shaded area on top of the other

Figure 5.14

A small increase in temperature can cause a large increase in the rate of reaction as there is a significant increase in the number of molecules with enough energy to undergo a successful collision.

Maxwell–Boltzmann distribution at different concentrations

If the concentration of the reactant molecules is increased, the shape of a Maxwell–Boltzmann distribution will again change.

The curve retains the **same basic shape**. This means that the most probable energy of the molecules remains the same so the peak should be higher but at the same energy value on the horizontal axis.

As there are more total reactant molecules at the same temperature, the overall area under the curve increases. This increases the number of reactant molecules which have enough energy to undergo a successful reaction. This leads to a higher rate of reaction.

Higher concentration of reactant molecules

Lower concentration of reactant molecules

Energy

The most probable energy of the molecules stays the same at different concentrations as the temperature has not changed.

This is the activation energy

Figure 5.15

If we compare the shaded areas under the distribution curves above the activation energy, it can be seen that at the higher concentration there are only a few more reactant molecules with enough energy to undergo a successful collision.

It does explain why the rate of reaction increases at higher concentrations but the effect is much less **significant** than is achieved when increasing the temperature. Small increases in temperature increase the rate of reaction much more significantly than increasing the concentration of the reactants.

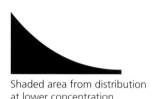
Shaded area from distribution at lower concentration

Shaded area from distribution at higher concentration

Setting one shaded area on top of the other

Figure 5.16

TEST YOURSELF 3

1 Explain how a catalyst increases the rate of reaction.

2 State which factor must be changed to alter the most probable energy of the particles in a Maxwell–Boltzmann distribution.

3 Explain why a small increase in temperature causes a larger increase in rate of reaction compared to an increase in concentration.

4 The diagram below shows a Maxwell–Boltzmann distribution for a sample of gas at 350 K.

   a) On the diagram, sketch the distribution for the same sample of gas at 330 K.

   b) With reference to the Maxwell–Boltzmann distribution, explain why an increase in temperature increases the rate of a chemical reaction.

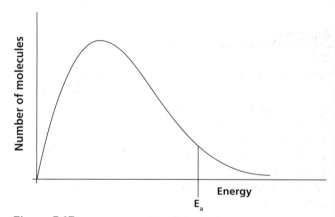

Figure 5.17

ACTIVITY

Investigating the effect of temperature on the reaction between sodium thiosulfate and hydrochloric acid.

25 cm^3 of sodium thiosulfate solution were measured into a conical flask. 10 cm^3 of hydrochloric acid were added from a test tube and the stop clock was started. The experimenter recorded the time taken for the sulfur precipitate, which was forming, to obscure a cross on the paper below the flask.

$$Na_2S_2O_3(aq) + 2HCl(aq) \rightarrow S(s) + SO_2(g) + H_2O(l) + 2NaCl(aq)$$

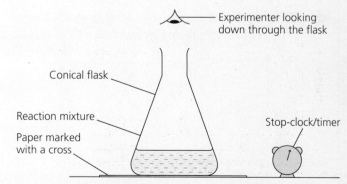

Figure 5.18 Investigating the effect of temperature on the rate of the reaction between sodium thiosulfate and hydrochloric acid.

Practice questions

1 Which one of the arrows on the enthalpy level diagram below represents the activation energy for the forward reaction? *(1)*

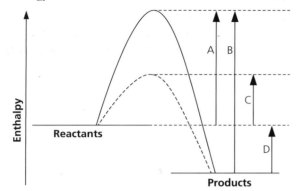

Figure 5.19

2 The diagram below shows the distribution of molecular energies in a gas at two different temperatures.

Which one of A, B, C or D is the most probable energy of the molecules in the gas at 450 K? *(1)*

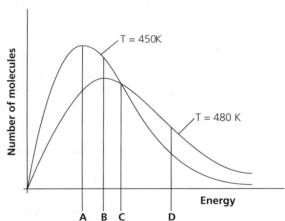

Figure 5.20

3 For the following enthalpy level diagram, four differences in enthalpy are labelled A, B, C and D.

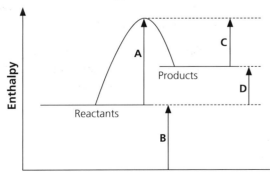

Figure 5.21

a) Which of the labelled differences in enthalpy are affected by the presence of a catalyst? *(1)*

b) Which one of the labelled differences in enthalpy is the enthalpy change for the reaction? *(1)*

c) Explain whether or not the reaction is exothermic or endothermic. *(1)*

4 In the Haber process nitrogen reacts with hydrogen according to the equation:

$$N_2(g) + 3H_2(g) \rightleftharpoons 2NH_3(g)$$
$$\Delta H = -92 \, kJ \, mol^{-1}$$

The reaction is carried out at a temperature of 450 °C and 20 MPa pressure. An iron catalyst is used.

a) The diagram below shows the distribution of molecular energies in the reaction mixture at 450 °C.

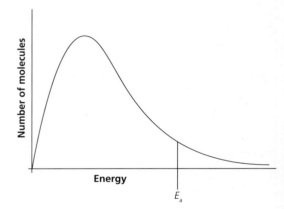

Figure 5.22

i) Sketch, on the diagram, the distribution of molecular energies at 400 °C. *(1)*

ii) With reference to the Maxwell–Boltzmann distribution, explain why the rate of reaction would be lower at 400 °C. *(2)*

b) With reference to the Maxwell–Boltzmann distribution, explain how a catalyst increases the rate of reaction. *(2)*

c) Explain whether the reaction is exothermic or endothermic. *(1)*

d) Explain why the curve starts at the origin (0,0). *(1)*

5 Hydrogen peroxide decomposes to form water and oxygen gas according to the equation:

$$2H_2O_2(aq) \rightarrow 2H_2O(l) + O_2(g)$$

The catalyst used in the laboratory is manganese(IV) oxide.

a) What is meant by the term catalyst? *(1)*

b) 50 cm³ of 0.080 mol dm⁻³ hydrogen peroxide were decomposed using manganese(IV) oxide as the catalyst at 20 °C.

The volume of oxygen gas was measured. The curve below shows how the total volume of oxygen collected changed with time under these conditions. Four points on the graph are labelled A, B, C, and D.

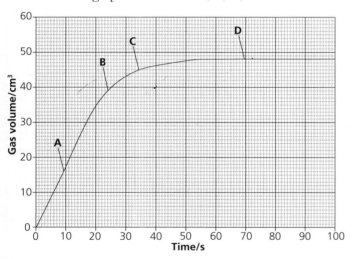

Figure 5.23

i) At which point (A, B, C or D) is the rate of reaction fastest? *(1)*

ii) At which point (A, B, C or D) is the rate of reaction zero? *(1)*

iii) Draw a curve on the figure above to show how the total volume of oxygen will change with time if the reaction is repeated at 20 °C using 50 cm³ of 0.040 mol dm⁻³ hydrogen peroxide solution.
Label this curve X. *(1)*

iv) Draw a curve on the figure above to show how the total volume of oxygen

will change with time if the reaction is repeated at 10 °C using 50 cm³ of 0.040 mol dm⁻³ hydrogen peroxide solution.
Label this curve Y. *(1)*

6 For the following Maxwell–Boltzmann distribution three values on the axes are labelled: n, E and E_a. This is the distribution at 150 °C.

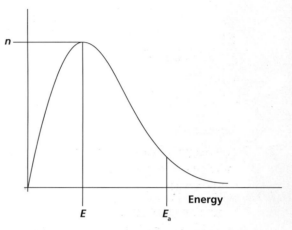

Figure 5.24

a) i) What label should be placed on the vertical axis? *(1)*

ii) What is E_a? *(1)*

b) i) Which of the values (n, E or E_a), if any, would decrease if the temperature was increased without changing the total number of molecules? *(1)*

ii) Which of the values (n, E or E_a), if any, would change if a catalyst was added? *(1)*

iii) State how, if at all, the values of E, E_a and n would change with an increase in the number of molecules at constant temperature. *(2)*

c) Explain how a catalyst increases the rate of a chemical reaction. *(1)*

d) Explain, in terms of particles how increasing the temperature increases the rate of a chemical reaction. *(2)*

6 Chemical equilibria and Le Chatelier's principle

PRIOR KNOWLEDGE

- At GCSE level most chemical reactions are not considered to be reversible.
  Most reactions are considered to go to completion and this is indicated using an arrow showing the direction of the reaction, for example:

  $CaCO_3(s) \rightarrow CaCO(s) + CO_2(g)$

- However some chemical reactions are reversible and these are indicated using a reversible reaction arrow.

  $NH_4Cl(s) \rightleftharpoons NH_3(g) + HCl(g)$

  $N_2(g) + 3H_2(g) \rightleftharpoons 2NH_3(g)$

- When the reactants start to react, the product(s) start to form and the product can also break down into the reactants again.

TEST YOURSELF ON PRIOR KNOWLEDGE 1

1 What is meant by a reversible reaction?
2 For the reaction: $2SO_2(g) + O_2(g) \rightleftharpoons 2SO_3(g)$
   a) Name the reactants and products
   b) What shows that the reaction is reversible?

Equilibrium

Irreversible reactions are reactions where the reactants convert to products and where the products cannot convert back to the reactants. For example, you cannot turn a baked cake back into its raw ingredients, or change the ashes from a piece of burnt newspaper, back into a newspaper.

Figure 6.1 Baking a cake is not reversible.

154

For the reaction: $CaCO_3(s) \rightarrow CaO(s) + CO_2(g)$

This equation means that 1 mole of calcium carbonate is broken down on heating to form 1 mole of calcium oxide and 1 mole of carbon dioxide. There is 1 mole of a substance on the left-hand side of the equation and 2 moles of substances on the right-hand side. Equations should be understood in moles.

Many chemical reactions are reversible. The reversible arrow is used to show that a reaction is reversible and it looks like this: $\rightleftharpoons$

A reversible reaction is one in which the reactants are converted into products but the products are then converted back into the reactants.

For example: $N_2(g) + 3H_2(g) \rightleftharpoons 2NH_3(g)$

If nitrogen and hydrogen are mixed they react to form some ammonia but some of the ammonia also starts to break down to reform nitrogen and hydrogen.

For example: $2SO_2(g) + O_2(g) \rightleftharpoons 2SO_3(g)$

If sulfur dioxide and oxygen are mixed they react to form some sulfur trioxide but some of the sulfur trioxide also starts to break down to re-form sulfur dioxide and oxygen.

For a general reaction: $A(g) + B(g) \rightleftharpoons AB(g)$

In general, a particle of A reacts with a particle of B and forms a particle of AB **and** a particle of AB breaks down to form a particle of A and a particle of B. Imagine particles of A and B moving around randomly. When they collide they may pair up if they collide with sufficient energy to form a pair AB. However a pair AB can also break up to release one particle of A and one particle of B.

The reaction starts with A and B only present and as the reaction starts some pairs of AB appear. The reaction is described as having reached equilibrium when the number of particles of A, B and AB remains constant. However the A, B and AB particles present are not always the same ones even though the number of each remains the same. They are constantly being formed and broken up. This is called a **dynamic equilibrium** in which you have a **steady state** because the rates of the forward reaction and the reverse reaction are the same (Figure 6.4).

Figure 6.2 Statue of Claude Louis Berthollet in Annecy, France. The world believed that all chemical reactions were irreversible until 1803 when French chemist Claude Louis Berthollet introduced the concept of reversible reactions.

Figure 6.3 These batteries can be recharged as the reaction that releases the electrical energy is reversible. Plugging the batteries into the mains via the charger converts electrical energy back into chemical energy, stored in the batteries.

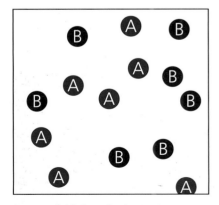

Initially only reactants
A and B

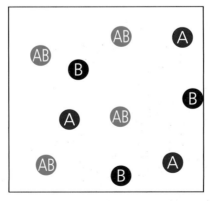

Forward reaction proceeds
making product AB

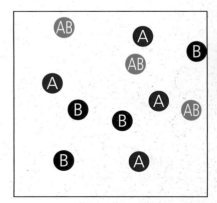

Some AB break down to form
A and B so steady state reached
(3AB, 4A and 4B)

Figure 6.4 Reversible reaction reaching dynamic equilibrium.

The definition of a dynamic equilibrium is a system where:

- the concentrations of the reactants and products remain constant, and
- where the forward and reverse reactions are proceeding at equal rates.

Homogeneous and heterogeneous reactions

A homogeneous reaction is one in which all the reactants and products are in the same physical state.

For example: $N_2(g) + 3H_2(g) \rightleftharpoons 2NH_3(g)$

All reactants and products are in the same state, i.e. in this example they are all gases.

When explaining why a reaction is considered to be homogeneous, it is important to explain that homogeneous means that all reactants and products are in the same state but also make it specific to the reaction by explaining for example that they are all gases.

Position of equilibrium

The position of equilibrium is a notional measure of how far the reaction is to the left-hand side or to the right-hand side. In the homogeneous A and B example:

$$A(g) + B(g) \rightleftharpoons AB(g)$$

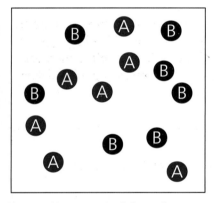

Equilibrium to the left totally as only reactants

Figure 6.5 Position of equilibrium.

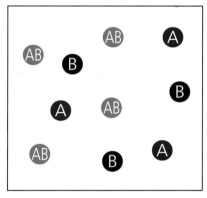

Equilibrium more to the right hand side as there are more AB than A and B

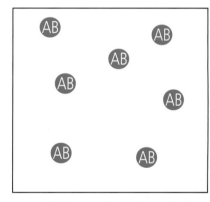

Equilibrium to the right totally as only products present

If the conditions of the reaction, such as temperature, pressure and concentration, are changed, this is likely to affect the position of equilibrium and therefore result in more reactants or more products formed.

TEST YOURSELF 2
1 State two features of a reaction at equilibrium.
2 Explain why the following equilibrium is described as a homogeneous equilibrium: $2SO_2(g) + O_2(g) \rightleftharpoons 2SO_3(g)$.
3 State two factors which could affect the position of equilibrium of a reaction.

Le Chatelier's principle

Le Chatelier's principle states that if a factor is changed which affects a system in equilibrium, the position of equilibrium will move in a direction so as to oppose the change.

The principle can also be stated as: When a system in equilibrium is disturbed, the position of equilibrium will move in a direction to reduce the effect of the disturbance.

Factors which may affect the position of equilibrium are changes in temperature, pressure or concentration of a particular reactant or product.

Changes in pressure

For the A and B homogeneous equilibrium: $A(g) + B(g) \rightleftharpoons AB(g)$

There are 2 moles of gas on the left-hand side (1 mole of $A(g)$ and 1 mole of $B(g)$) whereas there is only 1 mole of gas on the right-hand side (1 mole of $AB(g)$).

An increase in pressure at constant temperature shifts the position of equilibrium to the side with a smaller gas volume. Increasing the pressure on this equilibrium will move the position of equilibrium from **left to right**, which results in an increase in the concentration of AB. A smaller number of moles of gas has a smaller gas volume.

The reverse is true if pressure is decreased. A decrease in pressure, at constant temperature, shifts the position of equilibrium to the side with a larger gas volume. The position of equilibrium will move from **right to left** as there are 2 moles of gas on the left-hand side and 1 mole of gas on the right-hand side. A larger number of moles of gas has a greater volume.

> **TIP**
> At this stage it is only an idea of whether the position of equilibrium does not change, or moves from left to right or from right to left that is required.

> **EXAMPLE 1**
>
> In the Haber process, ammonia is manufactured and the following equilibrium is established.
>
> $N_2(g) + 3H_2(g) \rightleftharpoons 2NH_3(g)$
>
> 1 Use Le Chatelier's principle to explain how the equilibrium yield of ammonia is affected by increasing the pressure on this equilibrium system at constant temperature.
> 2 State why industry uses a pressure of 20 MPa despite a higher yield being obtained at higher pressures. Do not include references to safety.
>
> **Answers**
>
> 1 There are 4 moles of gas on the left-hand side and 2 moles of gas on the right-hand side
> An increase in pressure shifts the position of equilibrium to the side with a smaller gas volume.
> So increasing pressure moves the position of equilibrium from **left to right to oppose the increase in pressure**.
> Therefore the yield of ammonia increases.
> 2 Pressure is expensive due to the cost of electrical pumps to apply the pressure and requires expensive strong-walled vessels and expensive valves and other equipment to withstand the pressure.

> **TIP**
> State the number of moles **of gas**.

The question could be rephrased to 'Explain why a higher pressure is not used in this industrial process?' or a diagram or graph like the one shown below could be given for the Haber process at different temperatures and pressures. At any temperature an increase in pressure, increases the percentage of ammonia formed.

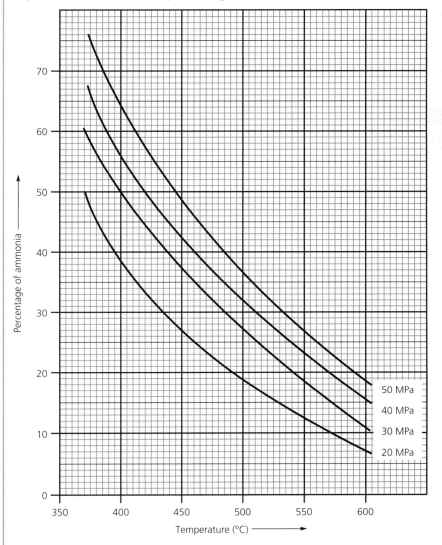

Figure 6.6 The equilibrium yield of ammonia varies with changes in temperature and pressure.

A pressure of 20 MPa will give a reasonable yield of ammonia as the position of equilibrium will move from left to right at high pressure as there are 4 moles of gas on the left-hand side and only 2 moles of gas on the right-hand side. A high pressure shifts the position of equilibrium to the side with a smaller gas volume so increasing pressure will move the position of equilibrium from left to right and increase the yield of ammonia.
If safety had been considered, it is important to note that higher pressures require safety controls due to the risk of explosion.

If there is the same amount of moles of gas on each side of the equilibrium reaction then a change in pressure will have no effect on the position of that equilibrium. For example:

$$2HI(g) \rightleftharpoons H_2(g) + I_2(g)$$

A change in pressure would have no effect on the position of this equilibrium as there are 2 moles of gas on each side of the equation.

The industrial production of ammonia is often called the Haber–Bosch process, illustrating the collaborative nature of science. Fritz Haber discovered the conditions for the production of ammonia, and Bosch, working with the German company BASF, up-scaled his method to factory size and developed the high pressure chemical plant needed (Figure 6.7). Both were awarded the Nobel Prize in 1913.

Changes in temperature

The position of equilibrium can be altered by changing the temperature but this depends on whether the reaction is exothermic or endothermic.

For the equilibrium: $N_2(g) + 3H_2(g) \rightleftharpoons 2NH_3(g)$; $\Delta H = -92\,kJ\,mol^{-1}$

The ΔH (change in enthalpy) given is for the forward reaction and it is negative, so the forward reaction is exothermic. Remember that this means that the reverse reaction is endothermic.

- For a reaction in equilibrium where the forward reaction is exothermic, an increase in temperature would shift the position of equilibrium in the direction of the reverse endothermic reaction so the position of equilibrium would move from right to left.
- For a reaction in equilibrium where the forward reaction is exothermic, a decrease in temperature would shift the position of equilibrium in the direction of the forward exothermic reaction and the position of equilibrium would move from left to right.
- For a reaction in equilibrium where the forward reaction is endothermic, an increase in temperature would shift the position of equilibrium in the direction of the forward endothermic reaction and the position of equilibrium would move from left to right.
- For a reaction in equilibrium where the forward reaction is endothermic, a decrease in temperature would shift the position of equilibrium in the direction of the reverse exothermic reaction and the position of equilibrium would move from right to left.

Figure 6.7 Factory manufacturing ammonia by Haber–Bosch process.

TIP

Some mark schemes insist on the endothermic reactions absorbing heat and again for a full answer include:
- a statement about the forward and reverse reaction
- which direction the position of equilibrium moves in terms of endothermic or exothermic
- which direction the position of equilibrium moves (left to right or right to left)
- and if required a statement about the yield.

EXAMPLE 2

Explain how the yield of ammonia would change if the temperature is increased.

$N_2(g) + 3H_2(g) \rightleftharpoons 2NH_3(g)$; $\Delta H = -92\,kJ\,mol^{-1}$

Answer

The forward reaction is exothermic and the reverse reaction is endothermic.

An increase in temperature shifts the position of equilibrium in the direction of the endothermic reaction as it absorbs heat.

An increase in temperature would move the position of equilibrium from **right to left** to oppose the increase in temperature

So, an increase in temperature would result in a decrease in the yield of ammonia.

TIP

Make sure you understand compromise temperature. In particular that even though a low temperature would give a better yield, the reaction would be too slow. Often the question is simply 'Explain why 450 °C is described as a compromise temperature' which can be answered in the same way as Example 3.

EXAMPLE 3

Using the graph on page 158, explain fully why a temperature of 450 °C is used in the Haber process even though a higher yield could be achieved at a lower temperature.

Answer

The forward reaction is exothermic and the reverse reaction is endothermic.

A low temperature would increase the yield of ammonia because the forward reaction is exothermic and a low temperature shifts the position of equilibrium in the direction of an exothermic reaction.

However if the temperature is too low the rate of reaction is very low.

450 °C is a **compromise temperature** between rate and yield.

Changes in concentration

If more of a reactant or a product is added to or removed from the equilibrium system at constant temperature and constant pressure, this alters the position of equilibrium. The equilibrium will adjust to replace any substance that has been removed or remove any substance that has been added.

In the usual example: $N_2(g) + 3H_2(g) \rightleftharpoons 2NH_3(g)$

Adding more nitrogen will push the position of equilibrium from left to right as there are more nitrogen molecules to react which will increase the yield of ammonia.

Removing the ammonia (by cooling it to condense it) will result in the position of equilibrium moving to the right-hand side to form more ammonia, which will increase the yield of ammonia and lower the concentration of nitrogen and hydrogen that are present at equilibrium.

In Figure 6.8, consider the AB example. In this example if all the AB were to be removed, the only reaction which could occur is the forward reaction, so more AB would be formed.

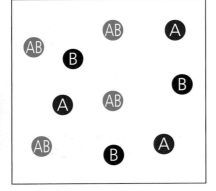

Equilibrium

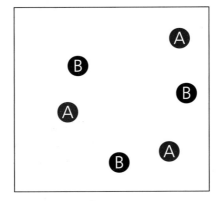

All AB removed

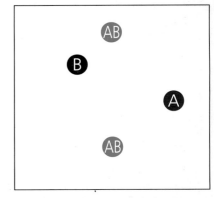

Forward reaction only occurs again and equilibrium established to form AB so less A and B than previously at equilibrium

Figure 6.8 Change in concentration.

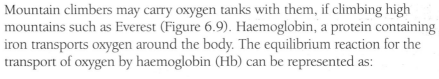

Mountain climbers may carry oxygen tanks with them, if climbing high mountains such as Everest (Figure 6.9). Haemoglobin, a protein containing iron transports oxygen around the body. The equilibrium reaction for the transport of oxygen by haemoglobin (Hb) can be represented as:

$$Hb(aq) + O_2(aq) \rightleftharpoons HbO_2(aq)$$

At high altitude there is a decrease in air pressure and there is less oxygen available. In accordance with Le Chatelier's principle the equilibrium shifts to the left, thus removing oxygen from haemoglobin. Without adequate oxygen fed to the body's cells a person tends to feel light-headed. Breathing oxygen from a tank shifts the equilibrium to the right.

Figure 6.9 Mountaineer with oxygen tank.

Another example to consider is the two different coloured Co(II) complexes, $[Co(H_2O)_6]^{2+}$ and $[CoCl_4]^{2-}$, which exist together in equilibrium in solution in the presence of chloride ions:

$$[Co(H_2O)_6]^{2+}(aq) + 4Cl^-(aq) \rightleftharpoons [CoCl_4]^{2-}(aq) + 6H_2O(l)$$

pink   blue

This **equilibrium** can be disturbed by changing the chloride ion **concentration** or by changing the **temperature**. The colour changes accompanying the shifts in equilibrium position are as predicted by **Le Chatelier's principle**.

At room temperature the cobalt and chloride ions are in equilibrium and the solution is a light pink colour. Heating the solution shifts the equilibrium and the solution turns blue. Cooling the solution shifts the solution in the opposite direction, intensifying the pink colour (Figure 6.10). This equilibrium is sensitive to the concentration of solutes, as well as temperature.

Figure 6.10 Cobalt chloride equilibrium. Experiment demonstrating the changes in equilibrium in cobalt chloride solutions that are heated (left) or cooled (right).

EXAMPLE 4

For the homogeneous equilibrium:

$$H_2(g) + I_2(g) \rightleftharpoons 2HI(g)$$

Use Le Chatelier's principle to explain how the concentration of hydrogen would change if the hydrogen iodide were removed by reacting with alkali at constant temperature and pressure.

Answer

Removing HI would move the position of equilibrium to the right-hand side to replace the HI.

The concentration of hydrogen would decrease.

EXAMPLE 5

For the following equilibrium:

$$CO_2(g) + 3H_2(g) \rightleftharpoons CH_3OH(g) + H_2O(g)$$

Use Le Chatelier's principle to explain how the equilibrium yield of CH_3OH would change if the concentration of carbon dioxide is increased at constant temperature and constant pressure.

Answer

The position of equilibrium moves from the left to the right to remove the carbon dioxide that has been added.

The equilibrium yield of CH_3OH increases.

EXAMPLE 6

When chlorine reacts with water the following equilibrium is established:

$$Cl_2(aq) + H_2O(l) \rightleftharpoons H^+(aq) + Cl^-(aq) + HOCl(aq)$$

Use Le Chatelier's principle to explain why the position of equilibrium moves to the right when sodium hydroxide solution is added.

Answer

Sodium hydroxide solution reacts with $H^+(aq)$ ions.

The $H^+(aq)$ ions are removed from the equilibrium.

The position of equilibrium shifts from the left to the right to replace the $H^+(aq)$ ions which have been removed.

Catalyst

A catalyst has no effect on the position of equilibrium but simply allows the reaction to get to equilibrium faster. A catalyst increases the rate of the forward and reverse reactions equally so equilibrium is attained much more quickly in the presence of a catalyst but it does not affect the position of equilibrium.

TEST YOURSELF 3

1 State Le Chatelier's principle.
2 Explain why a catalyst has no effect on the position of equilibrium.
3 In the following equilibrium:

$$CH_4(g) + 2H_2O(g) \rightleftharpoons CO_2(g) + 4H_2(g)$$

Use Le Chatelier's principle to state and explain how the yield of hydrogen changes when the total pressure is increased at constant pressure.
4 Dinitrogen tetroxide decomposes to nitrogen(IV) oxide according to the equilibrium:

$$N_2O_4(g) \rightleftharpoons 2NO_2(g); \Delta H = +58\,kJ\,mol^{-1}$$

   a) Use Le Chatelier's principle to state and explain how the concentration of NO_2 changes when the temperature is increased at constant pressure.
   b) Use Le Chatelier's principle to state and explain how the concentration of NO_2 changes when the pressure is increased at constant temperature.
5 In industry 450 °C is described as a compromise temperature in the production of sulfur trioxide from sulfur dioxide as part of the manufacture of sulfuric acid.

$$SO_2(g) + \tfrac{1}{2}O_2(g) \rightleftharpoons SO_3(g); \Delta H = -98\,kJ\,mol^{-1}$$

   a) Explain why 450 °C is described as a compromise temperature.
   b) Use Le Chatelier's principle to state and explain how the yield of SO_3 would change when the total pressure is increased at constant temperature.

Figure 6.11 Ethanol can be produced by fermentation of sugars using the enzyme zymase from yeast. This is carried out under anaerobic conditions and will be discussed in more detail in the alcohols section.

Figure 6.12 The platinum catalyst used to make hydrogen from the electrolysis of water is a large part of the cost. A new catalyst made from molybdenum and ground soybeans has just been developed and it is around 1500 times cheaper than platinum. This could make hydrogen the cheaper, clean fuel of the future.

Industrial examples

Methanol and ethanol are important chemicals and fuels. The importance of both of them is growing with the rise in the use of biofuels. Methanol can be used to synthesize other chemicals.

Ethanol can be produced by sugar fermentation and also by the hydration of ethene with steam. An equilibrium is established and the conditions used ensure the highest yield of ethanol is produced.

The equation for the hydration of ethene is:

$$C_2H_4(g) + H_2O(g) \rightleftharpoons CH_3CH_2OH(g); \Delta H = -42\,kJ\,mol^{-1}$$

A **hydration** reaction is one in which water is added to a compound.

The conditions below are used to increase the rate of the reaction and also to maximise the yield of ethanol:

1 Concentrated phosphoric acid is used as the catalyst (concentrated sulfuric acid can also be used).

- The presence of a catalyst allows equilibrium to be attained more rapidly. The catalyst increases the rate of reaction.
- A catalyst has no effect on the position of equilibrium so does not affect the yield of the product but it will ensure that equilibrium is attained more quickly so making the product more rapidly.

TIP

Some catalysts can increase the rate substantially so catalyst research is big business in the chemical industry. Development of more effective catalysts can increase the profit of an industrial process. The best catalysts will give the lowest activation energy but there must be a balance between the effectiveness of the catalyst and its cost.

2 A high pressure between 5 MPa and 10 MPa is used.

Using Le Chatelier's principle it is understandable that a high pressure will increase the equilibrium yield of ethanol as there are 2 moles of gas on the left and 1 mole of gas on the right so a high pressure will move the position of equilibrium to the right.

- A higher yield of ethanol would be obtained at higher pressures but as discussed previously high pressure is expensive to apply due to increased electrical pumping costs and this requires expensive strong-walled vessels and expensive valves and other equipment to withstand the pressure.

3 The temperature used is between 300 °C and 600 °C.

- The forward reaction is exothermic so an increase in temperature will shift the position of equilibrium in the direction of the reverse reaction, which is endothermic.
- A temperature between 300 °C and 600 °C is a compromise temperature between rate and yield.
- It increases the rate of reaction at the expense of a loss in some yield at equilibrium.

4 Excess ethene or excess steam can be used.

- Excess ethene (or steam) will move the position of equilibrium to the right to remove the excess reactant so increasing the yield of ethanol.
- Any unreacted ethene is recycled back into the reaction mixture.

5 The ethanol can also be condensed out of the reaction mixture.

- If a product is removed from an equilibrium reaction, the position of equilibrium will shift to replace it.
- In this example the position of equilibrium will move to the right to replace the ethanol which has been removed.

EXAMPLE 9

Methanol can be produced by the reaction of carbon monoxide with hydrogen.

$CO(g) + 2H_2(g) \rightleftharpoons CH_3OH(g)$; $\Delta H = -90\,kJ\,mol^{-1}$

The conditions used again maximise the rate of reaction and the equilibrium yield of methanol. The conditions are a catalyst which is a mixture of copper, zinc oxide and aluminium oxide, a temperature of 250 °C and a pressure between 5 and 10 MPa.

1 Explain why a high pressure increases the equilibrium yield of methanol.
2 Explain the effect, if any, of a catalyst on the equilibrium yield of methanol.
3 Explain why 250 °C is described as a compromise temperature.

Answers

1 There are 3 moles of gas on the left-hand side and 1 mole of gas on the right-hand side.
An increase in pressure shifts the position of equilibrium to the side with the fewer moles of gas to oppose the increase in pressure.
Therefore the position of equilibrium moves from the left to the right and the yield of methanol increases.
2 A catalyst has no effect on the yield of methanol.
3 As the forward reaction is exothermic so a temperature is chosen to give a compromise between yield and rate of reaction.

TEST YOURSELF 4

1 Ethene can be hydrated to form ethanol.
   a) Write an equation for this homogeneous equilibrium.
   b) What is the catalyst used in this reaction?
   c) State the conditions used to obtain a high yield of ethanol.
   d) Name the type of reaction.

2 In the equilibrium:

   $CO(g) + 2H_2(g) \rightleftharpoons CH_3OH(g)$

   a) Name the product.
   b) Use Le Chatelier's principle to state and explain how the yield of CH_3OH would change if the pressure is increased at constant temperature.
   c) The temperature used is described as a compromise temperature. Explain what is meant by compromise temperature.

3 In the following equilibrium:

   $2HBr(g) \rightleftharpoons H_2(g) + Br_2(g)$

   Explain why a change in pressure has no effect on the position of equilibrium.

Practice questions

1 For the following equilibrium:

$$CH_4(g) + 2H_2O(g) \rightleftharpoons CO_2(g) + 4H_2(g)$$

Which one of the following will decrease the yield of hydrogen?

A adding a catalyst

B increasing the pressure

C adding more steam

D decreasing the pressure   *(1)*

2 For the following four reactions in equilibrium:

A $N_2(g) + 3H_2(g) \rightleftharpoons 2NH_3(g)$;
$\Delta H = -92\,kJ\,mol^{-1}$

B $C_2H_4(g) + H_2O(g) \rightleftharpoons CH_3CH_2OH(g)$;
$\Delta H = -42\,kJ\,mol^{-1}$

C $2HI(g) \rightleftharpoons H_2(g) + I_2(g)$;
$\Delta H = +10\,kJ\,mol^{-1}$

D $N_2(g) + O_2(g) \rightleftharpoons 2NO(g)$;
$\Delta H = +180.5\,kJ\,mol^{-1}$

a) In which of the reactions, if any, will the position of equilibrium move from the left to the right when the temperature is increased?   *(1)*

b) In which of the reactions, if any, will the position of equilibrium not be affected when the pressure is increased?   *(1)*

c) In which of the reactions, if any, will the position of equilibrium move from the right to the left when the pressure is increased?   *(1)*

3 N_2O_4 decomposes into NO_2. The reaction is in equilibrium.

$$N_2O_4(g) \rightleftharpoons 2NO_2(g)$$

a) State two features of a reaction in equilibrium.   *(2)*

b) An increase in temperature moves the position of equilibrium to the right.

i) Explain whether the forward reaction is endothermic or exothermic.   *(2)*

ii) Explain in terms of particles how the rate of reaction changes with an increase in temperature.   *(2)*

c) Use Le Chatelier's principle to explain how the position of equilibrium changes if the total pressure is reduced.   *(3)*

4 In the following equilibrium, NO and O_2 are colourless gases and NO_2 is a brown gas.

$$2NO(g) + O_2(g) \rightleftharpoons 2NO_2(g);$$
$$\Delta H = -116\,kJ\,mol^{-1}$$

a) Use Le Chatelier's principle to explain why the amount of NO_2 increases when the pressure is increased at constant temperature.   *(3)*

b) Use Le Chatelier's principle to explain why the mixture of gases becomes darker in colour when the temperature is decreased.   *(3)*

5 Methanol (CH_3OH) may be manufactured from carbon monoxide by reacting it with hydrogen as shown in reaction 2 below. The hydrogen required may be produced from the reaction of methane with steam as shown in reaction 1.

(1) $CH_4(g) + H_2O(g) \rightleftharpoons CO(g) + 3H_2(g)$;
$\Delta H = +206\,kJ\,mol^{-1}$

(2) $CO(g) + 2H_2(g) \rightleftharpoons CH_3OH(g)$;
$\Delta H = -90\,kJ\,mol^{-1}$

a) Use Le Chatelier's principle to explain how an increase in pressure would affect the yield of hydrogen in reaction 1 at constant temperature.   *(3)*

b) Use Le Chatelier's principle to explain why an excess of hydrogen in reaction 2 would increase the yield of methanol at constant temperature and constant pressure.   *(2).*

6 The following equilibrium is established in the production of nitric acid from ammonia.

$$4NH_3(g) + 5O_2(g) \rightleftharpoons 4NO(g) + 6H_2O(g);$$
$$\Delta H = -905\,kJ\,mol^{-1}$$

a) Use Le Chatelier's principle to explain how the position of equilibrium is affected by an increase in pressure at constant temperature.   *(3)*

b) Use Le Chatelier's principle to explain how the position of equilibrium is affected by an increase in temperature at constant pressure.   *(3)*

c) A platinum/rhodium mixture is used as the catalyst in this reaction.

Explain how a catalyst affects the position of equilibrium. *(1)*

d) The catalyst is heated initially; explain why the catalyst remains hot during the reaction. *(1)*

7 In the production of sulfuric acid, SO_2 is converted to SO_3 in a reaction involving an equilibrium.

$$SO_2(g) + \tfrac{1}{2}O_2(g) \rightleftharpoons SO_3(g);$$
$$\Delta H = -98\,kJ\,mol^{-1}$$

a) Name the catalyst used in this process. *(1)*

b) Explain why a catalyst is used in this industrial process if it has no effect on the yield of SO_3. *(1)*

c) This reaction is carried out at 450 °C even though a higher yield of SO_3 could be obtained at a lower temperature.

i) Use Le Chatelier's principle to explain why a higher yield of SO_3 would be obtained at a lower temperature. *(3)*

ii) Explain why in industry a relatively high temperature is used despite a higher yield of SO_3 being achieved at a lower temperature. *(2)*

d) The reaction is carried out at 200 kPa pressure.

i) Use Le Chatelier's principle to explain why a high pressure would increase the yield of SO_3. *(3)*

ii) Explain why a higher pressure is not used despite a higher yield of SO_3 being achieved at a higher pressure. *(2)*

7

Equilibrium constant, K$_c$, for homogeneous systems

TEST YOURSELF ON PRIOR KNOWLEDGE 1

1 What is meant by *dynamic equilibrium*?

2 State Le Chatelier's principle.

3 For the equilibrium:

$$A(g) + 2B(g) \rightleftharpoons C(g); \Delta H = -52\,kJ\,mol^{-1}$$

   a) Explain how the yield of C would change if the temperature were increased.

   b) Explain how the yield of C would change if the total pressure were increased.

   c) Explain why a catalyst does not affect the position of equilibrium.

   d) Explain why this equilibrium is described as a homogeneous equilibrium.

The equilibrium law

In 1863, Norwegian chemists Cato Maximilian Guldberg and Peter Waage formulated what they called the law of mass action.

Today, this is called the law of chemical equilibrium, which states that the direction taken by a reaction is dependent not only on the mass of the various components of the reaction, but also upon the concentration.

When analysing the results of their experiments, Guldberg and Waage noticed that when they arranged the equilibrium concentrations into a specific form of ratio, the resulting value was the same no matter what combinations of initial concentrations were mixed at a fixed temperature. This value they called the equilibrium constant.

- K_c represents the equilibrium constant. The subscript letter after K shows what type of equilibrium is being expressed.
- K_c is an equilibrium constant calculated from concentrations of reactants and products (in $mol\,dm^{-3}$).
- K_c can be calculated for reactions in solution or homogeneous gaseous reactions, as the concentration of a solution or a gas can be calculated as the number of moles in a certain volume (in dm^3).
- All equilibrium constants are only constant at constant temperature. If the temperature remains constant, the equilibrium constant will not change. If any other factor is varied, such as pressure or concentration of reactants, the value of the equilibrium constant remains constant.

TIP

The temperature should be quoted when the value of any equilibrium constant is given but often it is stated that it is at a given temperature.

The equilibrium constant, K_c

- For the reaction:

$$aA + bB \rightleftharpoons cC + dD$$

$$K_c = \frac{[C]^c[D]^d}{[A]^a[B]^b}$$

where [C] represents the concentration of C in mol dm^{-3} in the equilibrium mixture and c is the balancing number for C in the equation for the reaction. The same applies to A, B and D.

- The concentrations of all products at equilibrium are on the top line of the expression raised to the power of their balancing numbers, and the concentrations of all reactants at equilibrium are on the bottom line of the expression, again raised to the power of their balancing numbers.

- Concentration is often calculated as the number of moles of a reactant or products divided by the volume (most often in dm^3).

$$\text{Units of } K_c = \frac{(\text{mol dm}^{-3})^{(c+d)}}{(\text{mol dm}^{-3})^{(a+b)}}$$

- The units are in terms of concentration in mol dm^3 but the overall power depends on the balancing numbers in the equation for the reaction.

Figure 7.1 The units of concentration are mol dm^{-3}. A more concentrated solution has a greater number of moles per dm^3.

Writing K_c expressions and calculation of units of K_c

A common question is to write an expression for the equilibrium constant, K_c, and to calculate the units of K_c.

TIP
The position of equilibrium may vary when external factors are changed but only changes in temperature will affect the value of the equilibrium constant.

EXAMPLE 1

Write an expression for K_c for the reaction:

$$PCl_5 \rightleftharpoons PCl_3 + Cl_2$$

and calculate its units.

Answer

$$K_c = \frac{[PCl_3][Cl_2]}{[PCl_5]}$$

$$\text{Units of } K_c = \frac{(\text{mol dm}^{-3})^2}{(\text{mol dm}^{-3})} = \text{mol dm}^{-3}.$$

Figure 7.2 Ethyl ethanoate ($CH_3COOCH_2CH_3$) is a colourless liquid with a characteristic sweet smell and is used in glues and nail polish removers and in decaffeinating tea and coffee. It is used in pear drops as a flavouring.

EXAMPLE 2

Write an expression for K$_c$ for the reaction:

$$CH_3COOH + CH_3CH_2OH \rightleftharpoons CH_3COOCH_2CH_3 + H_2O$$

and calculate its units.

Answer

$$K_c = \frac{[CH_3COOCH_2CH_3]\,[H_2O]}{[CH_3COOH]\,[CH_3CH_2OH]}$$

Units of $K_c = \dfrac{(mol\,dm^{-3})^2}{(mol\,dm^{-3})^2} =$ no units.

TIP

Remember that when the same term is multiplied, their powers are added and if the same term is divided then the powers are subtracted.

In the first example, concentration^2 divided by concentration = concentration$^{(2-1)}$ = concentration so the units are $mol\,dm^{-3}$.

In the second example, concentration^2 divided by concentration^2 = concentration$^{(2-2)}$ = concentration^0 = 1 (no units).

TIP

When fractions (or decimals) are used to balance the equation, the powers of the concentrations in the K$_c$ expression are these balancing numbers from the equation.

EXAMPLE 3

Write an expression for K$_c$ for the reaction:

$$SO_2(g) + \tfrac{1}{2}O_2(g) \rightleftharpoons SO_3(g)$$

and calculate its units.

Answer

$$K_c = \frac{[SO_3]}{[SO_2]\,[O_2]^{\frac{1}{2}}}$$

Units of $K_c = \dfrac{(mol\,dm^{-3})}{(mol\,dm^{-3})^{\frac{3}{2}}} = \dfrac{1}{(mol\,dm^{-3})^{\frac{1}{2}}} = (mol\,dm^{-3})^{-\frac{1}{2}} = mol^{-\frac{1}{2}}\,dm^{\frac{3}{2}}$.

No units of K$_c$

For the following reaction:

$$2HI \rightleftharpoons H_2 + I_2$$

The equilibrium constant, K$_c$, has no units.

$$K_c = \frac{[H_2]\,[I_2]}{[HI]^2}$$

Units of $K_c = \dfrac{(mol\,dm^{-3})^2}{(mol\,dm^{-3})^2} =$ no units

If K$_c$ has no units then it is because there are an equal number of moles on both sides of the equation and they cancel each other out in the K$_c$ expression (see also Example 2).

TEST YOURSELF 2

TEST YOURSELF 2

1 For the equilibrium:

$$2F + G \rightleftharpoons 2H$$

a) Write an expression for the equilibrium constant, K_c.
b) Deduce the units of K_c.

2 For the equilibrium:

$$CH_4 + 2H_2O \rightleftharpoons CO_2 + 4H_2$$

a) Write an expression for the equilibrium constant, K_c.
b) Deduce the units of K_c.

3 For the equilibrium:

$$H_2 + Br_2 \rightleftharpoons 2HBr$$

a) Write an expression for the equilibrium constant, K_c.
b) Explain why K_c for this equilibrium has no units.

Calculating K_c

Calculating a value for K_c may involve calculating and using equilibrium amounts, in moles or equilibrium concentrations. K_c is only constant at constant temperature which may be quoted in °C or K or simply stated as a constant temperature or given temperature or particular temperature.

Calculating equilibrium moles

The general format below is followed to calculate the amount, in moles, of each substance present at equilibrium (equilibrium moles).

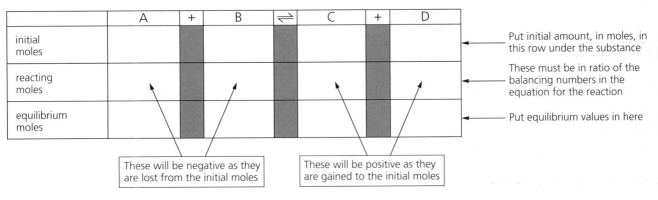

Figure 7.3

TIP
If the equilibrium moves from right to left then the reacting moles will be lost from the products (and be negative on the right) and gained by the reactants (and be positive on the left). This would be unusual.

EXAMPLE 4

A mixture was prepared using 2.0 mol of A and 1.5 mol of B. At a given temperature, the mixture was left to reach equilibrium according to the equation.

$$2A + B \rightleftharpoons 3C + D$$

The equilibrium mixture contained 0.74 mol of B.

Calculate the amount, in moles, of A, C and D in the equilibrium mixture.

Answer

Step 1

In this first step, create a table below the equation with three row headings as shown. Put the initial amounts, in moles, into the first row. There was no C or D present initially so they are put as 0 (zero).

	2A	+	B	$\rightleftharpoons$	3C	+	D
initial moles	2.0		1.5		0		0
reacting moles							
equilibrium moles							

Step 2

In this step you need to put in some information you have been given apart from initial amounts, in moles. In the example you are told that the equilibrium mixture contained 0.74 mol of B. This is placed below B in the equilibrium moles row.

	2A	+	B	$\rightleftharpoons$	3C	+	D
initial moles	2.0		1.5		0		0
reacting moles							
equilibrium moles			0.74				

Step 3

From this information you can now work out how much of B reacted to create this equilibrium. This is a simple subtraction. 1.5 − 0.74 = 0.76 mol of B reacted. This is filled into the reacting moles line below B. It is negative because this is how much has reacted.

	2A	+	B	$\rightleftharpoons$	3C	+	D
initial moles	2.0		1.5		0		0
reacting moles			−0.76				
equilibrium moles			0.74				

Step 4

All the other reacting moles are calculated from 0.76. The reacting moles must all be in the same ratio as the balancing numbers in the equation. In this example, 1.52 mol of A reacts (2A + B on the left) and 2.28 mol of C plus 0.76 mol of D are formed (3C + D). The moles of A reacting is a negative number and the moles of C and D formed are positive numbers.

	2A	+	B	$\rightleftharpoons$	3C	+	D
initial moles	2.0		1.5		0		0
reacting moles	−1.52		−0.76		+2.28		+0.76
equilibrium moles			0.74				

Step 5

The equilibrium moles for A, C and D are calculated by adding the initial moles and the reacting moles (taking into account the − and +).

For A, this is (2.0 − 1.52 = 0.48) mol. For C, this is (0 + 2.28) mol and for D, this is (0 + 0.76) mol.

	2A	+	B	$\rightleftharpoons$	3C	+	D
initial moles	2.0		1.5		0		0
reacting moles	−1.52		−0.76		+2.28		+0.76
equilibrium moles	0.48		0.74		2.28		0.76

The amount, in moles, of A, C and D in the equilibrium mixture at this temperature is:

A = 0.48 mol

C = 2.28 mol

D = 0.76 mol

TIP

This process seems long but once you get used to it, it becomes much faster. In these examples, the steps are colour-coded (**R**, **O**, **G**, **B**, **V**) in the table. (Yellow and indigo are omitted from the colours of the spectrum as they are harder to see in print!)

EXAMPLE 5

Hydrogen and nitrogen react to form ammonia. A mixture of 10.0 mol of nitrogen and 25.0 mol of hydrogen were allowed to react and come to equilibrium at temperature, T.

$N_2(g) + 3H_2(g) \rightleftharpoons 2NH_3(g)$

At temperature T, there were 2.8 mol of nitrogen in the equilibrium mixture. Calculate the amount, in moles, of hydrogen and ammonia in the equilibrium mixture.

	N_2	+	$3H_2$	$\rightleftharpoons$	$2NH_3$
initial moles	10.0		25.0		0
reacting moles	−7.2		−21.6		+14.4
equilibrium moles	2.8		3.4		14.4

Answer

Initial amounts in moles (given in the example)

10.0 mol of nitrogen are mixed with 25.0 mol of hydrogen

Amount in moles of one substance at equilibrium (given in the example)

2.8 mol of nitrogen are present in the equilibrium mixture

Reacting moles of this substance (calculated from given information)

7.2 mol of nitrogen has reacted (−7.2)

Reacting moles of other substances (calculated from balancing numbers in the equation)

7.2 mol of nitrogen reacts with 21.6 mol of hydrogen (1:3 ratio) and formed 14.4 mol of NH_3 (1:2 ratio)

Equilibrium moles of other substances (adding the initial moles and the reacting moles)

25.0 − 21.6 = 3.4 mol of H_2 remaining; 0 + 14.4 = 14.4 mol of NH_3 formed.

The amount, in moles, of H_2 and NH_3 in the equilibrium mixture at this temperature are:

H_2 = 3.4 mol

NH_3 = 14.4 mol

EXAMPLE 6

The following equilibrium was established in a closed container

$2P \rightleftharpoons R + 2S$

7.5 mol of P and 2.0 mol of S were placed in a closed container and allowed to come to equilibrium at a certain temperature.

5.04 mol of S were present in the equilibrium mixture. Calculate the amount, in moles, of P and R in the equilibrium mixture.

	$2P$	$\rightleftharpoons$	R	+	$2S$
initial moles	7.5		0		2
reacting moles	−3.04		+1.52		+3.04
equilibrium moles	4.46		1.52		5.04

Answer

Initial amounts in moles (given in the example)

7.5 mol of P and 2.0 mol of S are mixed.

Amount in moles of one substance at equilibrium (given in the example)

5.04 mol of S are present in the equilibrium mixture

Reacting moles of this substance (calculated from given information)

3.04 mol of S are formed (+3.04)

Reacting moles of other substances (calculated from balancing numbers in the equation)

3.04 mol of S formed from 3.04 mol of P (2:2 ratio) and also 1.52 mol of R formed ($\frac{3.04}{2}$) due to 2:1 ratio of S:R in equation.

Equilibrium moles of other substances (adding the initial moles and the reacting moles)

7.5 − 3.04 = 4.46 mol of P remaining; 0 + 1.52 = 1.52 mol of R formed.

The amount, in moles, of P and R in the equilibrium mixture at this temperature are:

P = 4.46 mol

R = 1.52 mol.

TIP

This example is slightly different as there was some of one of the products initially present but the same rules still apply to each step in the process.

Calculating equilibrium concentrations

The previous questions can be taken one stage further. If a volume is given in the example, the equilibrium moles are divided by the the volume in dm^3 to give the equilibrium concentration in $mol\,dm^{-3}$.

EXAMPLE 7

Sulfur dioxide reacts with oxygen at a high temperature to form sulfur trioxide.

$$2SO_2(g) + O_2(g) \rightleftharpoons 2SO_3(g)$$

1.5 mol of sulfur dioxide is mixed with 0.75 mol of oxygen at a high temperature in a container of volume 5.0 dm^3. 0.25 mol of oxygen are present at equilibrium.

1 Calculate the amount, in moles, of sulfur dioxide and sulfur trioxide in the equilibrium mixture.
2 Calculate the concentration of the reactants and products in $mol\,dm^{-3}$ in the equilibrium mixture.

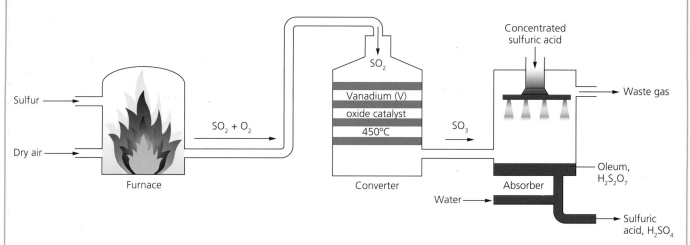

Figure 7.4 The equilibrum between sulfur dioxide and oxygen is an important part of the manufacture of sulfuric acid.

Answers

1

	$2SO_2$	+	O_2	$\rightleftharpoons$	$2SO_3$
initial moles	1.5		0.75		0
reacting moles	−1		−0.50		+1.0
equilibrium moles	0.5		0.25		1.0

Amounts, in moles, of SO_2 and SO_3 in the equilibrium mixture:
SO_2 = 0.5 mol

SO_3 = 1.0 mol.

2 The concentrations, in $mol\,dm^{-3}$, of the reactants and products in the equilibrium mixture are given in the table by dividing the amount, in moles, at equilibrium by the volume in dm^3.

	$2SO_2$	+	O_2	$\rightleftharpoons$	$2SO_3$
initial moles	1.5		0.75		0
reacting moles	−1		−0.50		+1.0
equilibrium moles	0.5		0.25		1.0
concentration at equilibrium $(mol\,dm^{-3})$	$\frac{0.5}{5} = 0.1$		$\frac{0.25}{5} = 0.05$		$\frac{1.0}{5} = 0.2$

The concentration of SO_2 in $mol\,dm^{-3}$ in the equilibrium mixture is 0.1 $mol\,dm^{-3}$
The concentration of O_2 in $mol\,dm^{-3}$ in the equilibrium mixture is 0.05 $mol\,dm^{-3}$
The concentration of SO_3 in $mol\,dm^{-3}$ in the equilibrium mixture is 0.2 $mol\,dm^{-3}$

TIP
Often one of these equilibrium concentrations is asked or all may be calculated and used to determine a value for the equilibrium constant, K_c.

TEST YOURSELF 3

1 1.0 mol of A and 1.0 mol of B were mixed and allowed to reach equilibrium at a certain temperature.

$$2A + 2B \rightleftharpoons C + D$$

0.2 mol of C were present in the equilibrium mixture.

a) Write an expression for the equilibrium constant, K_c, for this equilibrium and deduce its units.

b) Calculate the amount, in moles, of A, B and D present in the equilibrium mixture.

2 10.0 mol of PCl_5 were allowed to reach equilibrium in a container of volume $2.5 \, dm^3$ according to the equilibrium.

$$PCl_5 \rightleftharpoons PCl_3 + Cl_2$$

a) Write an expression for the equilibrium constant, K_c, for this equilibrium.

b) Deduce the units for K_c.

c) 4.0 mol of PCl_5 are present in the equilibrium mixture.

  i) Calculate the amount, in moles, of PCl_3 and Cl_2 in the equilibrium mixture.

  ii) Calculate the concentration of PCl_5, PCl_3 and Cl_2 in the equilibrium mixture.

3 For the equilibrium:

$$4NH_3 + 5O_2 \rightleftharpoons 4NO + 6H_2O$$

a) Write an expression for the equilibrium constant, K_c, for this equilibrium.

b) Deduce the units of K_c.

c) 2.00 mol of NH_3 and 2.50 mol of O_2 were mixed and allowed to reach equilibrium. 0.80 mol of NO were present in the equilibrium mixture.

  i) Calculate the amount, in moles, of NH_3, O_2 and H_2O present in the equilibrium mixture.

  ii) The total volume is $20.0 \, dm^3$. Calculate the equilibrium concentrations of all reactants and products.

The equilibrium concentrations may be determined by dividing the equilibrium moles by the volume (use V if volume is not given). The equilibrium concentration may then be substituted into the K_c expression to calculate a value for K_c.

For a K_c expression which has no units, the volumes will cancel out as there are equal number of moles on each side of the equilibrium equation.

TIP

It is important to note that in this question, no volume is given so $V\,dm^3$ is used as the volume and the equilibrium moles are divided by V to calculate the equilibrium concentrations in $mol\,dm^{-3}$.

TIP

As K_c has no units V cancels out in the equilibrium expression. It is acceptable to use equilibrium moles to calculate the value of a K_c which has no units but the use of V is better practice and appears in many mark schemes.

TIP

The answer has been given to 2 significant figures as all the initial data were given to 2 significant figures. This is good practice and you will notice this in mark schemes.

EXAMPLE 8

Nitrogen monoxide decomposes into nitrogen and oxygen according to the equilibrium.

$$2NO(g) \rightleftharpoons N_2(g) + O_2(g)$$

1 Write an expression for the equilibrium constant, K_c, for this equilibrium.
2 Explain why K_c for this reaction has no units.
3 5.0 mol of nitrogen monoxide were allowed to come to equilibrium in a sealed container at a particular temperature. 1.5 mol of nitrogen were in the equilibrium mixture.
   a) Calculate the amount, in moles, of nitrogen monoxide and oxygen in the equilibrium mixture.
   b) Calculate a value for K_c for this equilibrium at this temperature.

Answers

1 $K_c = \dfrac{[N_2][O_2]}{[NO]^2}$.
2 Equal numbers of moles on each side of the reaction so the concentrations cancel each other out in terms of units.
3 a)

	2NO	$\rightleftharpoons$	N$_2$	+	O$_2$
initial moles	5.0		0		0
reacting moles	−3.0		+1.5		+1.5
equilibrium moles	2.0		1.5		1.5
equilibrium concentration	$\dfrac{2.0}{V}$		$\dfrac{1.5}{V}$		$\dfrac{1.5}{V}$

b) $K_c = \dfrac{[N_2][O_2]}{[NO]^2} = \dfrac{\dfrac{1.5}{V} \times \dfrac{1.5}{V}}{\left(\dfrac{2.0}{V}\right)^2} = \dfrac{1.5^2}{2.0^2} = \dfrac{2.25}{4.0} = 0.56$ (no units).

EXAMPLE 9

A reacts with B to form C and D according to the equilibrium below:

$$A + 2B \rightleftharpoons C + D$$

0.25 mol of A were mixed with 0.80 mol of B in a container of volume $10\,dm^3$ and the mixture allowed to come to equilibrium at 500 K.

The equilibrium mixture at 500 K contained 0.20 mol of A.

1 Calculate the concentration, in $mol\,dm^{-3}$, of A, B, C and D in the equilibrium mixture.
2 Write an expression for the equilibrium constant, K_c, for this equilibrium.
3 Calculate a value for K_c for this equilibrium at 500 K and state its units, if any.

Answers

1

	A	+	2B	⇌	C	+	D
initial moles	0.25		0.80		0		0
reacting moles	−0.05		−0.10		+0.05		+0.05
equilibrium moles	0.20		0.70		0.05		0.05
equilibrium concentration (mol dm⁻³)	0.02		0.07		0.005		0.005

The equilibrium concentrations are determined by dividing the equilibrium moles by the volume which is given, in this example, as 10 dm³.

2 $K_c = \dfrac{[C][D]}{[A][B]^2}$

3 $K_c = \dfrac{0.005 \times 0.005}{0.02 \times (0.07)^2} = \dfrac{2.5 \times 10^{-5}}{9.8 \times 10^{-5}} = 0.26 \text{ mol}^{-1}\text{dm}^3$

TIP

If the calculation in part 3 of Example 9 for K_c were entered into a calculator as:

$0.005 \times 0.005 \div 0.02 \times (0.07)^2$

the answer would be 6.1×10^{-6}, which is incorrect. This calculates K_c as $\dfrac{0.005 \times 0.005}{0.02} \times (0.07)^2$ which is a completely different calculation. Try these calculations on your calculator to make sure you get the correct answer. The use of brackets is very important.

EXAMPLE 10

For the equilibrium:

$\frac{1}{2}N_2 \text{ (g)} + \frac{3}{2}H_2 \text{ (g)} \rightleftharpoons NH_3\text{(g)}$

1.00 mol of nitrogen and 3.00 mol of hydrogen were mixed in a container of volume 0.500 dm³. 0.240 mol of ammonia were present at equilibrium. Calculate a value for K_c and state its units, if any.

	$\frac{1}{2}N_2$	+	$\frac{3}{2}H_2$	⇌	NH_3
initial moles	1.00		3.00		0
reacting moles	−0.120		−0.360		+0.240
equilibrium moles	0.880		2.640		0.240
equilibrium concentration (mol dm⁻³)	1.760		5.280		0.480

TIP

The calculation of equilibrium moles is the same using fractional balancing numbers. $\frac{3}{2}$ of 0.240 is 0.360 (($0.240/3$) × 2 = 0.36). Half of 0.240 is 0.120. Equilibrium concentrations in mol dm⁻³ are worked out by dividing by the volume (0.500 dm³).

TIP

The power of $\frac{1}{2}$ is best accessed on your calculator by raising the number to the power of 0.5. Similarly a power of $\frac{3}{2}$ is a power of 1.5. The power of $\frac{1}{2}$ is the same as the square root of the number.

Answer

$[N_2] = \dfrac{0.880}{0.500} = 1.76 \text{ mol dm}^{-3}$

$[H_2] = \dfrac{2.64}{0.500} = 5.28 \text{ mol dm}^{-3}$

$[NH_3] = \dfrac{0.240}{0.500} = 0.480 \text{ mol dm}^{-3}$

$K_c = \dfrac{[NH_3]}{[N_2]^{\frac{1}{2}}[H_2]^{\frac{3}{2}}} = \dfrac{(0.480)}{(1.76)^{\frac{1}{2}}(5.28)^{\frac{3}{2}}} = \dfrac{0.480}{16.096} = 0.0298 \text{ mol}^{-1}\text{ dm}^3$

Calculating K_c from concentrations

Equilibrium concentrations of reactants and products may be given and the equilibrium constant, K_c, is calculated from these values. You will often be asked to deduce the units as well in this type of question.

The equilibrium constant for the reverse reaction is the reciprocal of the value for the forward reaction. Simply use $\frac{1}{K_c}$ to calculate the value of the equilibrium constant for the reverse reaction. The units of the equilibrium constant for the reverse reaction are again the reciprocal of the units of the equilibrium constant for the forward reaction. For example if the units of the equilibrium constant for the forward reaction are $mol^{-1}\,dm^3$, the units of the value of the equilibrium constant of the reverse reaction are $(mol^{-1}\,dm^3)^{-1} = mol\,dm^{-3}$.

EXAMPLE 11

Methane reacts with water as shown in the equilibrium below.

$CH_4(g) + H_2O(g) \rightleftharpoons CO(g) + 3H_2(g)$

The equilibrium concentrations, in $mol\,dm^{-3}$, of each gas at a particular temperature are given in the table below.

Gas	Equilibrium concentration ($mol\,dm^{-3}$)
CH_4	0.14
H_2O	0.55
CO	0.17
H_2	0.51

1 Write an expression for the equilibrium constant, K_c, for this equilibrium.
2 Calculate a value for K_c at this temperature and deduce its units. Give your answer to 2 significant figures.
3 Calculate a value for the equilibrium constant for the reverse reaction and deduce its units. Give your answer to 2 significant figures.

Answers

1 $K_c = \dfrac{[CO][H_2]^3}{[CH_4][H_2O]}$

2 $K_c = \dfrac{0.17 \times 0.51^3}{0.14 \times 0.55} = \dfrac{0.02255}{0.077} = 0.2929\ mol^2\,dm^{-6}$

To 2 significant figures, $K_c = 0.29\ mol^2\,dm^{-6}$

The units are deduced by $\dfrac{(mol\,dm^{-3})^4}{(mol\,dm^{-3})^2} = mol^2\,dm^{-6}$.

3 The value of the equilibrium constant for the reverse reaction

$\dfrac{1}{0.29} = 3.4$ (to 2 significant figures)

The units of the reverse equilibrium constant are $(mol^2\,dm^{-6})^{-1} = mol^{-2}\,dm^6$.

TEST YOURSELF 4

1 Sulfur dioxide and oxygen are in a homogeneous gaseous equilibrium with sulfur trioxide.

$2SO_2(g) + O_2(g) \rightleftharpoons 2SO_3(g)$

At 450 °C the equilibrium concentrations are:
$SO_2 = 0.240 \, mol \, dm^{-3}$

$O_2 = 1.47 \, mol \, dm^{-3}$

$SO_3 = 0.82 \, mol \, dm^{-3}$

a) Write an expression for the equilibrium constant, K_c.
b) Calculate a value of the equilibrium constant, K_c, at 450 °C and deduce its units.

2 2.0 mol of PCl_5 vapour are heated to 500 K in a vessel of volume 10 dm³. The equilibrium mixture contains 1.2 mol of chlorine.

$PCl_5(g) \rightleftharpoons PCl_3(g) + Cl_2(g)$

a) Write an expression for the equilibrium constant, K_c.
b) Calculate the amount, in moles, of PCl_5 and PCl_3 in the equilibrium mixture.
c) Calculate the equilibrium concentrations of PCl_5, PCl_3 and Cl_2.
d) Calculate a value of the equilibrium constant, K_c, at 500 K and deduce its units.

3 A reacts with B according to the equilibrium:

$2A(g) + B(g) \rightleftharpoons C(g)$

1.00 mol of A is mixed with 1.00 mol of B. The mixture is left to come to equilibrium. The equilibrium mixture contains 0.20 mol of A.

a) Calculate the amount, in moles, of B and C in the equilibrium mixture.
b) The volume of the container is 5.00 dm³. Calculate the equilibrium concentrations of A, B and C.
c) Write an expression for the equilibrium constant, K_c, for this equilibrium.
d) Calculate a value for K_c at 500 K and deduce its units.

Using K_c

Some calculations require you to use a given value of K_c at a certain temperature. There is a variety of these type of calculations.

In some examples, the K_c value may be given to you, and you are asked to calculate one of the equilibrium concentrations or amount, in moles, present at equilibrium from using the K_c value.

EXAMPLE 12

The following equilibrium was established in a closed container.

$2A(g) + B(g) \rightleftharpoons 2C(g)$

1 Write an expression for the equilibrium constant, K$_c$, for this equilibrium.
2 At 350 K, K$_c$ = 45.2 mol^{-1} dm^3 and the equilibrium concentration of A and B are 1.27 mol dm^{-3} and 0.240 mol dm^{-3} respectively. Calculate the equilibrium concentration of C at 350 K

Answer

1 $K_c = \dfrac{[C]^2}{[A]^2[B]}$

2 Rearranging the equilibrium expression gives: $[C]^2 = K_c \times [A]^2 \times [B]$.
The values given above are equilibrium concentrations so may be used directly in the expression.

$[C]^2 = 45.2 \times (1.27)^2 \times 0.240$

$[C]^2 = 17.497$

$[C] = \sqrt{17.497} = 4.18$ mol dm^{-3}

TIP

If you want to check your answer is correct, put it into the expression for K$_c$ and you should get the K$_c$ value you were given in the question.

$$K_c = \frac{(4.18)^2}{(1.27)^2 \times 0.24} = \frac{17.4724}{0.3871} = 45.14 \text{ mol}^{-1}\text{dm}^3$$

This is the value given. Often, as in this example, it may be close to the value given in the question due to rounding of answers.

TIP

Rearranging an expression like the expression for the equilibrium constant, K$_c$, is a common skill in maths and in the applied maths used in chemistry. Make sure you can do this. If a quantity is divided on one side of the expression, it will be multiplied on the other side when moved, for example:

$$K_c = \frac{[C]^2}{[A]^2[B]}$$

so $[C]^2 = K_c \times [A]^2 \times [B]$;

as the equation was divided by [A]^2 and [B] so when moved to the other side, it is multiplied by them. When you have values, practise rearranging a K$_c$

expression and check the values you have allowing you to obtain the value for the subject, for example, using this K$_c$ expression:

$$[B] = \frac{[C]^2}{[A]^2 K_c}$$

so $[B] = \dfrac{(4.183)^2}{(1.27)^2 \times 45.2} = 0.24$ mol dm^{-3}

which is the value you were given in the example. The most complex rearrangement comes when the quantity you want to make the subject of the expression is part of the denominator in a fraction.

EXAMPLE 13

The following equilibrium was established in a closed container.

$3D(g) + 2E(g) \rightleftharpoons F(g)$

At 550 K, $K_c = 96.2 \, mol^{-4} \, dm^{12}$. The equilibrium mixture contained 28.0 mol of D and 113.4 mol of F in a 140 dm^3 container.

1 Write an expression for the equilibrium constant, K_c, for this equilibrium.
2 Calculate the concentration, in mol dm^{-3}, of E in the equilibrium mixture.
3 Calculate the amount, in moles, of E present in the 140 dm^3 container.

Answers

1 $K_c = \dfrac{[F]}{[D]^3[E]^2}$

2 The equilibrium concentrations are calculated by dividing the amount, in moles, at equilibrium by the volume of the container in dm^3.

$$[D] = \frac{28.0}{140} = 0.2 \, mol \, dm^{-3}$$

$$[F] = \frac{113.4}{140} = 0.81 \, mol \, dm^{-3}$$

Rearranging the equilibrium expression for K_c to make [E] the subject.

$$[E]^2 = \frac{[F]}{[D]^3 K_c} = \frac{0.81}{(0.2)^3 \times 96.2} = \frac{0.81}{0.7696} = 1.0525$$

$$[E] = \sqrt{1.0525} = 1.03 \, mol \, dm^{-3}$$

3 If [E] = 1.03 mol dm^{-3}, this means that 1.03 mol of E are present in 1 dm^3 so to calculate the amount, in moles, present in 140 dm^3, the concentration is multiplied by 140.
Amount, in moles, in 140 dm^3 = 1.03 × 140 = 144 mol.
The answer is given to 3 significant figures as this is the lowest number of significant figures in the data given.

TIP

If you find you are prone to calculator errors in these types of calculations, it is best to calculate the numerator and denominator separately and then carry out the division.

TIP

Many of these calculations are carried throughout the calculation on a calculator. Make sure you give the answer to a specific number of significant figures asked for in the calculation or to the lowest level of precision (lowest number of significant figures) for the given data in the question.

TEST YOURSELF 5

1 For the equilibrium:
$A + 2B \rightleftharpoons C + D$

$K_c = 12.4 \, mol^{-1} \, dm^3$ at a certain temperature.

A and B are mixed and allowed to come to equilibrium at this temperature and the equilibrium concentrations are:

A: 1.20 mol dm^{-3}

B: 0.55 mol dm^{-3}

D: 2.10 mol dm^{-3}

Calculate the equilibrium concentration of C at this temperature.

2 For the same equilibrium in question 1:

A + 2B $\rightleftharpoons$ C + D at a different temperature

$K_c = 10.5 \, mol^{-1} \, dm^3$.

The equilibrium concentrations of A, C and D are:

A: $0.440 \, mol \, dm^{-3}$

C: $0.940 \, mol \, dm^{-3}$

D: $2.52 \, mol \, dm^{-3}$

Calculate the equilibrium concentration of B at this temperature.

3 For the equilibrium:

$2HI(g) \rightleftharpoons H_2(g) + I_2(g)$

a) Write an expression for the equilibrium constant, K_c.

b) Explain why K_c has no units.

c) At a particular temperature, an equilibrium mixture contains $0.0800 \, mol$ of H_2 and $0.0800 \, mol$ of I_2. K_c at this temperature is 7.45. The total volume is $4.0 \, dm^3$.
Calculate the concentration of HI, in $mol \, dm^{-3}$, in the equilibrium mixture at this temperature.

ACTIVITY

Determination of the value of K_c when ethanol reacts with ethanoic acid

The procedure involves setting up a known mixture of ethanoic acid, ethanol, water and dilute hydrochloric acid (catalyst) and leaving it for one week to reach equilibrium.

$CH_3COOH + CH_3CH_2OH \rightleftharpoons CH_3COOCH_2CH_3 + H_2O$

A control is also prepared containing only dilute hydrochloric acid. As soon as equilibrium has been established, the mixture is poured into excess water (this 'freezes' the equilibrium for a short period of time) and then rapidly titrated with standard sodium hydroxide solution, using phenolphthalein as an indicator. The results of this titration give a measure of the total concentration of acid present in the equilibrium mixture.

Titration of the control allows the concentration of dilute hydrochloric acid to be determined and from the two titration results, the equilibrium concentration of ethanoic acid can be determined. Provided the initial concentrations of ethanoic acid, ethanol, water

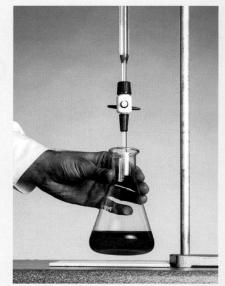

Figure 7.5 Titration of the acid with standard sodium hydroxide using phenolphthalein indicator.

and dilute hydrochloric acid are known, the equilibrium concentrations of each of these compounds can be calculated and hence a value of K_c can be found for the reaction under the conditions of the experiment.

Table 7.1 Initial volumes for each reagent to make equilibrium mixture and control.

Reagent	Density /g cm⁻³	Control flask Volume/cm³	Mixture flask Volume/cm³
Ethanoic acid	1.05	0.0	6.0
Ethanol	0.79	0.0	6.0
Water	1.00	18.0	6.0
Hydrochloric acid	1.00	2.0	2.0
Total volume		20.0	20.0

The titration was carried out using $0.10\,mol\,dm^{-3}$ NaOH.

Table 7.2 Titration results for the experiment.

	Initial burette reading/cm³	Final burette reading/cm³	Titre/cm³
Control flask	0.00	16.50	16.50
Titration of a sample from the mixture flask	0.00	34.70	34.70
Titration of a sample from the mixture flask	0.00	34.60	34.60
Average titre of mixture			34.65

1 Using the densities and volumes given for each reagent, calculate the initial amount, in moles, for each of ethanoic acid, ethanol and water in the equilibrium mixture. Assume that $2.0\,cm^3$ of the $1.0\,mol\,dm^{-3}$ hydrochloric acid catalyst adds an extra $2.0\,cm^3$ of water to the mixture.

2 Calculate the amount of hydrochloric acid, in moles, in the control flask using the titration result for the control.

3 Calculate the total amount of acid, in moles, in the equilibrium mixture using the average titre for titration of the equilibrium mixture.

4 Subtract the amount of hydrochloric acid from the total amount of acid in the equilibrium mixture to calculate the amount of ethanoic acid, in moles, remaining in the equilibrium mixture.

5 Calculate the amount of ethanoic acid, in moles, that has reacted as the equilibrium is established. This amount of moles is the same as the amount of ethyl ethanoate, in moles, and the amount of water, in moles, which have been formed at equilibrium. It is also the amount of alcohol, in moles, that has reacted as the equilibrium is established.

6 Calculate the amount of ethanol, in moles, that remains at equilibrium from the original amount, in moles, that was put into the flask.

7 Calculate the amount of water, in moles, at equilibrium. Do not forget to include the fact that the equivalent of $8.0\,cm^3$ of water were added initially.

8 Calculate the concentration in $mol\,dm^{-3}$ of ethanoic acid at equilibrium.

9 Calculate the concentration in $mol\,dm^{-3}$ of ethanol at equilibrium.

10 Calculate the concentration in $mol\,dm^{-3}$ of ethyl ethanoate at equilibrium.

11 Calculate the concentration in $mol\,dm^{-3}$ of water at equilibrium.

12 Use these data to calculate the value of K_c for the equilibrium mixture.

13 Given that the accepted value for this equilibrium constant is usually quoted as 4.0 at 298 K, comment on the result calculated.

○──

Effects of changes on concentrations, position of equilibrium and K_c

Many factors affect the position of equilibrium and concentrations or yields of certain substances in the equilibrium but only temperature affects the value of the equilibrium constant, K_c.

All equilibrium constants are constant at constant temperature and will vary when temperature is varied.

Effects of changes in temperature

Temperature is the only factor that affects the value of K_c for an equilibrium reaction. When temperature affects the position of equilibrium and the concentrations of reactants and products, K_c will change. When temperature changes, an increase in the concentration of the products (and decrease in the concentration of the reactants) will increase the value of K_c. When temperature changes, a decrease in the concentration of the products (and increase in the concentration of the reactants) will decrease the value of K_c. This is **only** when temperature changes.

When the forward reaction is **exothermic**, an **increase** in temperature shifts the equilibrium to the left as it absorbs heat. This lowers the concentrations of the products and increases the concentrations of the reactants and so K_c **decreases**.

Figure 7.6 The water gas shift reaction is the reaction of carbon monoxide and water vapour to form carbon dioxide and hydrogen (the mixture of carbon monoxide and hydrogen is known as water gas): $CO + H_2O \rightarrow CO_2 + H_2$. It is used industrially to manufacture hydrogen. The equilibrium constant is high at low temperature and the formation of products is favoured but drops quickly with increasing temperature, as shown in the graph above. Think about what temperature is used in industry for a high yield.

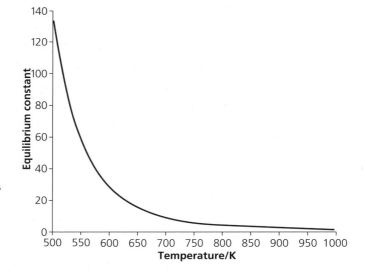

○ EXAMPLE 14

For the equilibrium:

$N_2(g) + 3H_2(g) \rightleftharpoons 2NH_3(g)$; $\Delta H = -92\,kJ\,mol^{-1}$

State the effect, if any, on the value of K_c of decreasing the temperature. All other factors are unchanged.

Answer

● The forward reaction is exothermic (ΔH is negative).

● A decrease in temperature moves the position of equilibrium in the direction of the forward exothermic reaction to oppose the decrease in temperature.

● The position of equilibrium shifts to the right increasing the concentration of the products and decreasing the concentrations of the reactants.

● K_c increases.

EXAMPLE 15

For the equilibrium:

$H_2(g) + I_2(g) \rightleftharpoons 2HI(g)$

K_c values are given for this equilibrium at different temperature in the table below.

Temperature /°C	K_c
200	74.5
500	50.0
1000	13.0

Explain whether the forward reaction is exothermic or endothermic.

Answer

- The value of K_c decreases with an increase in temperature.

- An increase in temperature moves the position of equilibrium to the left.

- The reverse reaction is endothermic to absorb the heat (on increasing temperature).

- So, the forward reaction is exothermic.

Effects of changes in pressure and volume

Changes in volume or total pressure in a gaseous homogeneous reaction may have an effect on the position of equilibrium and the concentrations of the reactants and products present in the equilibrium mixture. However, changes in volume or total pressure have **no effect** on the value of the equilibrium constant, K_c.

Although the equilibrium concentrations of the reactants and products may change, they will give the same value for K_c when put into the expression, as K_c is constant at constant temperature.

Remember that a decrease in volume or an increase in total pressure on the reaction will move the position of equilibrium to the side with the fewer moles of gas to oppose the increase in pressure.

EXAMPLE 16

For the equilibrium:

$2P(g) + Q(g) \rightleftharpoons R(g)$; $\Delta H = +57\,kJ\,mol^{-1}$

$K_c = 15.0$ at $500\,K$

1 State the effect, if any, on the equilibrium amount of R of using a container of larger volume. All other factors are unchanged.
2 State the effect, if any, on the value of K_c of using a container of larger volume. All other factors are unchanged.

Answers

1 The use of a container of larger volume decreases the pressure on the reaction. Using Le Chatelier's principle, the position of equilibrium will move to the side with the larger moles of gas. So the position of equilibrium will move to the left in this example and this will decrease the equilibrium amount of R.
The equilibrium amount of R decreases.
2 As there is no change in temperature, K_c does not change.

TIP

This point is particularly important in some reactions as removing a product can move the position of equilibrium to the right so increasing the yield of a product. For example in the Haber process, the reaction mixture is cooled and ammonia is condensed out of the mixture, the position of equilibrium moves to replace the ammonia.

Effects of changes in concentration

If the concentration of a reactant or product in the equilibrium mixture is changed, the position of equilibrium will change. The position of equilibrium will move to oppose any change in concentration.

If more of a reactant is added, the position of equilibrium will move to the right to oppose this change and decrease the concentration of the added reactant.

If a product is removed from the reaction mixture, the position of equilibrium will move to the right to replace the product which has been removed.

A change in concentration has **no effect** on the value of the equilibrium constant.

EXAMPLE 17

The following equilibrium is established in the production of the ester ethyl propanoate, $CH_3CH_2COOCH_2CH_3$.

$CH_3CH_2OH + CH_3CH_2COOH \rightleftharpoons CH_3CH_2COOCH_2CH_3 + H_2O$

1 State the effect, if any, on the equilibrium amount of the ester of increasing the concentration of propanoic acid (CH_3CH_2COOH). All other factors are unchanged.
2 State the effect, if any, on the value of K_c of increasing the concentration of propanoic acid (CH_3CH_2COOH). All other factors are unchanged.

Answers

1 Based on Le Chatelier's principle, the position of equilibrium will move to oppose the change so the position of equilibrium will move to the right to remove the added propanoic acid. The amount of the ester will increase.
2 As there is no change in temperature, K_c does not change.

TEST YOURSELF 6

1 State Le Chatelier's principle.
2 For the equilibrium:

$A(g) + 2B(g) \rightleftharpoons 2C(g)$; $\Delta H = +25 \, kJ \, mol^{-1}$

a) State the effect, if any, of increasing the temperature on the value of K_c for this equilibrium. All other factors are unchanged.
b) State the effect, if any, on the equilibrium amount of C on using a container of larger volume. All other factors are unchanged.
c) The concentration of A is increased.
   i) State the effect, if any, on the value of K_c for this equilibrium. All other factors are unchanged.
   ii) State the effect, if any, on the equilibrium amount of C. All other factors are unchanged.

3 For the equilibrium:

$2SO_2(g) + O_2(g) \rightleftharpoons 2SO_3(g)$; $\Delta H = -197 \, kJ \, mol^{-1}$

a) State the effect, if any, of increasing the total pressure on the equilibrium amount of SO_3. All other factors are unchanged.
b) State the effect, if any, of increasing the total pressure on the value of K_c. All other factors are unchanged.
c) State the effect, if any, of increasing the temperature on the value of K_c. All other factors are unchanged.

Practice questions

1 1.00 mol of propanoic acid and 1.00 mol of ethanol were mixed and allowed to reach equilibrium. At equilibrium 0.400 mol of the ester ethyl propanoate ($CH_3CH_2COOCH_2CH_3$) were formed.

$$CH_3CH_2COOH + CH_3CH_2OH \rightleftharpoons CH_3CH_2COOCH_2CH_3 + H_2O$$

The value of K_c to 2 decimal places at this temperature is:

A 0.27   **B** 0.44

C 0.67   **D** 2.25   *(1)*

2 For the following reaction in equilibrium:

$$N_2(g) + 3H_2(g) \rightleftharpoons 2NH_3(g)$$

Initially 1.0 mol of nitrogen was mixed with 2.5 mol of hydrogen. At equilibrium 1.0 mol of ammonia is formed.

What is the total number of moles of gas present in the equilibrium mixture?

A 2.0   **B** 2.5

C 3.0   **D** 3.5   *(1)*

3 For the equilibrium:

$$PCl_5(g) \rightleftharpoons PCl_3(g) + Cl_2(g);$$
$$K_c = 0.185 \, mol \, dm^{-3} \text{ at } 250\,°C$$

a) One equilibrium mixture at this temperature contains PCl_5 at a concentration of $0.220 \, mol \, dm^{-3}$ and PCl_3 at a concentration of $0.0120 \, mol \, dm^{-3}$. Calculate the concentration of Cl_2 in this equilibrium mixture. *(3)*

b) Another equilibrium mixture in a $2.00 \, dm^3$ vessel at the same temperature contains 0.150 mol of PCl_3 and 0.0900 mol of Cl_2. Calculate the number of moles of PCl_5 present at equilibrium. *(4)*

4 Sulfur dioxide reacts with oxygen according to the equilibrium:

$$SO_2(g) + \tfrac{1}{2}O_2(g) \rightleftharpoons SO_3(g); \Delta H = -98 \, kJ \, mol^{-1}$$

K_c for this reaction at 1000 K is $2.8 \times 10^2 \, mol^{-\frac{1}{2}} dm^{\frac{3}{2}}$.

a) Write an expression for the equilibrium constant, K_c, for this equilibrium. *(1)*

b) State the effect, if any, on the equilibrium amount of SO_3 of increasing the temperature. All other factors are unchanged. *(1)*

c) At 1000 K, equilibrium is established and the concentrations of SO_3 and O_2 are given below.

SO_3: $2.4 \, mol \, dm^{-3}$

O_2: $0.54 \, mol \, dm^{-3}$

Calculate the equilibrium concentration of SO_2 at 1000 K. *(3)*

d) Calculate the equilibrium constant, K_c, for the reaction:

$$SO_3(g) \rightleftharpoons SO_2(g) + \tfrac{1}{2}O_2(g) \text{ at } 1000\,K. \quad (2)$$

5 For the equilibrium:

$$2A(g) + B(g) \rightleftharpoons C(g)$$

a) Write an expression for the equilibrium constant, K_c, for this equilibrium. *(1)*

b) Deduce the units of K_c for this equilibrium. *(1)*

c) One equilibrium mixture at 400 °C contains the following concentrations of A, B and C.

A: $0.35 \, mol \, dm^{-3}$

B: $0.77 \, mol \, dm^{-3}$

C: $1.04 \, mol \, dm^{-3}$

i) Calculate a value for K_c at 400 °C. *(2)*

ii) The table below gives two other values of K_c for this reaction at different temperatures.

Temperature/°C	K_c
150	25.20
800	4.27

Explain whether the equilibrium reaction

$$2A(g) + B(g) \rightleftharpoons C(g)$$

is exothermic or endothermic. *(2)*

6 A mixture of 1.80 mol of hydrogen and 1.80 mol of iodine were allowed to come to equilibrium at 710 K in a 2.00 dm³ container.

$$H_2(g) + I_2(g) \rightleftharpoons 2HI(g)$$

The equilibrium mixture was found to contain 3.00 mol of hydrogen iodide.

Calculate the value of the equilibrium constant K_c and deduce its units. (4)

7 0.0420 mol of SO_2Cl_2 were placed in a 2.00 dm³ vessel, the vessel was sealed and its temperature raised to 375 °C. At equilibrium, the vessel contained 0.0345 mol of Cl_2.

$$SO_2Cl_2(g) \rightleftharpoons SO_2(g) + Cl_2(g)$$

a) Calculate the amount, in moles, of SO_2Cl_2 and SO_2 in the equilibrium mixture. (2)

b) Write an expression for the equilibrium constant, K_c, for this reaction. (1)

c) Calculate a value for K_c for this equilibrium at 375 °C. (3)

d) Deduce the units of K_c for this equilibrium. (1)

8 Hydrogen and carbon dioxide form an equilibrium mixture at 1000 °C.

$$H_2 + CO_2 \rightleftharpoons H_2O + CO$$

If 1.0 mole of H_2 and 1.0 mole of CO_2 are placed in a flask at 1000 °C and allowed to come to equilibrium. 0.2 mol of H_2 are present in the equilibrium mixture.

a) Write an expression for the equilibrium constant, K_c, for this equilibrium. (1)

b) Calculate the amount, in moles, of CO_2, H_2O and CO in the equilibrium mixture. (2)

c) Calculate a value for K_c for this equilibrium. (2)

8

Oxidation, reduction and redox equations

Redox reactions are reactions in which oxidation and reduction occur simultaneously and they are very important in everyday life.

In Figure 8.1 there are many redox reactions occurring. The electric currents from the batteries that power the computer and the games consoles are generated by redox reactions. The combustion of the natural gas in the fireplace is a redox reaction; even the respiration that all the family members are undergoing to keep alive is a redox reaction.

At this level, to explain redox in detail, you need to add one more factor to the three factors in the prior content. This factor is oxidation state.

Figure 8.1 What redox reactions can you identify in this photograph?

Oxidation state

Oxidation state is a numerical value of the degree of oxidation or reduction of an atom/element. Non-zero oxidation states are always given as a sign with a number, e.g. $+2$ or -3. The oxidation state in a compound is defined as the hypothetical charge on an atom assuming that the bonding is completely ionic.

The **oxidation state** in a compound is defined as the hypothetical charge on an atom assuming that the bonding is completely ionic.

> **TIP**
>
> It is important to write the oxidation state of an atom/element in a compound or an ion as a sign (either + or –) followed by a number. Elements have an oxidation state of 0 (zero). Both chlorine atoms in Cl_2 have an oxidation state of 0.

Oxidation state of ions

The oxidation state of a simple ion is the charge on the ion.

For example:

- oxidation state of iron in $Fe^{3+} = +3$
- oxidation state of oxygen in $O^{2-} = -2$
- oxidation state of sodium in $Na^+ = +1$
- oxidation state of chlorine in $Cl^- = -1$
- oxidation state of tin(II), $Sn^{2+} = +2$

General points on oxidation state

- The total sum of the oxidation states of all the elements in a compound must add up to zero.
- The total sum of the oxidation states of all the elements in a molecular ion must add up to the charge on the ion.
- The maximum oxidation state of an element is '+ group number', e.g. nitrogen is in Group 5 so its maximum oxidation state is $+5$.
- The minimum oxidation state of an element is 'its group number -8', e.g. nitrogen is in Group 5 so its minimum oxidation state is $5 - 8 = -3$.

Rules for determining oxidation state

There are some simple rules for determining the oxidation state of an element in a compound.

- Oxygen has an oxidation state of -2 in all compounds except in peroxides where it has an oxidation state of -1 and in OF_2 where it has an oxidation state of $+2$.
- Hydrogen has an oxidation state of $+1$ in all compounds except in hydrides where it has an oxidation state of -1.
- Group 1 elements have an oxidation state of $+1$ in all compounds.
- Group 2 elements have an oxidation state of $+2$ in all compounds.
- The oxidation states of transition elements and p block elements vary.
- In a simple binary compound (two elements), the more electronegative element has the negative oxidation state.
- d block elements where the charge on the ion is given, e.g. copper(II), Cu^{2+}, copper has an oxidation state of $+2$; iron(III), oxidation state of iron is $+3$; iron(II), oxidation state of iron is $+2$.

Iron in the oxidation states $+2$ and $+3$ is involved in a reaction causing cell damage and cancer in the brain. Recent research by Aleksandra Wandzilak from the AGH University of Science and Technology in Poland (2013) has shown that the higher the malignancy grade of a brain tumour, the higher the iron(II) to iron(III) ratio. It is thought that measuring the oxidation state of the iron in the tumour may be a way of determining the severity of the cancer.

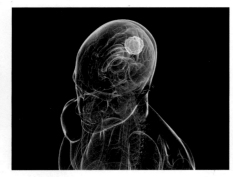

Figure 8.2 Analysing the oxidation state of iron in a brain tumour may be a future medical method of determining the grade of the tumour.

EXAMPLE 1

Give the oxidation state of sulfur in sodium sulfate, Na_2SO_4.

Answer

- Na: oxidation state = +1; 2 Na present so total for 2 Na = +2

- S: oxidation state = x; 1 S present so total for 1 S = x

- O: oxidation state = –2; 4 O present so total for 4 O = –8

The total of the oxidation states is zero as it is a compound.

Therefore: $+ 2 + x – 8 = 0$

Solving for x: $x = + 8 – 2 = +6$

So the oxidation state of S in Na_2SO_4 is +6.

EXAMPLE 2

Give the oxidation state of nitrogen in the ammonium ion, NH_4^+

Answer

- N: oxidation state = x; 1 N present so total for 1 N = x

- H: oxidation state = +1; 4 H present so total for 4 H = +4

The total of the oxidation states is +1 as it is an ion with a charge of +.

Therefore: $x + 4 = +1$

Solving for x: $x = +1 – 4 = –3$

Oxidation state of N in NH_4^+ is –3.

EXAMPLE 3

Give the oxidation state of manganese in $KMnO_4$

Answer

- K: oxidation state = + 1; 1 K present so total for 1 K = +1

- Mn: oxidation state = x; 1 Mn present so total for 1 Mn = x

- O: oxidation state = –2; 4 O present so total for 4 O = –8

The total of the oxidation states is zero as it is a compound.

Therefore: $+1 + x – 8 = 0$

Solving for x: $x = +8 – 1 = + 7$

Oxidation state of Mn in $KMnO_4$ is +7.

TIP
$KMnO_4$ is called potassium manganate(VII) where the (VII) represents the +7 oxidation state of manganese in the compound. The MnO_4^- ion is called the manganate(VII) ion. Potassium manganate(VII) is also called potassium permanganate. Sodium sulfate in Example 1 is sometimes called sodium sulfate(VI).

EXAMPLE 4

Give the oxidation state of chlorine in the perchlorate ion, ClO_4^-.

Answer

- Cl: oxidation state = x; 1 Cl present so total for 1 Cl = x

- O: oxidation state = –2; 4 O present so total for 4 O = –8

The total of the oxidation states is –1 as it is an ion with a charge of –.

Therefore: $x – 8 = –1$

Solving for x: $x = –1 + 8 = +7$

Oxidation state of Cl in ClO_4^- is +7.

TIP
The ClO_4^- ion is also called the chlorate(VII) ion.

EXAMPLE 5

Determine the oxidation state of sulfur in sodium disulfate, $Na_2S_2O_7$

Answer

● Na: oxidation state = +1; 2 Na present so total for 2 Na = +2

● S: oxidation state = x; 2 S present so total for 2 S = $2x$

● O: oxidation state = –2; 7 O present so total for 7 O = –14

The total of the oxidation states is zero as it is a compound.

Therefore: $+2 + 2x – 14 = 0$

Solving for x: $2x = +14 – 2 = +12$; $x = +6$

Oxidation state of S in $Na_2S_2O_7$ is +6.

This is the first example where there are two atoms of the unknown oxidation state. Both sulfur atoms have an oxidation state of +6.

EXAMPLE 6

Give the oxidation state of iron in magnetite, Fe_3O_4.

Answer

● Fe: oxidation state = x; 3 Fe present so total for 3 Fe = $3x$

● O: oxidation state = –2; 4 O present so total for 4 O = –8

The total of the oxidation states is zero as it is a compound.

Therefore: $3x – 8 = 0$

Solving for x: $3x = +8$; $x = 2\frac{2}{3}$

Oxidation state of Fe in Fe_3O_4 is $+2\frac{2}{3}$.

The oxidation state of iron in Fe_3O_4 is $+2\frac{2}{3}$ (this is due to one Fe atom having an oxidation state of +2 and two Fe atoms having oxidation states of +3 which gives an average oxidation state for the atoms of iron in this compound of $+2\frac{2}{3}$).

TIP
When you get an oxidation state which is not a whole number, it is because it is an average of the oxidation states of all the atoms of that element in the compound or ion.

TIP
This would give you the same answer if you took the nitrate ion, NO_3^-, out of the formula and worked out the oxidation state of the nitrogen in the nitrate ion. If nitrogen is x and each oxygen –2; $x – 6 = –1$; so $x = +5$. The full chemical name of copper(II) nitrate is copper(II) nitrate(V) where (V) represents the +5 oxidation state of the nitrogen.

EXAMPLE 7

Give the oxidation state of nitrogen in magnesium nitrate, $Mg(NO_3)_2$

Answer

● Mg: Oxidation state = +2; 1 Mg present so total for 1 Mg = +2

● N: Oxidation state = x; 2 N present so total for 2 N = $2x$

● O: Oxidation state = –2; 6 O present so total for 6 O = –12

The total of the oxidation states is zero as it is a compound.

Therefore: $+2 + 2x – 12 = 0$

Solving for x: $2x = +12 – 2 = +10$; $x = +5$

Oxidation state of N in $Mg(NO_3)_2$ is +5.

TIP

The ammonium ion, NH_4^+, can be counted as a unit which leaves the vanadate ions as VO_3^-. The removal of the NH_4^+ leaves the vanadate with a charge of –1 so the oxidation state of vanadium can be worked out from VO_3^-, which will give the same answer of +5 for the oxidation state of V.

TEST YOURSELF 2

1 Give the oxidation states of nitrogen in the following compounds.
   a) HNO_3 **b)** NO_2 **c)** N_2O **d)** NH_3 **e)** Mg_3N_2
2 Give the oxidation states of sulfur in the following compounds and ions.
   a) SO_3^{2-} **b)** $S_2O_3^{2-}$ **c)** H_2SO_4 **d)** H_2S **e)** SF_6
3 In which of the following d block compounds does the d block element have an oxidation state of +6?
   $KMnO_4$ $Na_2Cr_2O_7$ $BaFeO_4$ K_2MnO_4 $BaNiO_3$
4 Give the oxidation state of phosphorus in the following:
   a) P_4 **b)** H_3PO_3 **c)** P_4O_{10} **d)** HPO_4^- **e)** Na_3P
5 Plutonium displays colourful oxidation states in aqueous solution, as shown in Figure 8.3.
   Give the oxidation state of plutonium in each of the five aqueous solutions.

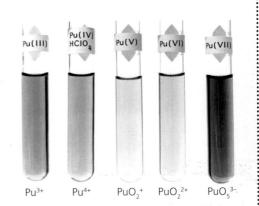

Figure 8.3 The oxidation states of plutonium in different aqueous solutions. These oxidation states are not stable, so plutonium solutions may spontaneously change oxidation states and colours!

Proper names and formulae of ions and compounds

The correct chemical name of an ion (or part of a compound) depends on its oxidation state. For compounds containing d block and p block elements the oxidation state of the p or d block element is often written after the name as a roman numeral in brackets.

Formulae of molecular ions

All molecular ions which end in –te contain d or p block element atoms and some oxygen atoms. For example:

- the carbonate ion contains one carbon atom and some oxygen atoms
- the manganate ion contains one manganese atom and some oxygen atoms
- the dichromate ion contains two chromium atoms and some oxygen atoms
- the nitrite ion contains one nitrogen atom and some oxygen atoms.

Some ions contain hydrogen as well, for example:

- the hydrogen carbonate ion contains one hydrogen atom, one carbon atom and some oxygen atoms
- the dihydrogen phosphate ion contains two hydrogen atoms, one phosphorus atom and some oxygen atoms
- the hydrogen phosphate ion contains one hydrogen atom, one phosphorus atom and some oxygen atoms.

Determining the number of oxygen atoms in an ion or compound

EXAMPLE 9

Calculate the value of x in potassium stannate (IV) K_4SnO_x.

Answer

In potassium stannate(IV), K_4SnO_x, the tin has the +4 oxidation state.

The number of oxygen atoms in the compound may be determined if the formula is given as K_4SnO_x where x represents the number of oxygen atoms. Each potassium is +1, the tin is +4 and each oxygen is –2 so the total for oxygen is $-2x$. As it is a compound the overall total for the oxidation states is zero.

Therefore: $+4 + 4 - 2x = 0$

Solving for x: $-2x = -4 - 4 = -8$; $x = 4$

Formula of potassium stannate(IV) is K_4SnO_4.

TEST YOURSELF 3

1 Give the oxidation state of iron in FeO_4^{2-}.
2 The chlorate(VII) ion is ClO_4^-. What is the formula of sodium chlorate(VII)?
3 What is the full chemical name of Mg_2SnO_4?
4 Deduce the value of x in the following molecular ions.
   a) chlorate(V) ion, ClO_x^-
   b) bromate(I) ion, BrO_x^-
   c) sulfate(IV) ion, SO_x^{2-}
   d) dihyrogenphosphate(V) ion, $H_2PO_x^-$
   e) manganate(VII) ion, MnO_x^-

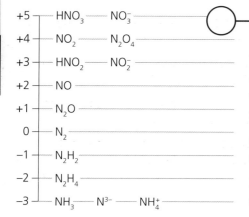

Oxidation and reduction

Many p and d block elements show great variation in the oxidation states of the compounds and ions which they form. It is important to be able to work out the oxidation state to determine if an element is being oxidised or reduced.

- An increase in oxidation state is oxidation.
- A decrease in oxidation state is reduction.

Figure 8.4 shows the different oxidation states of nitrogen. Nitrogen is used as an example but the skills apply to any element in any redox reaction.

Figure 8.4

TIP

The ending –ATE with an oxidation state may be used for negatively charged molecular ions, for example sulfate(IV) and nitrate(V). There are a few exceptions such as hydroxide, OH^- and cyanide, CN^-. The ending –ATE without an oxidation state usually indicates the highest oxidation state of the p or d block element in the molecular ion, for example sulfate is sulfate(VI), nitrate is nitrate(V).

TIP

Again note that the maximum oxidation state of nitrogen is +5 (which is + the group number) and the minimum oxidation state is the group number − 8 = 5 − 8 = −3.

EXAMPLE 10

In the following reaction:

$N_2H_4 + 2H_2O_2 \rightarrow N_2 + 4H_2O$

explain whether nitrogen is oxidised or reduced.

Answer

- In N_2H_4, nitrogen has the oxidation state −2.
- In N_2, nitrogen has the oxidation state 0.
- Nitrogen is oxidised as it shows an increase in oxidation state.

TIP

In Example 10 oxygen is reduced as it shows a decrease in oxidation number from −1 (in H_2O_2 – a peroxide) to −2 in H_2O.

EXAMPLE 11

The following reaction is a redox reaction of nitrogen monoxide, which occurs in a catalytic converter.

$2NO + 2CO \rightarrow N_2 + 2CO_2$

Explain the redox reaction in terms of change in oxidation state.

Answer

- Nitrogen is reduced (shows a decrease in oxidation state) from +2 (in NO) to 0 (in N_2).
- Carbon is oxidised (shows an increase in oxidation state) from +2 (in CO) to +4 (in CO_2).
- A redox reaction is one in which both oxidation and reduction occur simultaneously.

EXAMPLE 12

Consider the following reactions of sulfuric acid:

A $NaCl + H_2SO_4 \rightarrow NaHSO_4 + HCl$
B $2KOH + H_2SO_4 \rightarrow K_2SO_4 + 2H_2O$
C $2HBr + H_2SO_4 \rightarrow Br_2 + SO_2 + 2H_2O$

Which one of the reactions is a redox reaction? Explain your answer.

Answer

- Reaction C is a redox reaction.
- Sulfur is reduced (shows a decrease in oxidation state) from +6 (in H_2SO_4) to +4 (in SO_2)
- Bromine is oxidised (shows an increase in oxidation state) from −1 (in HBr) to 0 (in Br_2)
- A redox reaction is one in which both oxidation and reduction occur simultaneously.
- There is no change in oxidation state for any of the elements in the other reactions.

TEST YOURSELF 4

1 Explain in terms of oxidation states why the following reactions are described as redox reactions.

a) $4NH_3 + 5O_2 \rightarrow 4NO + 6H_2O$

b) $6HI + H_2SO_4 \rightarrow 3I_2 + S + 4H_2O$

c) $Cl_2 + 2NaBr \rightarrow 2NaCl + Br_2$

2 State which element is oxidised in the following reactions. Explain your answer in terms of oxidation state.

a) $CuO + H_2 \rightarrow Cu + H_2O$

b) $Al_2O_3 + 6Na \rightarrow 2Al + 3Na_2O$

3 Explain why the following reaction is not a redox reaction.

$CaCO_3 \rightarrow CaO + CO_2$

Redox

A redox reaction is one in which an oxidation and reduction reaction occur.

In the reaction $6HI + H_2SO_4 \rightarrow 3I_2 + S + 4H_2O$

- I is oxidised as it shows an increase in oxidation state from -1 (in HI) to 0 (in I_2).
- S is reduced as it shows a decrease in oxidation state from $+6$ (in H_2SO_4) to 0 (in S).
- This is a redox reaction as both oxidation and reduction are occurring together in the same reaction.

In the reaction $Cl_2 + H_2O \rightarrow HCl + HOCl$

- Cl is oxidised as it shows an increase in oxidation state from 0 (in Cl_2) to $+1$ (in HOCl).
- Cl is also reduced as it shows a decrease in oxidation state from 0 (in Cl_2) to -1 (in HCl).
- This is a redox reaction as chlorine (Cl) is oxidised and reduced in the same reaction.

If you are asked to explain a redox reaction, you should calculate the oxidation states and then state the definition of a redox reaction.

Oxidising agents and reducing agents

- An oxidising agent is a chemical which causes oxidation in another species.
- Oxidising agents accept electrons (they cause another substance to lose electrons thus oxidising it).
- A reducing agent is a chemical that causes a reduction in another species.
- Reducing agents lose electrons (they cause another substance to gain electrons thus reducing it).

In the body strong oxidising agents such as hydrogen peroxide are formed, and cause damaging chemical changes to cell DNA. Healthy cells can convert the altered DNA back to its normal form, however ageing, cancer and heart disease slow down this repair mechanism. Antioxidants such as vitamins C and E remove oxidising agents from our body, and as a result slow down the alteration of DNA.

Figure 8.5 Five portions of fruit and vegetables per day are thought to slow down damage caused by oxidising agents in the body. Do you eat your five a day?

EXAMPLE 13

Explain the reaction of chlorine with sodium bromide according to the following equation:

$$Cl_2 + 2NaBr \rightarrow 2NaCl + Br_2$$

Answer

In this reaction:

- Br is oxidised from –1 (in NaBr) to 0 (in Br_2).
- Cl is reduced from 0 (in Cl_2) to –1 (in NaCl).
- Chlorine causes the oxidation of bromide ions to bromine.
- Chlorine is the oxidising agent (or oxidant).

EXAMPLE 14

Explain the reaction of hydrogen peroxide with iodide ions. The equation is:

$$H_2O_2 + 2I^- + 2H^+ \rightarrow 2H_2O + I_2$$

Answer

In this reaction:

- I is oxidised from –1 (in I^-) to 0 (in I_2).
- O is reduced from –1 (in H_2O_2) to –2 (in H_2O).
- Hydrogen peroxide causes the oxidation of iodide ions to iodine.
- Hydrogen peroxide is an oxidising agent (or oxidant).

EXAMPLE 15

Explain the reaction of zinc with ammonium vanadate(V) in acid conditions as follows:

$$3Zn + 2NH_4VO_3 + 6H_2SO_4 \rightarrow 3ZnSO_4 + 2VSO_4 + (NH_4)_2SO_4 + 6H_2O$$

Answer

In this reaction:

- Zn is oxidised from 0 (in Zn) to +2 (in $ZnSO_4$).
- V is reduced from +5 (NH_4VO_3) to +2 (in VSO_4).
- Zinc causes the reduction of vanadium from the +5 oxidation state to +2.
- Zinc is a reducing agent (or reductant).

TEST YOURSELF 5

1 Ammonia can act as a reducing agent. When passed over heated copper(II) oxide, the following reaction occurs.

$$2NH_3 + 3CuO \rightarrow 3Cu + N_2 + 3H_2O$$

Give the oxidation states of nitrogen and copper in the reactants and products and use them to explain this redox change.

2 The equation for the reaction of silver with nitric acid is:

$$3Ag + 4HNO_3 \rightarrow 3AgNO_3 + 2H_2O + NO$$

Give the oxidation states of the underlined elements in the following species from the equation and use them to explain the redox reaction taking place.

$\underline{A}g \qquad H\underline{N}O_3 \qquad Ag\underline{N}O_3 \qquad \underline{N}O$

3 Chlorine reacts with water according to the following equation:

$$Cl_2 + H_2O \rightarrow HCl + HOCl$$

Give the oxidation states of chlorine in the following molecules.

$Cl_2 \qquad HCl \qquad HOCl$

4 In the following reaction:

$$5SO_2 + 2MnO_4^- + 2H_2O \rightarrow 5SO_4^{2-} + 2Mn^{2+} + 4H^+$$

a) Give the oxidation state of the underlined elements in the species below.
$\underline{S}O_2 \quad Mn\underline{O}_4^- \quad \underline{S}O_4^{2-} \quad \underline{M}n^{2+}$

b) Name the oxidising agent in this reaction.

Half equations

Half equations include electrons and help to show the oxidation and reduction processes in a chemical reaction.

Basics of half equations

Reduction is gaining electrons. In a reduction half equation the electrons should be placed on the left. For example:

$$Sn^{2+} + 2e^- \rightarrow Sn$$

In this example tin(II) ions are gaining electrons as shown by $+2e^-$ on the left-hand side of the equation. Tin(II) ions are reduced to tin.

Oxidation is losing electrons. In an oxidation half equation the electrons should be placed on the right. For example:

$$Mg \rightarrow Mg^{2+} + 2e^-$$

In this example magnesium atoms are losing electrons as shown by the $+2e^-$ on the right-hand side of the equation. Magnesium atoms are oxidised to magnesium ions.

The total change in the oxidation state is the same as the number of electrons required in the equation. If nitrogen in nitrate(V), NO_3^- (oxidation state of N is $+5$), is reduced to nitrogen(II) oxide, NO (oxidation state of N is $+2$), the total change in oxidation state is 3 so $+3e^-$ are required on the left-hand side with NO_3^-, as it is a reduction.

If oxygen atoms are added or removed from a compound or ion, H_2O and hydrogen ions, H^+ are used to balance this change in oxygen content. For example in the reduction of NO_3^- to NO, two oxygen atoms are removed so 2 H_2O are required on the right-hand side with 4 H^+ on the left-hand side.

TIP

If two or more atoms are reduced, the total oxidation state change is calculated from (the oxidation state change × the number of atoms undergoing oxidation or reduction).

The reduction of NO_3^- to NO is written:

$$NO_3^- + 4H^+ + 3e^- \rightarrow NO + 2H_2O$$

Finally for all half equations the total of the charges on the left-hand side and the right-hand side should be the same. In this example $NO_3^- + 4H^+ + 3e^-$ on the left-hand side give a total of -1 and $+4$ and $-3 = 0$ and there are no charges on the right-hand side so the total charge is zero as well.

> **TIP**
> The total of the charges on both sides of the equation is useful as a final check to make sure any half equation, or even an ionic equation, is correct.

Writing half equations

EXAMPLE 16

Write a half equation for the reduction of chlorine. (This would be similar for any halogen being reduced.)

Answer

The following points explain the process of writing a half equation for the reduction of chlorine:

- Chlorine as an element is diatomic so must be written Cl_2. This is the reactant in the half equation and should appear on the left-hand side of the arrow.

$$Cl_2 \rightarrow$$

- Two chloride ions are formed and each of these has a charge of -1 and should appear on the right-hand side of the arrow.

$$Cl_2 \rightarrow 2Cl^-$$

- The oxidation state of each chlorine atom in Cl_2 is 0.
- The oxidation state of each chloride ion is -1.
- The reaction is a reduction (decrease in oxidation state) so the electrons are placed on the left-hand side.

$$Cl_2 + e^- \rightarrow 2Cl^-$$

- Two electrons are required for the total change in oxidation state;
  $2e^-$ are placed on the left as electrons are gained as it is a reduction half equation.

$$Cl_2 + 2e^- \rightarrow 2Cl^-$$

- The overall charge on each side of the equation should be the same $(2^-) \rightarrow (2^-)$.
- The process is a reduction because:
  - Chlorine gains electrons.
  - The oxidation state of the chlorine decreases.

EXAMPLE 17

Write a half equation for the oxidation of magnesium. (This would be similar for any metal being oxidised to a 2+ ion.)

Answer

The following points explain the process of writing a half equation for the oxidation of magnesium:

- Magnesium is the reactant in the half equation and should appear on the left-hand side of the arrow.

$$Mg \rightarrow$$

- Mg^{2+} ions are formed and should appear on the right-hand side of the arrow.

$$Mg \rightarrow Mg^{2+}$$

- The oxidation state of magnesium is 0.

- The oxidation state of the Mg^{2+} ion is $+2$.
- The reaction is an oxidation (increase in oxidation state) so the electrons are placed on the right-hand side.

$$Mg \rightarrow Mg^{2+} + e^-$$

- Two electrons are required for the total change in oxidation state;
  $2e^-$ are placed on the right as electrons are lost as it is an oxidation half equation.

$$Mg \rightarrow Mg^{2+} + 2e^-$$

- The overall charge on each side of the equation should be the same:

$$(0) \rightarrow (2+ \text{ and } 2- = 0)$$

- The process is an oxidation because:
  - Magnesium loses electrons.
  - The oxidation state of the magnesium increases.

TIP

Sulfite is often called sulfate(IV) and sulfate is often called sulfate(VI).

TIP

Check that you can calculate the oxidation states of sulfur in both the sulfite and sulfate ions. Go back to the section on calculating oxidation states if you have problems.

TIP

Iodate(V) is often simply called iodate.

TIP

This is the first oxygen containing half equation where the atoms being oxidised or reduced have to be balanced. It is important to watch for this in a half equation.

EXAMPLE 18

Sulfite, SO_3^{2-}, may be oxidised to sulfate, SO_4^{2-}. Write a half equation for this reaction.

Answer

- SO_3^{2-} is on the left and SO_4^{2-} is on the right

  $$SO_3^{2-} \rightarrow SO_4^{2-}$$

- The sulfur atoms balance so no balancing numbers are needed for sulfite and sulfate ions.

- Calculate the oxidation state of the element being oxidised or reduced.
  - SO_3^{2-}; oxidation state of sulfur is +4
  - SO_4^{2-}; oxidation state of sulfur is +6
  - The reaction is an oxidation (increase in oxidation state) so the electrons are placed on the right-hand side.

  $$SO_3^{2-} \rightarrow SO_4^{2-} + e^-$$

- Two electrons are required for the total change in oxidation state; $2e^-$ are placed on the right as electrons are lost because it is an oxidation half equation.

  $$SO_3^{2-} \rightarrow SO_4^{2-} + 2e^-$$

- One oxygen is gained from left to right (sulfite has three and sulfate has four). H_2O is placed on the left-hand side and $2H^+$ on the right-hand side.

  $$SO_3^{2-} + H_2O \rightarrow SO_4^{2-} + 2H^+ + 2e^-$$

- The overall charge on each side of the equation should be the same:

  $$(2-) \rightarrow (2- \text{ and } 2+ \text{ and } 2- = 2-)$$

- The process is an oxidation because:
  - The sulfite ions are losing electrons.
  - The oxidation state of the sulfur increases (+4 to +6).

EXAMPLE 19

Write a half equation for the reduction of iodate(V), IO_3^-, to iodine, I_2

Answer

- IO_3^- is on the left and I_2 is on the right.

  $$IO_3^- \rightarrow I_2$$

- The iodine atoms do not balance as there are 2 I atoms on the right and only one on the left. A '2' is needed in front of the IO_3^-

  $$2IO_3^- \rightarrow I_2$$

- Calculate the oxidation state of the element being oxidised or reduced.
  - IO_3^-: oxidation state of iodine is +5.
  - I_2: oxidation state of iodine is 0.
  - The reaction is a reduction (decrease in oxidation state) so the electrons are placed on the left-hand side.

  $$2IO_3^- + e^- \rightarrow I_2$$

- 10 electrons are required for the total change in oxidation state (two iodine changing from +5 to 0); $10e^-$ are placed on the left as electrons are gained as it is a reduction half equation.

$$2IO_3^- + 10e^- \rightarrow I_2$$

- Six oxygen are lost from left to right (each of the two iodate ions has 3 and iodine has 0). $6H_2O$ are placed on the right-hand side and $12H^+$ are on the left-hand side.

$$2IO_3^- + 12H^+ + 10e^- \rightarrow I_2 + 6H_2O$$

- The overall charge on each side of the equation should be the same $(2- \text{ and } 12+ \text{ and } 10- = 0) \rightarrow (0)$

- The process is a reduction because:
  - The iodate(V) ions are gaining electrons.
  - The oxidation state of the iodine decreases (+5 to 0).

EXAMPLE 20

Write a half equation for the reduction of manganate(VII) ions, MnO_4^-, to manganese(II), Mn^{2+}.

Answer

- MnO_4^- is on the left and Mn^{2+} is on the right.
- The manganese atoms balance.
- Calculate the oxidation state of the element being oxidised or reduced.
  - MnO_4^-: oxidation state of manganese is +7
  - Mn^{2+}: oxidation state of manganese is +2
  - The reaction is a reduction (decrease in oxidation state) so the electrons are placed on the left-hand side.

- Five electrons are required for the total change in oxidation state; $5e^-$ are placed on the left as electrons are gained as it is a reduction half equation.

- Four oxygen are lost from left to right. $4H_2O$ are placed on the right-hand side and $8H^+$ on the left-hand side.

$$MnO_4^- + 8H^+ + 5e^- \rightarrow Mn^{2+} + 4H_2O$$

- The overall charge on each side of the equation should be the same:

$(- \text{ and } 8+ \text{ and } 5- = 2+) \rightarrow (2+)$

- The process is a reduction because:
  - The manganate(VII) ions are gaining electrons.
  - The oxidation state of the manganese decreases (+7 to +2).

EXAMPLE 21

Write a half equation for the reduction of dichromate(VI), $Cr_2O_7^{2-}$, to chromium(III), Cr^{3+}.

Answer

- $Cr_2O_7^{2-}$ is on the left and Cr^{3+} is on the right.

- The chromium atoms do not balance so a '2' is placed before the Cr^{3+} on the right-hand side

- Calculate the oxidation state of the element being oxidised or reduced.
  - $Cr_2O_7^{2-}$: oxidation state of chromium is +6.
  - Cr^{3+}: oxidation state of chromium is +3.
  - The reaction is a reduction (decrease in oxidation state) so the electrons are placed on the left-hand side.

- Six electrons are required for the total change in oxidation state (two chromium changing from +6 to +3; $6e^-$ are placed on the left as electrons are gained as it is a reduction half equation).

- Seven oxygen are lost from left to right. $7H_2O$ are placed on the right-hand side and $14H^+$ on the left-hand side.

$$Cr_2O_7^{2-} + 14H^+ + 6e^- \rightarrow 2Cr^{3+} + 7H_2O$$

- The overall charge on each side of the equation should be the same:

$$(2- \text{ and } 14+ \text{ and } 6- = 6+) \rightarrow (6+)$$

- The process is a reduction because:
  - The dichromate(VI) ions are gaining electrons.
  - The oxidation state of the chromium decreases (+6 to +3).

TIP

Sometimes you may be asked to write an oxidation or reduction half equation from a balanced symbol equation for a redox reaction. Simply work out which element is being oxidised or reduced and write the half equation as shown from the species in the equation.

EXAMPLE 22

Concentrated nitric acid reacts with copper according to the equation:

$$Cu + 4HNO_3 \rightarrow Cu(NO_3)_2 + 2NO_2 + O_2$$

Write a half equation for the reduction occurring in this reaction.

Answer

- Nitrogen in HNO_3 is reduced from +5 to +4 in NO_2.

$$HNO_3 \rightarrow NO_2$$

- e^- is added to the left-hand side as it is a reduction.

$$HNO_3 + e^- \rightarrow NO_2$$

- One oxygen is removed from HNO_3 but also a hydrogen so H_2O is added to the right-hand side and H^+ to the left.

$$HNO_3 + H^+ + e^- \rightarrow NO_2 + H_2O$$

- The overall charge on each side of the equation should be the same $(+ \text{ and } - = 0) \rightarrow (0)$

TEST YOURSELF 6

1 Write half equations for the following processes:
   a) reduction of bromine to bromide ions.
   b) oxidation of zinc to zinc(II) ions.
   c) oxidation of iron to iron(III) ions.

2 Write half equations for the following processes:
   a) reduction of dioxovanadium(V), VO_2^+, to oxovanadium(IV), VO^{2+}
   b) reduction of oxovanadium(IV), VO^{2+}, to vanadium(III), V^{3+}
   c) reduction of sulfate(VI), SO_4^{2-}, to sulfur dioxide, SO_2
   d) oxidation of chromium(III), Cr^{3+}, to chromate, CrO_4^{2-}
   e) reduction of nitrate(V), NO_3^-, to nitrogen monoxide, NO
   f) oxidation of chlorine, Cl_2, to hypochlorite, OCl^-.

Organic oxidation and reduction reactions

Organic chemicals such as primary and secondary alcohols are oxidised by reagents such as acidified potassium dichromate solution.

TIP
Classification of alcohols and their oxidation is covered in the alcohols topic.

TIP
Aldehydes, ketones and carboxylic acids will also be covered in the alcohols topic.

- Primary alcohols (such as ethanol and propan-1-ol) are oxidised to aldehydes which may be further oxidised to carboxylic acids. Ethanol, CH_3CH_2OH, may be oxidised to ethanal, CH_3CHO, which may be further oxidised to ethanoic acid, CH_3COOH. Under the correct conditions ethanol can be oxidised directly to ethanoic acid.
- Secondary alcohols (such as propan-2-ol) may be oxidised to ketones. Propan-2-ol, $CH_3CH(OH)CH_3$, can be oxidised to propanone, CH_3COCH_3.

Aldehydes, ketones and carboxylic acid can be reduced using lithal ($LiAlH_4$).

- Butanoic acid, $CH_3CH_2CH_2COOH$, can be reduced to butanal, $CH_3CH_2CH_2CHO$, which can be further reduced to butan-1-ol, $CH_2CH_2CH_2CH_2OH$.
- Pentan-2-one, $CH_2COCH_2CH_2CH_3$ can be reduced to pentan-2-ol, $CH_3CH(OH)CH_2CH_2CH_3$.

Simple oxidation reactions

Equations for the oxidation of organic chemicals may be written simply using [O] to represent the oxidising agent.

- Oxidation in organic chemistry may be the addition of oxygen atoms and/or the removal of hydrogen atoms.
- [O] is placed as a reactant in the equation to represent the oxidising agent.
- If hydrogen atoms are removed, H_2O is placed on the right-hand side as a product.
- The [O] is balanced to reflect the number of oxygen atoms added or the number of hydrogen atoms removed as H_2O.

EXAMPLE 23

Write an equation for the oxidation of ethanol, CH_3CH_2OH, to ethanal, CH_3CHO, using [O] to represent the oxidising agent.

Answer
- Write the reactant and product with [O] on the reactant side.

$CH_3CH_2OH + [O] \rightarrow CH_3CHO$

- Examine the reactants and products for addition of oxygen atoms and removal of hydrogen atoms.
- In this example, no O atoms are added but 2 H atoms are removed.
- To remove 2 H atoms, H_2O is placed on the right-hand side.

$CH_3CH_2OH + [O] \rightarrow CH_3CHO + H_2O$

- The equation is now balanced as there are 2 O atoms on the left and the right including [O] on the left-hand side. There are 6 H atoms on both sides of the equation.

EXAMPLE 24

Write an equation for the oxidation of propan-1-ol, $CH_3CH_2CH_2OH$, to propanoic acid, CH_3CH_2COOH, using [O] to represent the oxidising agent.

Answer

● Write the reactant and product with [O] on the reactant side.

$CH_3CH_2CH_2OH + [O] \rightarrow CH_3CH_2COOH$

● Examine the reactants and products for addition of oxygen atoms and removal of hydrogen atoms.

● In this example, 1 O atom is added and 2 H atoms are also removed.

● To remove 2 H atoms, H_2O is placed on the right-hand side.

● 2[O] are needed on the left to balance the oxygen atoms (providing the oxygen atom for the carboxylic acid and the oxygen atom to remove the 2 H atoms as water).

$CH_3CH_2CH_2OH + 2[O] \rightarrow CH_3CH_2COOH + H_2O$

● The equation is now balanced as there are 3 O atoms on the left and the right including 2[O] on the left-hand side. There are 9 H atoms on both sides of the equation.

Simple reduction reactions

Equations for the reduction of organic chemicals may be written simply using [H] to represent the reducing agent.

● Reduction in organic chemistry may be the removal of oxygen atoms and/or the addition of hydrogen atoms.
● [H] is placed as a reactant in the equation to represent the reducing agent.
● If oxygen atoms are removed, H_2O is placed on the right-hand side as a product.
● The [H] is balanced to reflect the number of hydrogen atoms added or the number of oxygen atoms removed as H_2O.

EXAMPLE 25

Write an equation for the reduction of ethanoic acid, CH_3COOH, to ethanol, CH_3CH_2OH, using [H] to represent the reducing agent.

Answer

● Write the reactant and product with [H] on the reactant side.

$CH_3COOH + [H] \rightarrow CH_3CH_2OH$

● Examine the reactants and products for removal of oxygen atoms and addition of hydrogen atoms.

● In this example, 1 O atom is removed and 2 H atoms are removed.

● To remove 1 O atom, H_2O is placed on the right-hand side.

● 4[H] are required on the left-hand side to provide the 2 H atoms to be added and also 2 H atoms to form the H_2O.

$CH_3COOH + 4[H] \rightarrow CH_3CH_2OH + H_2O$

● The equation is now balanced as there are 2 O atoms on the left and the right. There are 8 H atoms on both sides of the equation, including the 4[H] on the left-hand side.

Organic half equations

Organic oxidation and reduction reactions may also be written as half equations involving electrons, but this is less common. The same procedure is followed as for the half equations for s, p and d block elements except the oxidation state of the elements is not determined.

EXAMPLE 26

Write a half equation for the oxidation of ethanol, CH_3CH_2OH, to ethanoic acid, CH_3COOH.

Answer

- CH_3CH_2OH is on the left and CH_3COOH is on the right.
- There is one oxygen extra on the right, therefore H_2O is placed on the left of the reaction.
- Two hydrogens are removed so $2H^+$ are needed on the right.
- Another $2H^+$ are also needed on the right-hand side to balance the H_2O on the left giving a total of $4H^+$ on the right.
- $4e^-$ are placed on the right-hand side to balance the charge.

  $$CH_3CH_2OH + H_2O \rightarrow CH_3COOH + 4H^+ + 4e^-$$

- The process is an oxidation reaction because electrons are lost.

TEST YOURSELF 7

1 Write oxidation equations for the following organic reactions using [O] to represent the oxidising agent.
   a) propan-2-ol, $CH_3CH(OH)CH_3$, to propanone, CH_3COCH_3.
   b) methanol, CH_3OH, to methanoic acid, $HCOOH$.
   c) butan-1-ol, $CH_3CH_2CH_2CH_2OH$, to butanal, $CH_3CH_2CH_2CHO$.
2 Write reduction equations for the following organic reactions using [H] to represent the reducing agent.
   a) butanone, $CH_3COCH_2CH_3$, to butan-2-ol, $CH_3CH(OH)CH_2CH_3$.
   b) ethanal, CH_3CHO, to ethanol, CH_3CH_2OH.
   c) propanoic acid, CH_3CH_2COOH, to propanal, CH_3CH_2CHO.
3 Write a half equation including electrons to show the oxidation of pentan-2-ol, $CH_3CH(OH)CH_2CH_2CH_3$, to pentan-2-one, $CH_3COCH_2CH_2CH_3$.

ACTIVITY

Redox reactions in photochromic glass

Sunglasses can be made from photochromic glass. When bright light strikes photochromic glass it darkens. Photochromic glass contains small amounts of silver(I) chloride, and copper(I) chloride, evenly distributed throughout the glass.

Figure 8.6 Photochromic sunglasses darken in sunlight due to redox reactions.

As sunlight passes through the glass, the silver(I) chloride is separated into ions. The chloride ions are then oxidised to chlorine atoms, and silver ions are reduced to silver atoms, which cluster together and block the transmittance of light, resulting in the darkening of the lenses.

What is the formula for a) silver(I) chloride, b) copper(I) chloride?

1 Write an equation for the separation of silver(I) chloride into its ions.
2 Calculate the maximum mass of silver ions that can be formed when 0.287 g of silver(I) chloride decomposes.

3 Write a half equation for the oxidation of chloride ions to chlorine atoms, and explain, using oxidation states, why this is an oxidation reaction.

4 Write a half equation for the reduction of silver ions into silver atoms, and explain in terms of electrons, why this is a reduction.

The darkening reaction is reversible, which allows the lenses to become transparent again. The presence of copper(I) chloride reverses the darkening process. When the lenses are removed from the light, copper(I) ions react with chlorine atoms, removing them and forming copper(II) ions and chloride ions. The copper(II) ions are then reduced back to copper(I) ions by the silver. The silver and chlorine atoms are converted to their original oxidised and reduced states and the glass becomes transparent.

5 Suggest what might happen to the chlorine atoms, if they were not removed by the copper(I) ions.

6 Write an equation for the reaction of copper(I) ions with chlorine atoms. Explain the redox in this reaction, in terms of oxidation states.

7 Write an equation for the reaction of copper(II) ions with silver. Explain the redox in this reaction, in terms of oxidation states.

Copper(I) chloride, similar to that used in photochromic glass, can be prepared in the laboratory by reacting copper metal with copper(II) ions as shown in Figure 8.7. This preparation must be carried out in a fume cupboard.

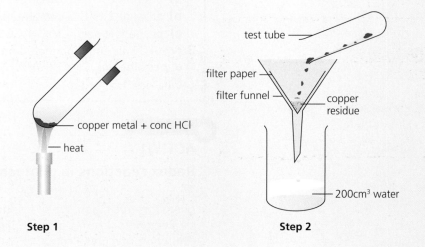

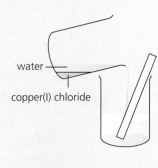

Figure 8.7 Preparation of copper(I) chloride.

Step 1 Warm 0.5 g of copper(II) oxide with 5 cm³ of concentrated hydrochloric acid in a test tube, for 1 minute. Add 1.0 g of copper turnings and boil gently for five minutes.

Step 2 Filter the solution into 200 cm³ of deionised water. Allow the precipitate of copper(I) chloride to settle.

Step 3 Decant the copper(I) chloride.

8 State and explain the safety precautions which should be observed in this preparation.

9 The overall equation for Step 1 is:

$$CuO + Cu + 2HCl \rightarrow 2CuCl + H_2O$$

Explain, using oxidation states why this is a redox reaction.

10 Suggest why copper(I) chloride forms as a precipitate in Step 3.

11 Explain what is meant by the term 'decant', in Step 3.

Combining half equations

Half equations include electrons. A half equation is half a redox reaction and involves the oxidation or reduction of one particular species.

Examples of half equations are:

$$Ni \rightarrow Ni^{2+} + 2e^-$$

$$Cr_2O_7^{2-} + 14H^+ + 6e^- \rightarrow 2Cr^{3+} + 7H_2O$$

Ionic equations do not include electrons. An ionic equation is the reaction between two ionic species transferring electrons.

Examples of ionic equations are:

$$Mg + 2H^+ \rightarrow Mg^{2+} + H_2$$

$$Cl_2 + 2I^- \rightarrow 2Cl^- + I_2$$

Often two half equations are given to you and you are asked to write the ionic equation. This is simply a matter of multiplying the half equations by a number which gives the same number of electrons in the oxidation half equation and in the reduction half equation.

When the equations are added together to make an ionic equation, there will be the same number of electrons on both sides of the ionic equation so they can be cancelled out. If H_2O or H^+ appear on both sides of the equations these can be cancelled down. Some H_2O and H^+ will remain as they will not cancel out completely.

There are two ways in which half equations are presented at this level.

1 You may be given one half equation which is a reduction and the other will be written as an oxidation.

2 You may be given two half equations which are both written as reductions. In this case, one equation needs to be reversed to make it an oxidation before you can add the equations together.

TIP
Half equations involve electrons and only one species is either oxidised or reduced.

TIP
Ionic equations do not involve electrons; two species are involved where one is oxidised and the other is reduced.

Adding half equations

Make sure you have one oxidation half equation (electrons on the right) and one reduction half equation (electrons on the left). For example:

$$Al \rightarrow Al^{3+} + 3e^- \quad \text{OXIDATION}$$

$$F_2 + 2e^- \rightarrow 2F^- \quad \text{REDUCTION}$$

Make sure the number of electrons are the same in both the oxidation and reduction half equations. To do this the oxidation equation needs to be multiplied by 2. The reduction equation needs to be multiplied by 3. This will give both equations six electrons.

$$2Al \rightarrow 2Al^{3+} + 6e^-$$

$$3F_2 + 6e^- \rightarrow 6F^-$$

To add them, simple write down all the species from the left-hand side of both half equations, then put an arrow and finally write down all the species from the right-hand side of both half equations.

$$Al + 3F_2 + 6e^- \rightarrow 2Al^{3+} + 6e^- + 6F^-$$

The next step is to cancel out the electrons on both sides of the equation.

$$Al + 3F_2 \rightarrow 2Al^{3+} + 6F^-$$

This is the ionic equation for the reaction between aluminium and fluorine.

TIP

Always check any ionic equation for charges on the left-hand side and right-hand side – the total charges should be the same on both sides. 0 on the left-hand side and (6+ and 6– = 0) on the right-hand side.

EXAMPLE 27

Write an ionic equation for the reaction in which iron(II) ions are oxidised by acidified potassium managanate(VII).

Answer

The two half equations are:

$Fe^{2+} \rightarrow Fe^{3+} + e^-$ OXIDATION

$MnO_4^- + 8H^+ + 5e^- \rightarrow Mn^{2+} + 4H_2O$ REDUCTION

Note there are five electrons on the left-hand side of the second equation and only 1 electron on the right-hand side of the first equation.

In order to write a complete ionic equation, the first equation must be multiplied by 5 and then the two equations are simply added together.

$5Fe^{2+} \rightarrow 5Fe^{3+} + 5e^-$

$MnO_4^- + 8H^+ + 5e^- \rightarrow Mn^{2+} + 4H_2O$

$MnO_4^- + 8H^+ + 5e^- + 5Fe^{2+} \rightarrow Mn^{2+} + 4H_2O + 5Fe^{3+} + 5e^-$

The $5e^-$ on each side can be cancelled so that the overall equation reads:

$MnO_4^- + 8H^+ + 5Fe^{2+} \rightarrow Mn^{2+} + 4H_2O + 5Fe^{3+}$

Sometimes the half equations are given as two reductions (as is the trend in chemistry), so one equation must be reversed to enable the electrons to be eliminated. The reaction will indicate which two species are reacting.

EXAMPLE 28

In this example, the same ionic equation is required as in Example 27 but with both half equations being given as reductions.

$Fe^{3+} + e^- \rightarrow Fe^{2+};$ REDUCTION

$MnO_4^- + 8H^+ + 5e^- \rightarrow Mn^{2+} + 4H_2O;$ REDUCTION

You are asked to write an ionic equation for the reaction between iron(II) ions, Fe^{2+}, and manganate(VII) ions, MnO_4^-. The reaction requires the first half equation to be reversed to:

$Fe^{2+} \rightarrow Fe^{3+} + e^-$

and then multiplied by 5 as before.

The final equation is the same when the electrons have been eliminated.

$MnO_4^- + 8H^+ + 5Fe^{2+} \rightarrow Mn^{2+} + 4H_2O + 5Fe^{3+}$

EXAMPLE 29

Write an ionic equation for the reaction between nitrate(III) ions, NO_2^- and dichromate(VI) ions, $Cr_2O_7^{2-}$

Answer

$NO_2^- + H_2O \rightarrow NO_3^- + 2H^+ + 2e^-;$ OXIDATION

$Cr_2O_7^{2-} + 14H^+ + 6e^- \rightarrow 2Cr^{3+} + 7H_2O;$ REDUCTION

The equations given are an oxidation (first equation) and a reduction (second equation) so they can be combined directly once the electrons have been balanced.

$3NO_2^- + 3H_2O \rightarrow 3NO_3^- + 6H^+ + 6e^-$

$Cr_2O_7^{2-} + 14H^+ + 6e^- \rightarrow 2Cr^{3+} + 7H_2O$

$Cr_2O_7^{2-} + 14H^+ + 3NO_2^- + 3H_2O + 6e^- \rightarrow 2Cr^{3+} + 7H_2O + 3NO_3^- + 6H^+ + 6e^-$

The electrons, water and H^+ need to be cancelled down to remove them from both sides of the equation.

$Cr_2O_7^{2-} + 8H^+ + 3NO_2^- \rightarrow 2Cr^{3+} + 4H_2O + 3NO_3^-$

EXAMPLE 30

Write an ionic equation for the reaction between sulfur dioxide and acidified potassium dichromate(VI).

Answer

$SO_4^{2-} + 4H^+ + 2e^- \rightarrow SO_2 + 2H_2O;$ REDUCTION

$Cr_2O_7^{2-} + 14H^+ + 6e^- \rightarrow 2Cr^{3+} + 7H_2O;$ REDUCTION

The SO_2 reacts with the $Cr_2O_7^{2-}$ so the first equation needs to be reversed to create an oxidation equation.

$SO_2 + 2H_2O \rightarrow SO_4^{2-} + 4H^+ + 2e^-;$ OXIDATION

$Cr_2O_7^{2-} + 14H^+ + 6e^- \rightarrow 2Cr^{3+} + 7H_2O;$ REDUCTION

The oxidation equation needs to be multiplied by 3.

$3SO_2 + 6H_2O \rightarrow 3SO_4^{2-} + 12H^+ + 6e^-$

$Cr_2O_7^{2-} + 14H^+ + 6e^- \rightarrow 2Cr^{3+} + 7H_2O$

$3SO_2 + 6H_2O + Cr_2O_7^{2-} + 14H^+ + 6e^- \rightarrow 3SO_4^{2-} + 12H^+ + 6e^- + 2Cr^{3+} + 7H_2O$

Cancel out the electrons and cancel down the H⁺ and H₂O.

$$3SO_2 + Cr_2O_7^{2-} + 2H^+ \rightarrow 3SO_4^{2-} + 2Cr^{3+} + H_2O$$

To check any half equation or ionic equation is correct, check the charges on each side. The above equation has a total charge of 0 on the left and 0 on the right.

TEST YOURSELF 8

1 Using the following half equations, write the equation for the reaction between hydrogen peroxide and hydrazine (N_2H_4).

$$N_2H_4 \rightarrow N_2 + 4H^+ + 4e^-$$

$$H_2O_2 + 2H^+ + 2e^- \rightarrow 2H_2O$$

2 Hydrogen peroxide oxidises iodide ions in solution to iodine. Use the half equations below to write an overall equation for this reaction.

$$H_2O_2 + 2H^+ + 2e^- \rightarrow 2H_2O$$

$$I_2 + 2e^- \rightarrow 2I^-$$

3 Given the following half equations.

$$ClO^- + H_2O + 2e^- \rightarrow Cl^- + 2OH^-$$

$$2HBrO + 2H^+ + 2e^- \rightarrow Br_2 + 2H_2O$$

Write an overall equation for the reaction between chloride ions and hydrobromous acid, HBrO.

4 Using the half equations below, write a balanced ionic equation for the reaction between manganate(VII) ions and ethanedioate ions?

$$MnO_4^- + 8H^+ + 5e^- \rightarrow Mn^{2+} + 4H_2O$$

$$C_2O_4^{2-} \rightarrow 2CO_2 + 2e^-$$

Practice questions

1 What is the oxidation state of nitrogen in HNO_2?

    A +2   **B** +3

    C +4   **D** +5   *(1)*

2 Which of the following reactions does not involve a change in oxidation state?

    A $Mg + \frac{1}{2}O_2 \rightarrow MgO$

    B $CuSO_4 + 2NaOH \rightarrow Cu(OH)_2 + Na_2SO_4$

    C $H_2 + I_2 \rightarrow 2HI$

    D $2HBr + H_2SO_4 \rightarrow Br_2 + SO_2 + 2H_2O$ *(1)*

3 What is the formula of potassium bromate(V)?

    A KBr   **B** KBrO

    C $KBrO_2$ **D** $KBrO_3$ *(1)*

4 Phosphorus forms compounds in a variety of oxidation states.

    a) Give the oxidation state of phosphorus in the following:

        H_3PO_4 P_4O_{10} P_4 H_3PO_3 *(4)*

    b) P_4 reacts with oxygen to form P_4O_{10}. Write an equation for this reaction.   *(1)*

    c) P_4O_{10} reacts with water to form H_3PO_4. Write an equation for this reaction and explain why this reaction is not a redox reaction.   *(2)*

    d) Write a half equation for the conversion of H_3PO_4 to H_3PO_3.   *(1)*

5 a) Balance the following half equations by inserting the correct balancing numbers.

    i) $SO_2 + \underline{} H_2O \rightarrow SO_4^{2-} + \underline{} H^+ + \underline{} e^-$ *(1)*

    ii) $MnO_4^{2-} + \underline{} H^+ + \underline{} e^- \rightarrow Mn^{2+} + \underline{} H_2O$ *(1)*

    b) Write an ionic equation for the reaction between sulfur dioxide and manganate(VI) ions.   *(1)*

6 Dioxovanadium(V) chloride, VO_2Cl reacts with sodium iodide in the presence of sulfuric acid according to the equation:

$$2VO_2Cl + 2H_2SO_4 + 2NaI \rightarrow VOCl_2 + VOSO_4 + Na_2SO_4 + I_2 + 2H_2O$$

a) Vanadium is reduced from the $+5$ to the $+4$ oxidation state. Give evidence from the equation for this statement. (2)

b) State which element is oxidised in this reaction and give the change in oxidation states. (2)

7 State which elements in the following reactions are reduced and which are oxidised.

a) $Mg + 2HCl \rightarrow MgCl_2 + H_2$ (2)

b) $Cl_2 + H_2O \rightarrow HOCl + HCl$ (2)

c) $3SO_2 + K_2Cr_2O_7 + H_2SO_4 \rightarrow K_2SO_4 + Cr_2(SO_4)_3 + H_2O$ (2)

8 Xenon compounds are rare but the equation below shows a reaction between hydrogenxenate and hydroxide ions:

$$2OH^- + 2HXeO_4^- \rightarrow Xe + 2H_2O + XeO_6^{4-} + O_2$$

a) Give the oxidation states of xenon in each of the following:

$HXeO_4^-$ Xe XeO_6^{4-} (3)

b) Explain why this is considered to be a redox reaction. (1)

9 The following reactions occur when concentrated sulfuric acid is added to a sample of solid sodium iodide.

A $NaI + H_2SO_4 \rightarrow NaHSO_4 + HI$

B $2HI + H_2SO_4 \rightarrow I_2 + SO_2 + 2H_2O$

C $6HI + H_2SO_4 \rightarrow 3I_2 + S + 4H_2O$

D $8HI + H_2SO_4 \rightarrow 4I_2 + H_2S + 4H_2O$

a) Explain why reaction A is not a redox reaction. (1)

b) Combine reactions A and B to write an overall equation for the reaction of sodium iodide with sulfuric acid to form iodine. (1)

c) Give the change in oxidation state for reaction B for the element being oxidised and the element being reduced. (2)

d) Suggest why both a grey-black solid and a yellow solid are observed during the reaction. (2)

e) Using reaction D, write a half equation for the conversion of H_2SO_4 to H_2S. (1)

f) Name the reducing agent in reaction B. (1)

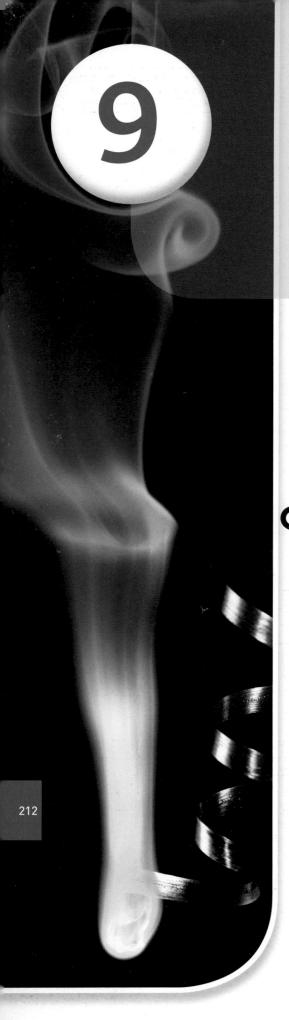

Periodicity and Group 2

In this chapter, we will look at periodicity – how properties change across a period. Then, by contrast, we will look at how properties change down a group – Group 2.

The Group 2 metals are known as the '**alkaline earth metals**' as their oxides and hydroxides are alkaline. 'Earth' is a term used by the first chemists to describe non-metallic substances which were insoluble in water. The white suspension formed when calcium reacts with water was described as an 'earthy' residue.

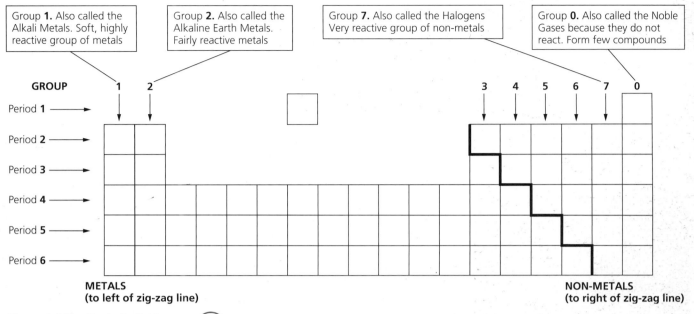

Figure 9.1 The Periodic Table.

Blocks of the Periodic Table

The Periodic Table can be divided into blocks depending on which sub-shell the outer (highest energy level) electrons are located.

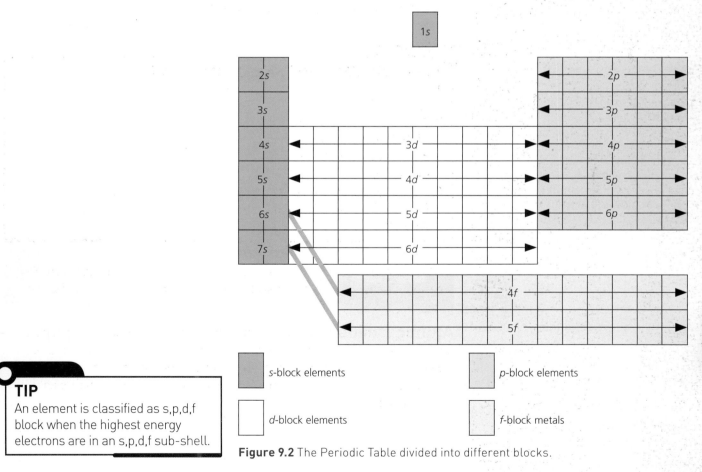

s-block elements

p-block elements

d-block elements

f-block metals

Figure 9.2 The Periodic Table divided into different blocks.

TIP

An element is classified as s,p,d,f block when the highest energy electrons are in an s,p,d,f sub-shell.

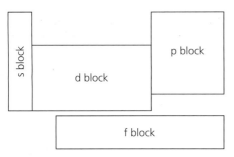

Figure 9.3 s, p, d f blocks of the Periodic Table.

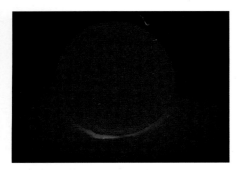

Figure 9.4 A pellet of plutonium-238, illuminated by the glow of its own radioactivity. In nature, plutonium only occurs in minute quantities that arise from the decay of uranium-238, the heaviest naturally occurring element.

Figure 9.5 Successive elements across Period 3. There are immense changes in the chemical and physical properties. Sodium, magnesium and aluminium are metals and have metallic properties whereas phosphorus, sulfur, chlorine and argon are non-metals and have non-metallic properties.

Figure 9.6 Atomic radius (nm) of the atoms of Period 3.

- The electron configuration of a lithium atom is $1s^2\ 2s^1$ so the highest energy level electrons are in an s sub-shell; the **electron configuration** of a magnesium atom is $1s^2\ 2s^2\ 2p^6\ 3s^2$ so the highest energy level electrons are also in an s sub-shell. Groups 1 and 2 form the **s block**.
- The electron configuration of a boron atom is $1s^2\ 2s^2\ 2p^1$ so the highest energy level electrons are in a p sub-shell; the electron configuration of a chlorine atom is $1s^2\ 2s^2\ 2p^6\ 3s^2\ 3p^5$ the highest energy level electrons are in a p sub-shell – Groups 3 to 0 form the **p block**.
- The elements in the middle of the Periodic Table (Sc to Zn in Period 4) have their highest energy electrons in the 3d sub-shell the **d block**.
- Figure 9.2 shows the highest energy levels for the different blocks of the Periodic Table including the **f block** elements. This block fits between s block and d block at Period 6 and 7. It consists of metal elements whose highest energy electrons are in an f sub-shell.

The f block elements are also referred to as the lanthanide series and the actinide series. The elements of the lanthanide series can be found naturally on Earth. Only one element in the series is radioactive. The actinide series are all radioactive and some are not found in nature. Some of the elements with higher atomic numbers have only been made by bombarding atoms with fast neutrons. Plutonium was the first actinide to be identified (Figure 9.3).

Major trends across Period 3

Atomic radius

The atomic radius **decreases** across Period 3.

Figure 9.6 shows the atomic radius of the Period 3 atoms. The atomic radius is given in nanometres (nm).

Na	Mg	Al	Si	P	S	Cl	Ar
0.18	0.16	0.14	0.11	0.11	0.10	0.09	0.09

The decreasing atomic radius is due to the increasing nuclear charge across the period, which draws outer electrons (which are in the same level) closer to the nucleus.

Melting point

The graph in Figure 9.7 shows the melting points of the Period 3 elements (Na to Ar) with atomic numbers 11 to 18.

Figure 9.7 Melting points (°C) of the elements of Period 3.

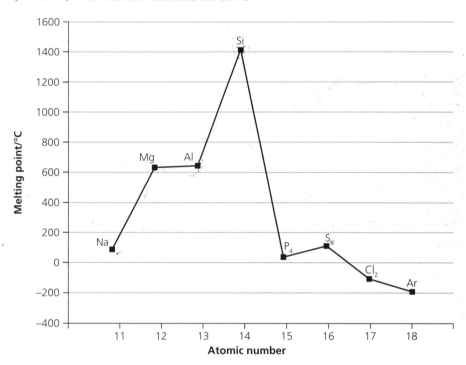

1 The melting points of metallic elements increase.

From Na to Al, the number of outer level electrons increases so more electrons can be delocalised leading to a greater attraction between positive ions and delocalised electrons. The size of the ions also increases and this again leads to a greater attraction between the smaller ions and the delocalised electrons.

2 The melting point of silicon is very high.

Silicon has a giant covalent structure and a lot of energy is required to break the many strong covalent bonds.

3 The melting points of phosphorus (P_4), sulfur (S_8), chlorine (Cl_2) and argon (Ar) are low.

- Phosphorus, sulfur and chlorine are simple covalent molecules and little energy is required to overcome the weak van der Waals' forces between the molecules.
- Argon exists as atoms and very little energy is required to overcome the very weak van der Waals' forces between the atoms.

4 The melting points of the elements after silicon increases from phosphorus to sulfur then decreases again.

- Phosphorus (44 °C), sulfur (115 °C) and chlorine (−101 °C)
- Phosphorus exists as P_4 molecules, sulfur as S_8 molecules and chlorine as Cl_2 (Figure 9.9 on page 216.)

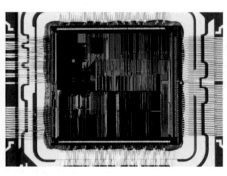

Figure 9.8 A macrophotograph of a 486 microprocessor silicon chip used in some personal computers. Microscopic electronic components have been imprinted onto the surface of the silicon by various deposition and etching processes. Silicon is used as it is a semiconductor with high thermal conductivity and high melting point.

P₄
molecule

S₈
molecule

Cl₂
molecule

Ar atom

Figure 9.9

- P_4 (60 electrons per molecule; each P atom has 15 electrons)
- S_8 (128 electrons per molecule; each S atom has 16 electrons)
- Cl_2 (34 electrons per molecule; each Cl atom has 17 electrons)

The attractions between the molecules are van der Waals' forces as the molecules are non-polar. If a molecule has more electrons, the extent of the induced dipole attractions will be greater thus causing greater van der Waals' forces of attraction between the molecules which increases the energy required to melt the substance.

- Argon has a very low melting point ($-189\,°C$). This is due to the very weak van der Waals' forces between the atoms.

> **TIP**
> The trends and explanations used are similar when comparing boiling points of the elements in Period 3.

First ionisation energy

The first ionisation energy is the energy required to remove one mole of electrons from one mole of gaseous atoms to form one mole of gaseous 1^+ ions. For example for sodium, it is the energy required to cause the following change:

$$Na(g) \rightarrow Na^+(g) + e^-$$

The first ionisation energy increases across Period 3. The graph in Figure 9.11 shows the values of the first ionisation energy of the elements in Period 3.

Figure 9.10 When burned, sulfur melts to a red liquid and emits a blue flame that is best observed in the dark. It has a low melting point and can be easily melted in a Bunsen flame.

Figure 9.11 The first ionisation energies of the elements of Period 3.

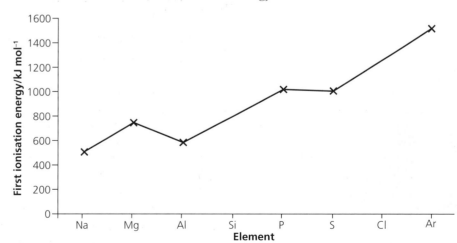

First ionisation energy increases across a period

The general increase in first ionisation energy across a period is caused by the increase in nuclear charge with no increased shielding. The atomic radius is decreasing. The outer electron is closer to the nucleus with a greater nuclear charge holding it so more energy is required to remove it.

First ionisation energies in Group 3 and Group 6

The atoms of Group 3 elements, such as aluminium, show a lower than expected first ionisation energy. This is due to the division of the energy level into sub-shells. Aluminium has the electron configuration $1s^2 2s^2 2p^6 3s^2 3p^1$. The $3p^1$ electron is further from the nucleus and has additional shielding from the $3s^2$ inner electrons so it requires less energy to ionise it.

The atoms of Group 6 elements, such as sulfur, also show a lower than expected first ionisation energy. This is due to the pairing of electrons in the p sub-shell. The $3p^4$ electron configuration of sulfur ($1s^2 2s^2 2p^6 3s^2 3p^4$) means that two electrons are paired in a p orbital in this sub-shell. The additional repulsion between these two electrons lowers the energy required to remove one of the electrons and this decreases the first ionisation energy.

First ionisation energies in Group 1 and Group 0

Atoms of Group 1 elements have the lowest first ionisation energy in every period as they have the greatest atomic radius and the lowest nuclear charge in a particular period.

Atoms of Group 0 elements have the highest first ionisation energy in every period for the opposite reasons; they have the smallest atomic radius and the highest nuclear charge in a period.

TIP

Many of the trends seen in Period 3 are also seen in Period 2. Questions involving Period 2 (usually Li to C) are often seen. The patterns are not seen in the molecules of elements in Groups 5, 6 and 7 as they are all diatomic in Period 2.

TEST YOURSELF 2

1 State the trend in atomic radius from sodium to argon.
2 First ionisation energy generally increases from sodium to argon. Explain why.
3 Explain why sulfur has a higher melting point than phosphorus.
4 State the element in Period 3 which has:
   a) the highest melting point
   b) the lowest first ionisation energy
   c) the largest atomic radius.
5 State the block of the Periodic Table to which the following elements belong:
   a) nitrogen **b)** molybdenum **c)** barium.

ACTIVITY

Trends in first ionisation energy across a period

1 Use the data from dynamic learning to plot a graph of first ionisation energy (*y*-axis) against atomic number (*x*-axis). Draw lines from one point to the next to show a pattern of peaks and troughs. Label each point with the symbol of its corresponding element. (For later parts of this activity, ensure that your y-axis scale goes up to $4600\,\text{kJ}\,\text{mol}^{-1}$.)
2 What do you notice about the position of the first ionisation energy of Group 1 elements on the graph? Explain your answer.
3 What do you notice about the position of the first ionisation energy of Group 0 elements on the graph? Explain your answer.
4 Explain the general trend in first ionisation energy across Period 2.
5 Is the general trend in first ionisation energy across Period 3 the same or different? Explain your answer.
6 Using your graph, compare the values of first ionisation energy for elements in the same group. State and explain any trends observed in the ionisation energy within a group.

7 The **second ionisation energy** is the energy required to remove one mole of electrons from one mole of gaseous ions with one positive charge to form one mole of gaseous 2+ ions. For example for sodium, it is the energy required to cause the following change:

$$Na^+(g) \rightarrow Na^{2+}(g) + e^-$$

a) On the same graph paper, plot the values for second ionisation energy for each element from sodium to argon. Use the data in Table 9.1. Join each point with a red pen.

Table 9.1 Second ionisation energy values for the elements of Period 3.

Element	Na	Mg	Al	Si	P	S	Cl	Ar
Second ionisation energy /kJ mol^{-1}	4562	1450	1816	1577	1907	2252	2298	2665

b) Compare the trend in second ionisation energy across the period with the trend in first ionisation energy.

c) Suggest why the second ionisation energy of each element is greater than the corresponding first ionisation energy.

d) Write an equation for the second ionisation energy of aluminium.

Group 2

Burning a strip of magnesium in a Bunsen flame may have been one of the first experiments you tried when you started your study of Chemistry. The thin strips of the shiny silver metal burn with an extremely bright white light as shown in Figure 9.12.

Group 2, the alkaline earth metals, are a group of reactive metals. In this section you will study the reactivity of the metals with water, but first it is important to understand how properties such as atomic radius, ionisation energy and melting point change down the group.

Figure 9.13 The Group 2 element magnesium has been used for over 100 years to provide a bright flash of white light to aid the taking of photographs.

Figure 9.12 Magnesium burns with a bright white light.

Trends down Group 2

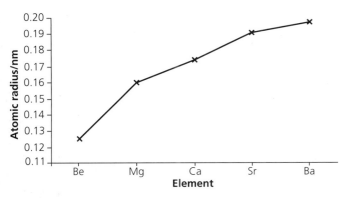

Figure 9.14 The trend in atomic radius down Group 2.

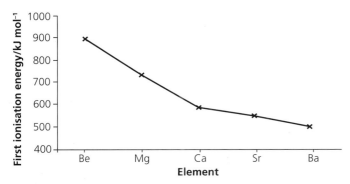

Figure 9.15 The trend in first ionisation energy down Group 2.

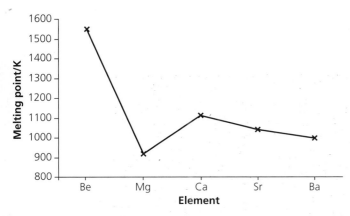

Figure 9.16 The melting points of the Group 2 elements.

Atomic radius

Atomic radius is measured as the distance between the centre of the nucleus to the outer electron. The outer electron is pulled in closer to the nucleus when the attraction between the outer electron and the nucleus increases, therefore the atomic radius becomes smaller. Account must also be taken of the effect of increasing number of electrons which repel each other.

As Group 2 is descended, the outer electron is less tightly held as it is shielded from the attractive force of the increasing nuclear charge due to the electrons filling in new energy levels. These electrons also provide extra repulsion and the outer electron becomes further away from the nucleus, therefore the atomic radius increases (Figure 9.14).

Ionisation energy

First ionisation energy decreases down the group. The energy required to remove the outer electron from an atom depends on the attraction between the outer electron and the nucleus. The less strongly the electron is held the less energy is required to remove it. As Group 2 is descended, the atomic size increases, the outer electron is less tightly held as it is shielded from the attractive force of the increasing nuclear charge due to the electrons filling in new energy levels, therefore it is easier to remove and the ionisation energy decreases (Figure 9.15).

Melting point

In general, the melting points of the Group 2 metals decrease as the group is descended (Figure 9.16).

Group 2 elements exist as giant metallic structures; a regular arrangement of positive ions surrounded by a sea of delocalised electrons. The metallic bond is the attraction between the layers of regularly arranged positive ions and the delocalised electrons. When a metal is melted, energy is supplied to break the metallic bond; the greater the strength of the metallic bond, the higher the melting point.

The strength of the metallic bond is proportional to the number of delocalised electrons per atom, the charge of the metal ion and the size of the ion. The ions produced from the metallic atoms of Group 2 elements all have a +2 charge and the same number of delocalised electrons per atom. However, as the group is descended the size/atomic radius of the atom increases. The attraction between the ion and the delocalised electrons decreases. The strength of the metallic bond decreases and the melting point decreases.

TIP
The melting point of magnesium is lower than expected. At this level however, you are not expected to explain this. The focus should be on the general trend which is a decrease in melting point down the group.

219

TEST YOURSELF 3

1 Write the electron configurations of the atoms of the Group 2 elements Mg to Ba.

2 a) Explain the meaning of the term ionisation energy.

 b) Explain why the ionisation energy of strontium is less than the ionisation energy of magnesium.

3 Explain why the atomic radius of calcium is smaller than that of strontium.

4 a) Draw a labelled diagram of the structure of a metal such as calcium.

 b) Describe the bonding in calcium.

 c) State and explain the general trend in melting points of the Group 2 metals.

5 Transition metals generally have higher melting points than Group 2 metals.

 a) Compare the strength of the metallic bond in the Group 2 metals and the transition metals.

 b) Explain your answer to part 5a.

Reactions of Group 2 metals with water

Beryllium does not react with water. Magnesium reacts very, very slowly with water but reacts readily with steam. Magnesium hydroxide is formed when magnesium reacts with water but the oxide of magnesium is formed when magnesium reacts with steam.

$$Mg + 2H_2O_{(l)} \rightarrow Mg(OH)_2 + H_2$$

$$Mg + H_2O_{(g)} \rightarrow MgO + H_2O$$

Magnesium oxide is formed rather than magnesium hydroxide when magnesium reacts with steam as the hydroxide is not stable at higher temperatures. It thermally decomposes to give magnesium oxide and water.

Calcium strontium and barium all react with cold water to produce the metal hydroxide and hydrogen. The **reactivity** of the reaction **increases as the group is descended**.

$$Ca + 2H_2O \rightarrow Ca(OH)_2 + H_2$$

$$Sr + 2H_2O \rightarrow Sr(OH)_2 + H_2$$

$$Ba + 2H_2O \rightarrow Ba(OH)_2 + H_2$$

Calcium reacts with cold water thus releasing bubbles of hydrogen in an exothermic reaction. A white precipitate of calcium hydroxide is formed and the remaining solution is slightly alkaline. This is because calcium hydroxide is slightly soluble in water. The reactions of strontium and barium are similar; strontium is more reactive than calcium and barium is even more reactive but the amount of precipitate becomes less because the hydroxides of the Group 2 metals become more soluble as the group is descended.

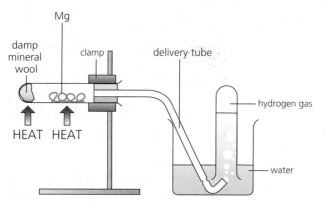

Figure 9.17 A Magnesium can be heated in steam using the apparatus shown.

The solubility of the Group 2 hydroxides in water

Table 9.2 The solubility of the Group 2 hydroxides in water.

Solution of Group 2 metal ion	Observations on adding sodium hydroxide solution	
Mg^{2+}	thick white precipitate of magnesium hydroxide	solubility of hydroxides increases
Ca^{2+}	white precipitate of calcium hydroxide	
Sr^{2+}	thin white precipitate of strontium hydroxide	
Ba^{2+}	very thin white precipitate of barium hydroxide	

The relative solubility of the Group 2 hydroxides can be found by adding a solution of sodium hydroxide to solutions of the Group 2 ions and observing the precipitate (Table 9.2). The Group 2 hydroxides become **more soluble** as the **group is descended**. Magnesium hydroxide is said to be **sparingly soluble** as the solution is slightly alkaline indicating that some hydroxide ions are dissolved in the water.

ACTIVITY

Group 2 hydroxides

The hydroxides of the Group 2 metals, magnesium to barium, are sparingly soluble in water; the solubility rises with atomic number. Table 9.3 shows the solubility of some Group 2 hydroxides.

Table 9.3 Solubility data for the Group 2 hydroxides.

Group 2 hydroxide	Solubility g/100cm³ water at 20°C
Magnesium hydroxide	0.014
Calcium hydroxide	0.173
Strontium hydroxide	1.770
Barium hydroxide	3.890

1 Suggest how you could determine the solubility of barium hydroxide in g/100cm³ water at 20°C. Give full practical details.

2 Use the data given in the table to calculate the concentration, in mol dm⁻³, of a saturated solution of barium hydroxide at 20°C.

Figure 9.18 Barium hydroxide is a white ionic solid.

In an experiment a solution of barium hydroxide was titrated with ethanoic acid using thymol blue indicator. Thymol blue is yellow in acid and blue in alkali. The ethanoic acid was added from a burette to the conical flask containing 25.0 cm³ of a barium hydroxide solution and a few drops of thymol blue.

3 Describe the colour change at the end-point of this titration.

4 Thymol blue is an acid. Suggest how the average titre would change if a few cm³, rather than a few drops, of the indicator were used by mistake in this titration.

5 State and explain two safety precautions that you would follow during this practical.

6 Suggest one reason why repeating a titration can improve its reliability.

Barium hydroxide reacts with dilute nitric acid to give a solution of barium nitrate.

7 Write a balanced symbol equation for the reaction of barium hydroxide and nitric acid.

Figure 9.19 Barium nitrate is mixed with thermite to form Thermate-TH₃, an explosive used in military thermite grenades.

The solubility of the Group 2 sulfates in water

The Group 2 sulfates, unlike the hydroxides, become **less soluble** in water as the **group is descended** (Table 9.4). The relative solubility of the Group 2 sulfates can be found by adding a solution of sodium sulfate to solutions of the Group 2 ions and observing the precipitate. Calcium sulfate is said to be sparingly soluble and barium sulfate is insoluble.

Table 9.4 Solubility data for the Group 2 sulfates.

Solution of Group 2 metal ion	Observation on addition of sodium sulfate solution	
Mg^{2+}	no precipitate	↑
Ca^{2+}	thin white precipitate	solubility of sulfates increases
Sr^{2+}	white precipitate	
Ba^{2+}	thick white precipitate	

The use of barium salts as a test for a sulfate
The thick white precipitate which forms when barium ions are added to sulfate ions can be used as a test for sulfate ions (Figure 9.20).

Method
To 1 cm³ of the unknown solution

- add 1 cm³ of dilute hydrochloric acid
- add 1 cm³ of aqueous barium chloride solution
- if a thick white precipitate is formed then sulfate ions are present.

The hydrochloric acid removes any other ions such as carbonate ions which may affect the test.

The simplest ionic equation for the reaction is:

$$Ba^{2+}(aq) + SO_4^{2-}(aq) \rightarrow BaSO_4(s)$$

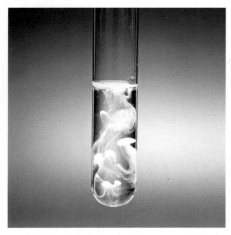

Figure 9.20 A white precipitate forms when acidified barium chloride solution is added to a solution of sodium sulfate.

Selected uses of the Group 2 elements and their compounds

Magnesium

Magnesium is used in the extraction of titanium. Titanium is an extremely useful metal. It has applications in the aerospace, marine and motor vehicle industries due to its extremely high corrosion resistance. It is very difficult to extract titanium from its ores e.g. TiO_2. In theory carbon can be used to extract titanium from its oxide but the titanium forms a brittle carbide. The production process requires several steps, one of which involves the Group 2 metal magnesium.

Titanium oxide is initially converted to the chloride which is subsequently reduced to titanium by reaction with magnesium, as described by this balanced symbol equation.

$$TiCl_4 + 2Mg \rightarrow Ti + 2MgCl_2$$

Magnesium hydroxide

Magnesium hydroxide is **sparingly soluble** and is sold as a suspension in water. In this form it is known as 'milk of magnesia' (Figure 9.21). It is taken to alleviate constipation and, as an antacid, to neutralise excess acid in the gut. It is used as the solutions are only slightly alkaline, due to its low solubility, and do not irritate the oesophagus.

Figure 9.21 A suspension of magnesium hydroxide in water.

Calcium hydroxide

Calcium hydroxide in solid form is known as 'slaked lime' and is used to neutralise acidic soil (Figure 9.22). The pH of soil affects its physical, chemical and biological properties, which in turn affects crop yield. As the pH of soil decreases, soluble aluminium and manganese ion levels increase to toxic proportions thus restricting root growth, and causing curling and black spot on plant leaves. Low pH levels in soil restrict the growth of bacteria, which function in conjunction with leguminous plants to fix nitrogen from the air.

Figure 9.22 Commercially available calcium hydroxide is used to alter the pH of soil.

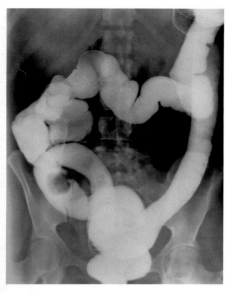

Reducing the acidity of the soil prevents all of these effects and increases the availability of other important nutrients, such as phosphorus. In addition the soil texture improves. It becomes finer thus allowing plants with smaller seedlings an improved opportunity to break through the soil surface.

Calcium oxide

Calcium oxide is used to neutralise sulfur dioxide, which may be produced as a by-product of the combustion fossil fuels in order to prevent the formation of acid rain. Calcium carbonate can also be used (see p.274).

Barium sulfate

Barium sulfate can be eaten as part of a 'barium meal'. Barium is good at absorbing X-rays and so when the barium sulfate gets to the gut the outline of the gut can be located using X-rays (Figure 9.23). Although barium ions are very toxic, this technique is harmless because barium sulfate is insoluble and is not absorbed into the blood.

Figure 9.23 The gut shows up on this X-ray as the patient has ingested a barium meal.

Practice questions

1 Which one of the following has the highest second ionisation energy?

    A aluminium   **B** chlorine

    C magnesium   **D** sodium   *(1)*

2 Which of the elements is a d block element?

    A bromine   **B** calcium

    C hydrogen   **D** silver   *(1)*

3 The electron configuration of an element is:

    $1s^2\ 2s^2\ 2p^6\ 3s^2\ 3p^6\ 3d^{10}\ 4s^2\ 4p^6\ 4d^{10}\ 5s^2\ 5p^5$

    a) In which block of the Periodic Table is the element located?   *(1)*

    b) State the atomic number and identity of the element.   *(2)*

4 The elements in Period 2 show similar trends to the elements in Period 3 of the Periodic Table.

    a) State the name of the element in Period 2 which has the highest first ionisation energy.   *(1)*

    b) State which element in Period 2 has the highest melting point.   *(1)*

    c) Write an equation for the first ionisation of beryllium.   *(1)*

    d) Explain why the first ionisation energy of boron is lower than expected.   *(3)*

5 a) State the trend in reactivity of the Group 2 elements with water.   *(1)*

    b) Under what conditions will magnesium react rapidly with H_2O?   *(1)*

    c) Write a balanced symbol equation for the reaction of barium with water.   *(1)*

6 a) State how the solubilities of the hydroxides of the elements Mg to Ba change as Group 2 is descended.   *(1)*

    b) State how the solubilities of the sulfates of the elements Mg to Ba change as Group 2 is descended.   *(1)*

    c) Describe a test to indicate the presence of sulfate ions in aqueous solution.   *(2)*

    d) Explain why barium sulfate is used as a 'barium meal', when barium salts are known to be toxic.   *(1)*

    e) What property of barium sulfate makes it useful when taking X-rays of the human gut?   *(1)*

7 The melting points and ionisation energies of the Group 2 metals generally decrease as the group is descended.

a) Explain fully the trend in melting point. *(3)*

b) Why is the first ionisation energy of magnesium larger than the first ionisation energy of calcium? *(2)*

c) Write an equation showing

i) the first ionisation energy of calcium *(1)*

ii) the second ionisation energy of barium. *(1)*

8 Group 2 metals and their compounds are used commercially in a variety of processes and applications.

a) State and explain the use of magnesium hydroxide in medicine. *(2)*

b) Calcium carbonate is an insoluble compound that can be added to water in lakes to reduce the acidity. Write an equation for the reaction which occurs in the lake. *(1)*

c) Magnesium is used in excess to produce titanium from titanium(IV) chloride. The excess magnesium is removed by reacting it with sulfuric acid.

i) Write a balanced equation for the reaction of:

- magnesium with titanium(IV) chloride *(1)*

- magnesium with sulfuric acid. *(1)*

ii) Use your knowledge of Group 2 sulfates to explain why the magnesium sulfate is easy to remove from the titanium metal. *(1)*

10 Halogens

TEST YOURSELF ON PRIOR KNOWLEDGE 1

1 State the names of the elements in Group 7.
2 Write the formula of the following compounds.
  a) sodium iodide   **b)** calcium chloride
  c) aluminium fluoride   **d)** iron(III) chloride
  e) lithium bromide   **f)** zinc chloride
3 Chlorine reacts with a solution of potassium iodide forming potassium chloride and iodine. Write an equation for this reaction.

The physical properties of the halogens

The elements in Group 7 are fluorine, chlorine, bromine, iodine and astatine. Group 7 remains the most common number used for the halogens but Group 17 is also used as the halogens are the 17th group across including the d block elements.

- Fluorine is a poisonous yellow gas.
- Chlorine is a poisonous, dense, yellow-green gas (from the Greek 'chloros' meaning green).
- Bromine is a caustic and toxic red-brown volatile liquid (from the Greek 'bromos' meaning stench). It forms a red-brown vapour.
- Iodine is a shiny, grey-black solid, which sublimes to form a violet vapour on gentle heating. (Sublimation is the change of state from a solid directly to a gas on heating.) Iodine is named from the Greek work 'iodos' meaning purple or violet.
- Astatine is the rarest naturally occurring element, it has no stable isotopes; the longest lived isotope has a half-life of just 8.1 hours, hence no one has ever obtained enough of it to be visible to the naked eye. A mass large enough to be seen would be immediately vaporised by the heat generated by its own radioactivity. Following the trend within the group, astatine may be dark, and it is likely to have a much higher melting point than iodine.

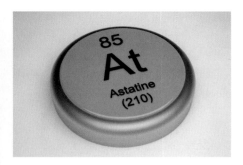

Figure 10.1 Astatine is one of the most mysterious chemical elements; no one knows what it looks like.

Diatomicity and bond enthalpy

The halogens exist as diatomic molecules, F_2, Cl_2, Br_2 and I_2, due to the formation of a single covalent bond between the atoms. The bonding between the atoms is important, as the bond enthalpy must be overcome in order to allow the halogen atoms to react.

The bond between the halogen atoms is due to the attraction of the electrons by the two nuclei. As the atomic number increases, atomic radius increases and so the distance between the electrons and the nucleus also increases. This would lead us to suspect that the mean bond enthalpy should decrease as the atomic number increases. This is true for chlorine, bromine and iodine. However fluorine rather than having the highest bond enthalpy, has a value close to that of iodine. The reason for this is that in a molecule of fluorine the atoms are so close together that the lone pairs of electrons around each atom repel each other and so the bond is weakened.

Trends in Group 7

It is important that we know the colour and physical state of these elements at room temperature and the general trends in Group 7.

The following trends all occur as the atomic number increases in Group 7 (i.e. going down the group).

1 The colour of the elements becomes darker.

2 The atomic radius increases.

- On moving down the group there are more energy levels of electrons.
- The outer energy levels of electrons are further from the nucleus.
- This increases the size of the atoms down the group.

3 The reactivity decreases.

- The outer electrons are further from the nucleus as the group is descended as an extra energy level is added each time.
- There is an increase in shielding.
- The atoms gain electrons more easily as there is a weaker attraction between the nucleus and the incoming electron. Smaller halogen atoms are more reactive.

4 The boiling points of the elements increase.

Figure 10.2 Fluorine is incredibly reactive, and is normally found as fluoride ions in ores, such as the purple fluorite, CaF_2.

- The size of diatomic molecules increases down the group.
- The larger molecules have more electrons.
- More electrons lead to greater induced dipole-dipole forces.
- Greater induced dipole-dipole attractions mean greater van der Waals' forces between the molecules.
- More energy is needed to overcome the greater van der Waals' forces between the molecules to change them from a liquid into a gas.
- Figure 10.3 shows the boiling points of the halogens.
- The increase in boiling points (and melting points) of the halogens going down the group explains why fluorine and chlorine are gases, bromine is a liquid and iodine is a solid.

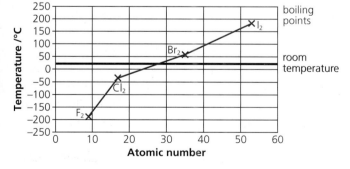

Figure 10.3 Boiling points of the halogens.

227

5 The electronegativity decreases. The values are given in the following table:

Element	F	Cl	Br	I	At
Electronegativity	4.0	3.0	2.8	2.5	2.2

- For the smaller atoms, the bonding electrons in a covalent bond are closer to the positive nucleus.
- There is less shielding.
- So the positive nucleus attracts the bonding pair of electrons more strongly and so the element has a greater electronegativity value.

TEST YOURSELF 2

1 Copy and complete the table below.

Halogen	State at room temperature and pressure	Colour
fluorine		
iodine		
bromine		
chlorine		

2 Iodine sublimes on heating.
  a) Explain what is meant by sublimation.
  b) What would be observed when iodine sublimes?
3 Fluorine has a higher electronegativity value than chlorine.
  a) Define what is meant by the term electronegativity.
  b) Explain why the electronegativity of fluorine is higher than chlorine.
4 Explain why the halogens should be used in a fume cupboard.
5 Explain why iodine has a higher boiling point than bromine.

ACTIVITY

Extraction of iodine from seaweed

Iodine is an element essential in the body for the production of thyroid hormones that regulate growth and metabolism. Diets deficient in iodine increase the risk of retarded brain development in children, mental slowness, high cholesterol, lethargy, fatigue, depression, weight gain, and goitre – a swelling of the thyroid gland in the neck. Of all foods, seaweed like kelp and nori are the best sources of natural iodine.

Iodine in seaweed can be extracted by heating it in air to an ash, in which the iodine is present as iodide. The iodide is dissolved out of the residue with boiling water, the solution is filtered and the iodide is oxidised to iodine using acidified hydrogen peroxide.

$$2I^-(aq) + 2H^+(aq) + H_2O_2(aq) \rightarrow 2H_2O(l) + I_2(aq)$$

TIP

Iodine has a very low solubility in water but is soluble in a solution containing iodide ions. Iodine reacts with iodide ions to form the brown triiodide ion $I_2 + I^- \rightleftharpoons I_3^-$.

The iodine colours the solution at this stage. The mixture is transferred into a separating funnel and shaken with cyclohexane. Two layers form, the top layer is purple. The layers are separated and the purple solution is left in a fume cupboard to allow the liquid (solvent) to evaporate. The crystals formed are weighed.

1. The heated seaweed contained 0.01% iodine by mass. In a 1 kg sample, calculate the mass of iodine present and use this to find the moles of iodine present.
2. Write the two half-equations for the redox equation between iodide and hydrogen peroxide.
3. Using oxidation states, explain why the iodide is oxidised.
4. What colour is the solution formed after oxidation?
5. Explain why the top layer in the separating funnel is purple.
6. Identify the crystals formed and state their colour.
7. The crystals in the evaporating basin are warmed slightly after weighing. Describe and explain what is observed.

Redox reactions of halogens

Chlorine reacting with cold, dilute, aqueous sodium hydroxide

Chlorine reacts with cold, dilute, aqueous sodium hydroxide solution forming sodium chlorate(I) and sodium chloride.

This reaction may be applied to other halogens. The chlorate(I) ion, ClO^-, is a halate(I) ion and there is a pattern in their formulae.

All halate(I) ions have similar formulae:

- chlorate(I) is ClO^-
- bromate(I) is BrO^-
- iodate(I) is IO^-

TIP

The oxidation number of the halogen atoms in halate(I) salts is +1 hence halate(I).

The formulae of sodium halate(I) salts are:

- sodium chlorate(I) is $NaClO$
- sodium bromate(I) is $NaBrO$
- sodium iodate(I) is $NaIO$.

Chlorine reacts with cold, dilute, aqueous sodium hydroxide to produce chlorate(I) ions, ClO^-.

Ionic equation:

$$2OH^- + Cl_2 \rightarrow Cl^- + ClO^- + H_2O$$

Symbol equation:

$$2NaOH + Cl_2 \rightarrow NaCl + NaClO + H_2O$$

Figure 10.4 The reaction between cold dilute aqueous sodium hydroxide is used to produce bleach. Bleach is a solution of sodium chorate(I).

Figure 10.5 Placing a capful of bleach into the water containing freshly cut flowers, helps to preserve the flowers. The bleach kills any fungi and bacteria and keeps the water clear. It may however slightly discolour the stems of the flowers.

Observations:

- yellow-green gas forms a colourless solution

This is a redox reaction as chlorine is oxidised from the oxidation state of 0 in Cl_2 to $+1$ in NaClO and chlorine is also reduced from 0 (zero) in Cl_2 to -1 in NaCl.

The solution containing sodium chlorate(I) is used as bleach. The chlorate(I) ions are responsible for the bleaching action of the solution. The sodium chlorate(I) kills bacteria and other micro-organisms.

Chlorine reacting with water

Chlorine is slightly soluble in water and results in a pale green solution. Some chlorine reacts with the water to form a mixture of hydrochloric acid (HCl) and chloric(I) acid (HOCl). HOCl is responsible for the strong germicidal and bleaching power of wet chlorine. HClO is also called hypochlorous acid. The reaction is reversible.

$$Cl_2(g) + H_2O(l) \rightleftharpoons HCl + HClO$$

This equation explains the use of chlorine to sterilise drinking water and water in swimming pools. HClO kills micro-organisms. It is safe to use because a very low concentration of chlorine is used. Also as HClO kills micro-organisms, the position of equilibrium moves from left to right so very little chlorine remains once HClO has done its job. In addition, the benefits of using chlorine in drinking water and in water in swimming pools far outweigh the potential health risks to humans if it were not present. In higher concentrations, chlorine is toxic.

However in bright sunlight, chlorine reacts with water to produce a solution containing chloride ions and oxygen gas.

$$2Cl_2(g) + 2H_2O(l) \rightarrow 4HCl(aq) + O_2(g)$$

This equation may be written:

$$2Cl_2(g) + 2H_2O(l) \rightarrow 4H^+(aq) + 4Cl^-(aq) + O_2(g)$$

The ultraviolet radiation from the sun breaks down the HClO into HCl and O_2.

$$2HClO(aq) \rightarrow 2HCl(aq) + O_2(g)$$

Society assesses the advantages and disadvantages of adding chemicals like chlorine and sodium fluoride to water supplies. Sodium fluoride and other fluorine containing compounds may be added to drinking water. This is

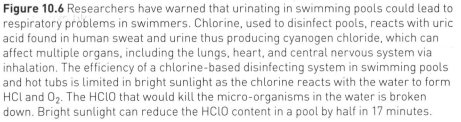

Figure 10.6 Researchers have warned that urinating in swimming pools could lead to respiratory problems in swimmers. Chlorine, used to disinfect pools, reacts with uric acid found in human sweat and urine thus producing cyanogen chloride, which can affect multiple organs, including the lungs, heart, and central nervous system via inhalation. The efficiency of a chlorine-based disinfecting system in swimming pools and hot tubs is limited in bright sunlight as the chlorine reacts with the water to form HCl and O_2. The HClO that would kill the micro-organisms in the water is broken down. Bright sunlight can reduce the HClO content in a pool by half in 17 minutes.

because fluoride ions can help to prevent tooth decay as they strengthen enamel. Some people object to this as there may be other health risks associated with fluoride and also because it can be viewed as mass medication.

Reaction with other halides in solution

Some halogens react with other halide ions in solution. The reactions can be explained by examining the oxidising ability of the halogens and the reducing ability of the halide ions.

For many of these reactions it is important to understand the colours of the halogens in aqueous solution.

Chlorine and bromine react with water so they appear to be partially soluble in water.

A solution of chlorine in water is called chlorine water. It is a pale green solution depending on the concentration, though can appear colourless at low concentrations of chlorine.

A solution of bromine in water is called bromine water. It is an orange solution.

Iodine is only very slightly soluble in water but forms a pale yellow solution. Iodine dissolves in a solution containing iodide ions to form a brown solution.

Figure 10.7 Halogens in solution. Left, chlorine in water; middle, bromine in water; right, iodine in a solution of potassium iodide.

Oxidising ability of the halogens

The oxidising power of the halogens reduces as atomic number increases.

Oxidising agents (also called oxidants) cause an oxidation to occur so they cause another atom, molecule or ion to lose electrons. This means that an oxidising agent gains electrons and becomes reduced.

Oxidising agents readily accept electrons. The ease with which an oxidising agent gains electrons determines how effective it is as an oxidising agent.

- A substance which gains electrons easily will be a good oxidising agent.
- The halogen atoms are gaining electrons into their outer energy level to complete the energy level.
- The electron which is gained by a fluorine atom completes an energy level closer to the nucleus than the electron which completes the outer energy level in chlorine.
- The electron which fluorine gains has a stronger attraction to the nucleus as it is **closer** to the nucleus than for a chlorine atom.
- Also the electron being gained by a fluorine atom is not subjected to as much shielding by inner electrons as there are fewer electrons between the electron gained and the nucleus.

The oxidising ability of the halogens decreases from fluorine to chlorine to bromine to iodine.

> **TIP**
> Bromine water is used in organic chemistry to test for the presence of C=C in a molecule.

Figure 10.8 Garlic is widely used to flavour food but varies in strength from species to species and crop to crop. A sensor to determine the strength of garlic for the food industry has been developed by UK scientists. An electrochemical sensor suspends garlic purée particles in a solution of bromide ions. The bromine produced reacts with the sulfur-sulfur bonds in garlic to regenerate bromide – this results in an increase in current, which is shown on the trace produced.

Displacement reactions

From our knowledge of oxidising powers of the halogens we would expect fluorine to displace all other halides from a solution of a halide compound. We would also expect chlorine to displace bromide and iodide from solutions of iodide compounds and bromine to displace iodide in solution.

231

TIP

The reactions of fluorine are not examined experimentally as fluorine is too dangerous to be used in the laboratory. The reactions of fluorine would follow the pattern for the other halogens.

TIP

You may be asked to write individual half equations for the conversion of molecules into ions or from ions or molecules. You may also have to identify these reactions as oxidation or reduction processes.

EXAMPLE 1

Write the simplest ionic equation for the reaction of chlorine with sodium bromide solution.

Write half-equations for the oxidation and reduction reactions occurring.

Answer

The equation for this reaction is:

$$Cl_2 + 2NaBr \rightarrow 2NaCl + Br_2$$

The Na^+ ion is a spectator ion so is not included in the ionic equation.

$$Cl_2 + 2Br^- \rightarrow 2Cl^- + Br_2$$

The half equation for the conversion of chlorine molecules into chloride ions is:

$$Cl_2 + 2e^- \rightarrow 2Cl^-$$

This is a reduction half equation as chlorine is gaining electrons.

The half equation for the conversion of bromide ions into bromine molecules is:

$$2Br^- \rightarrow Br_2 + 2e^-$$

This is an oxidation half equation as the **bromide ions** are losing electrons.

In this reaction the colourless solution changes to orange because bromine is produced in the solution and bromine water is an orange colour.

EXAMPLE 2

The equation below shows the simplest ionic equation for the reaction of chlorine with sodium iodide solution.

$$Cl_2(aq) + 2I^-(aq) \rightarrow 2Cl^-(aq) + I_2(aq)$$

Write a half equation for the conversion of chlorine into chloride ions.

Write a half equation for the conversion of iodide ions into iodine.

Answer

The half equation for the conversion of chlorine into chloride ions is:

$$Cl_2 + 2e^- \rightarrow 2Cl^-$$

This is a reduction half equation as chlorine is gaining electrons.

The half equation for the conversion of iodide ions into iodine is:

$$2Br^- \rightarrow Br_2 + 2e^-$$

This is an oxidation half equation as the **iodide ions** are losing electrons.

In this reaction the colourless solution changes to brown (as iodine is produced in a solution containing iodide ions).

Figure 10.9 Chlorine gas bubbling through an aqueous solution of potassium iodide in a test tube. The reaction is turning brown due to displacement of iodine. A colourless solution of potassium iodide is seen on the left in a glass bottle.

Bromine reacting with aqueous iodide ions

This reaction is not used as extensively as the other two above as the colour change in the solution is not as clear because it simply darkens from orange to brown.

$$Br_2(aq) + 2I^-(aq) \rightarrow 2Br^-(aq) + I_2(aq)$$

Half equations:

$$Br_2 + 2e^- \rightarrow 2Br^- \quad \text{REDUCTION}$$
$$2I^- \rightarrow I_2 + 2e^- \quad \text{OXIDATION}$$

Reducing ability of the halides

The reducing power of the **halide ions** increases as the atomic number increases.

Reducing agents (also called reductants) cause a reduction to occur so they cause another atom, molecule or ion to gain electrons. This means that a reducing agent loses electrons and becomes oxidised.

Reducing agents readily donate electrons. The ease with which a reducing agent loses electrons determines how effective it is as a reducing agent.

- A substance which loses electrons easily will be a good reducing agent.
- Halide ions are losing electrons from their outer energy level.
- The electron which is lost by an iodide ion comes from an energy level **further** from the nucleus than the electron that would be lost by a bromide ion.
- The electron which an iodide ion loses has a weaker attraction to the nucleus as it is further from the nucleus compared to the other halide ions.
- Also the electron being lost by an iodide ion is subjected to greater shielding by inner electrons than other halide ions as there are more electrons between the electron being lost and the nucleus.

The reducing ability of the halide ions increases from fluoride to chloride to bromide to iodide.

Fluoride and chloride ions have little reducing ability as they are not as able to donate electrons as bromide and iodide ions.

Reaction of solid halides with concentrated sulfuric acid

Solid halide salts, usually sodium or potassium halides, react with concentrated sulfuric acid, depending on the reducing ability of the halide ions and the hydrogen halides. Group 1 halides have been used to represent the standard reactions.

1 Reaction of concentrated H_2SO_4 with NaF(s)

$$NaF + H_2SO_4 \rightarrow NaHSO_4 + HF \quad \text{(NOT REDOX)}$$

Products names: sodium hydrogen sulfate and hydrogen fluoride

Observations: misty white fumes (HF)

TIP
The sodium and potassium are interchangeable throughout this and questions will vary in which solid salt is given. Many of the reactions produce misty white fumes of the hydrogen halide gas. The hydrogen halide gases fume in moist air reacting with the moisture in the air so appearing as misty fumes.

The simplest ionic equation for this reaction is:

$$F^- + H_2SO_4 \rightarrow HSO_4^- + HF$$

Hydrogen fluoride is released as a gas.

2 Reaction of concentrated H_2SO_4 with NaCl(s)

$$NaCl + H_2SO_4 \rightarrow NaHSO_4 + HCl \quad \text{(NOT REDOX)}$$

Products names: sodium hydrogen sulfate and hydrogen chloride

Observations: misty white fumes (HCl)

The simplest ionic equation for this reaction is:

$$Cl^- + H_2SO_4 \rightarrow HSO_4^- + HCl$$

Hydrogen chloride is released as a gas.

3 Reaction of concentrated H_2SO_4 with KBr(s)

$$KBr + H_2SO_4 \rightarrow KHSO_4 + HBr; \text{ NOT REDOX}$$

$$2HBr + H_2SO_4 \rightarrow Br_2 + SO_2 + 2H_2O; \text{ REDOX}$$

Products names: potassium hydrogen sulfate, hydrogen bromide, bromine, water, sulfur dioxide

Observations: misty white fumes (HBr), red-brown vapour (Br_2)

Redox: In the second equation the Br is oxidised from -1 (in HBr) to 0 (in Br_2) and the S is reduced from $+6$ (in H_2SO_4) to $+4$ (in SO_2).

Note that the bromine is the oxidation product in this reaction and the sulfur dioxide is the reduction product.

Ionic equations

The simplest ionic reaction for the first reaction is:

$$Br^- + H_2SO_4 \rightarrow HSO_4^- + HBr$$

The simplest ionic equation for the second reaction is:

$$2Br^- + SO_4^{2-} + 4H^+ \rightarrow Br_2 + SO_2 + 2H_2O$$

The bromide ion is able to reduce the sulfur in sulfuric acid from the oxidation state of $+6$ in SO_4^{2-} to $+4$ in SO_2.

Overall equation

The equation may be written as an overall equation:

$$2KBr + 3H_2SO_4 \rightarrow 2KHSO_4 + Br_2 + SO_2 + 2H_2O$$

For this reaction the observations would be red-brown gas released.

You may be asked to determine the ionic equation from this overall reaction:

$$2Br^- + 3SO_4^{2-} + 6H^+ \rightarrow 2HSO_4^- + Br_2 + SO_2 + 2H_2O$$

$$\text{or } 2Br^- + 3H_2SO_4 \rightarrow 2HSO_4^- + Br_2 + SO_2 + 2H_2O$$

TIP

Fluoride ions are not strong enough reducing agents to reduce the sulfur in sulfuric acid.

TIP

Chloride ions are also not strong enough reducing agents to reduce the sulfur in sulfuric acid.

Figure 10.10 Bromine gas forming as the product of a reaction between concentrated sulfuric acid (in bottle) and sodium bromide (in test tube).

4 Reaction of concentrated H_2SO_4 with KI(s)

$$KI + H_2SO_4 \rightarrow KHSO_4 + HI; \text{ NOT REDOX} \quad \leftarrow overall$$

$$2HI + H_2SO_4 \rightarrow I_2 + SO_2 + 2H_2O; \text{ REDOX}$$

$$6HI + H_2SO_4 \rightarrow 3I_2 + S + 4H_2O; \text{ REDOX}$$

$$8HI + H_2SO_4 \rightarrow 4I_2 + H_2S + 4H_2O; \text{ REDOX}$$

Products names: sodium hydrogen sulfate, hydrogen iodide, iodine, water, sulfur dioxide, sulfur, hydrogen sulfide.

Observations: misty white fumes (HI), purple vapour (I_2), yellow solid formed (S), rotten egg smell (H_2S), black solid formed (I_2).

Redox: in the second equation the I is oxidised from -1 (in HI) to 0 (in I_2) and the S is reduced from $+6$ (in H_2SO_4) to $+4$ (in SO_2).

In the third equation, the I is oxidised from -1 (in HI) to 0 (in I_2) and the S is reduced from $+6$ (in H_2SO_4) to 0 (in S)

In the fourth equation, the I is oxidised from -1 (in HI) to 0 (in I_2) and the S is reduced from $+6$ (in H_2SO_4) to -2 (in H_2S)

The iodine is the oxidation product and the sulfur dioxide, sulfur and hydrogen sulfide are the reduction products.

Ionic equations

The simplest ionic equation for the first reaction is:

$$I^- + H_2SO_4 \rightarrow HSO_4^- + HI$$

The simplest ionic equation for the second reaction is:

$$2I^- + SO_4^{2-} + 4H^+ \rightarrow I_2 + SO_2 + 2H_2O$$

The iodide ion is able to reduce the sulfur in sulfuric acid from the oxidation state of $+6$ to $+4$ in SO_2.

The simplest ionic equation for the third reaction is:

$$6I^- + 8H^+ + SO_4^{2-} \rightarrow 3I_2 + S + 4H_2O$$

The simplest ionic equation for the fourth reaction is:

$$8I^- + 10H^+ + SO_4^{2-} \rightarrow 4I_2 + H_2S + 4H_2O$$

The redox reactions occur because of the relative reducing powers of the halides in the hydrogen halides.

Chloride ions are not as powerful a reducing agent as bromide ions, which in turn are not as powerful a reducing agent as iodide ions. Hence HCl cannot reduce sulfuric acid; the chloride ions are simply displaced by the acid.

Bromide ions in hydrogen bromide are better reducing agents than chloride ions so some of the sulfuric acid is reduced to SO_2 and H_2O. The products are a mixture of bromine and hydrogen bromide.

Iodine ions in hydrogen iodide are the best of the three as a reducing agent and reduce the sulfur in sulfuric acid to SO_2, even reducing it further to S and H_2S. The product is mainly iodine.

Using concentrated sulfuric acid to test for halide ions

- When concentrated sulfuric acid is added to a solid bromide compound, misty white fumes (of hydrogen bromide) are released along with some brown gas (bromine).
- When concentrated sulfuric acid is added to a solid iodide compound, misty white fumes (of hydrogen iodide) are released, a black solid is formed (solid I_2), purple fumes are observed (I_2 gas) and there is a rotten egg smell (H_2S gas).

This is not normally used as a test for halide ions as concentrated sulfuric acid is corrosive and toxic gases are formed.

Figure 10.11 A halogen bulb, contains a halogen such as iodine or bromine. As tungsten evaporates from the filament, it usually condenses on the inner surface of the bulb. The halogen combines with this tungsten deposit on the glass to produce tungsten halides, which evaporate easily. When the tungsten halide reaches the filament, the intense heat of the filament causes the halide to break down, releasing tungsten back to the filament. This process, known as the halogen cycle, extends the life of the filament and keeps the inner surface of the bulb clean, which lets halogen bulbs stay close to full brightness as they age.

TEST YOURSELF 3

1 Chlorine is used to treat drinking water.
  Write an equation for the reaction of chlorine with cold water.
2 a) Write an equation for the reaction of chlorine with cold, dilute, aqueous sodium hydroxide solution.
  b) State the name of the product in which the chlorine has an oxidation state of +1.
  c) State a use for the solution formed in this reaction.
3 Potassium iodide reacts with concentrated sulfuric acid. The overall equation for the reaction is given below:

  $$8KI + 9H_2SO_4 \rightarrow 4I_2 + 8KHSO_4 + H_2S + 4H_2O$$

  a) Give the changes in oxidation state of iodine and sulfur in this reaction.
  b) Which product would lead to a rotten egg smell?
  c) Explain why iodide ions can reduce sulfur in this way but chloride ions cannot.
4 Chlorine reacts with sodium iodide solution.
  a) Write the simplest ionic equation for this reaction.
  b) Write a half equation for the conversion of iodide ions to iodine.
  c) Explain why the colour of the solution changes to brown.

Identification of halide ions

Halide ions (Cl⁻, Br⁻, I⁻) can be identified and distinguished from one another by the use of silver(I) ions in solution. Silver(I) nitrate ($AgNO_3$) solution is most often used.

The solution of the halide ions is first acidified using dilute nitric acid (HNO_3) and then silver nitrate solution is added.

The dilute nitric acid removes other ions which would react with the silver ions in the silver nitrate solution, for example carbonate ions (CO_3^{2-}), sulfite ions (SO_3^{2-}) or hydroxide ions (OH^-). These ions would form precipitates with the silver(I) ions and interfere with the test.

Silver nitrate solution does not form a precipitate with fluoride ions in solution as silver fluoride is soluble in water. Silver nitrate solution cannot be used to test for fluoride ions.

The other three silver halides are insoluble in water and the precipitates differ in colour. A precipitate is a solid appearing in a solution and precipitate is usually abbreviated to 'ppt'.

$$Ag^+(aq) + Cl^-(aq) \rightarrow AgCl(s) \text{ white precipitate}$$

$$Ag^+(aq) + Br^-(aq) \rightarrow AgBr(s) \text{ cream precipitate}$$

$$Ag^+(aq) + I^-(aq) \rightarrow AgI(s) \text{ yellow precipitate}$$

The equations given above are the simplest ionic equations but full equations may be written, for example

$$AgNO_3 (aq) + NaCl (aq) \rightarrow AgCl(s) + NaNO_3(aq)$$

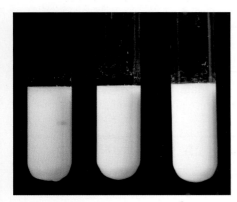

Figure 10.12 From left to right, these precipitates are silver chloride (AgCl), silver bromide (AgBr) and silver iodide (AgI).

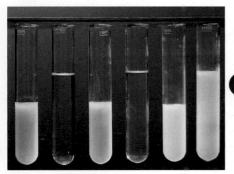

Figure 10.13 Test tubes (from left). 1, 3 and 5 show the white AgCl ppt, the cream AgBr ppt and the yellow AgI ppt. Test tubes 2, 4 and 6 show the result of adding excess concentrated ammonia solution. AgCl and AgBr dissolve but it has no effect on AgI. The yellow ppt remains.

TIP
It is important to be able to identify the precipitates formed in these tests. When the word 'identify' is used in a question, either the name or chemical formula is accepted.

A further test can be carried out with ammonia solution to confirm the identity of the halide ion.

- Silver chloride will redissolve in dilute and concentrated ammonia solution forming a colourless solution.
- Silver bromide does not redissolve in dilute ammonia solution but does redissolve in concentrated ammonia solution forming a colourless solution.
- Silver iodide does not redissolve in dilute or concentrated ammonia solution.

The redissolution of AgCl in dilute ammonia occurs because of the formation of a complex. Ammonia molecules have a lone pair of electrons, which can form a coordinate bond with an unoccupied orbital of the silver(I) ion. The complex ion is soluble and so the silver chloride redissolves. Silver iodide does not redissolve in either dilute ammonia solution or concentrated ammonia solution. The explanation of this is beyond the specification at this level though silver iodide is mostly a covalent compound based on the electronegativity values of silver and iodine so the more covalent a compound the less likely it is to form a complex ion with ammonia.

The dissolution of silver chloride is explained by the equation:

$$AgCl(s) + 2NH_3(aq) \rightarrow [Ag(NH_3)_2]^+(aq) + Cl^-(aq)$$

white ppt   colourless solution

Silver halides darken when exposed to sunlight forming a black deposit of silver.

$$2AgCl(s) \rightarrow 2Ag(s) + Cl_2(aq)$$

This plays an important part in photography as photographic plates darken when exposed to light.

TIP

A solution containing a mixture of sodium chloride and sodium iodide would form a cream precipitate with silver nitrate solution (resulting from a mixture of the yellow precipitate and the white precipitate). Adding dilute ammonia solution leaves a yellow precipitate as the white precipitate of silver chloride will dissolve.

Identification tests

This section deals with identification of the common anions and cations found in ionic compounds.

Many of the tests rely on the formation of a precipitate (abbreviated as ppt). A ppt is a solid which forms when two solutions are mixed. It results because a combination of ions from the solutions gives an insoluble compound which appears as a ppt. The formation of the ppt and its colour are vital clues to the identity of the ions present in a compound.

- It is vital that you know how to carry out the test practically and use the correct term, e.g. if you are expecting to see a ppt then a solution must be added to a solution; this is the most frequent mistake when describing tests for ions, e.g. to a solution of the solid, sodium hydroxide solution was added dropwise and then until it is in excess.
- You must also recognise the test so that you know what you are testing for, e.g. if you add silver nitrate solution to a solution of an unknown ionic compound, you should be aware that you are testing for the presence of a halide (chloride, bromide or iodide) ion and what the results you get mean in terms of the ion present.
- When writing equations for the reactions remember that the precipitate is the only product of the ionic equation:

EXAMPLE 4

When silver(I) nitrate solution is added to sodium chloride solution, a white precipitate of silver(I) chloride is formed.

Answer

Equation: $AgNO_3 + NaCl \rightarrow AgCl + NaNO_3$

Ionic equation: $Ag^+ + Cl^- \rightarrow AgCl$

If state symbols are required, the precipitate has (s) after it and the rest are (aq).

EXAMPLE 5

When sodium hydroxide solution is added to a solution of magnesium chloride, a white precipitate of magnesium hydroxide is formed.

Answer

Equation: $MgCl_2 + 2NaOH \rightarrow Mg(OH)_2 + 2NaCl$

Ionic equation: $Mg^{2+} + 2OH^- \rightarrow Mg(OH)_2$

The tables below give details of all the anion and cation tests you are required to know for AS.

Some ions require more than one test to be carried out. A white ppt is observed when magnesium chloride solution is added to a solution of a compound. This could indicate carbonate ions or hydroxide ions being present. The addition of acid to the substance would differentiate between them as only the carbonate would produce a gas.

Test and testing for	How to carry out the test	Typical observations	Deductions from observations (Detail of reactions)
Acidified silver(I) nitrate solution (often just referred to as acidified silver nitrate solution) Testing for halide ions	Dissolve a spatula measure of the sample in dilute nitric acid and add a few cm³ of silver nitrate solution Add dilute/concentrated ammonia solution	white ppt which redissolves in dilute ammonia solution to form a colourless solution	chloride ion/Cl^- present. White ppt is AgCl
		cream ppt which does not redissolve in dilute ammonia solution but redissolves in concentrated ammonia solution to form a colourless solution	bromide ion/Br^- present. Cream ppt is AgBr
		yellow ppt which does not redissolve in either dilute ammonia solution or concentrated ammonia solution	iodide ions/I^- present. Yellow ppt is AgI
		no ppt formed	no chloride, bromide or iodide ions present

Test and testing for	How to carry out the test	Typical observations	Deductions from observations (Detail of reactions)
Acidified barium chloride solution Testing for sulfate ions (This test may also be used to identify the presence of barium ions – adding a solution containing sulfate ions to a solution containing barium ions gives the same result)	Dissolve a spatula measure of the sample in dilute nitric acid (hydrochloric acid also works here) and add a few cm³ of barium chloride solution	white ppt	sulfate ions (SO_4^{2-}) present. White ppt is $BaSO_4$
		no ppt formed	no sulfate ions (SO_4^{2-}) present

Test and testing for	How to carry out the test	Typical observations	Deductions from observations (Detail of reactions)
Dilute nitric acid Testing for carbonate ions (hydrogen carbonate ions also react with acid producing carbon dioxide)	Place a few cm³ of dilute nitric acid in a test tube and add a spatula measure of the sample	effervescence; solid disappears gas evolved can be passed through limewater in another test tube – limewater changes from colourless to milky	carbonate ions/CO_3^{2-} present (or hydrogen carbonate ions/HCO_3^- present). Carbon dioxide released from reaction of a carbonate ion or hydrogen carbonate ion with acid
		no effervescence/no gas produced	no carbonate ions/CO_3^{2-} (or hydrogen carbonate ions/HCO_3^-) present

Test and testing for	How to carry out the test	Typical observations	Deductions from observations (Detail of reactions)
Magnesium nitrate solution (or magnesium chloride solution). Testing for carbonate ions (rules out hydrogen carbonate ions)	Dissolve a spatula measure of the sample in deionised water and add a few cm³ of magnesium nitrate solution (or magnesium chloride solution)	white ppt	carbonate ions/CO_3^{2-} ions present. White ppt is magnesium carbonate/$MgCO_3$
		no ppt formed	no carbonate ions/CO_3^{2-} present

Test and testing for	How to carry out the test	Typical observations	Deductions from observations (Detail of reactions)
Sodium hydroxide solution Testing for some metal cations (This test may also be used to test for the presence of the hydroxide ion – add a solution of magnesium nitrate to a solution suspected to contain hydroxide ions and a white ppt will form)	Dissolve a spatula measure of the sample in deionised water and add a few drops of sodium hydroxide solution; then add about 5 cm³ of sodium hydroxide solution (adding until in excess)	white ppt which does not dissolve in excess sodium hydroxide solution	magnesium ions/Mg^{2+} or calcium ions/Ca^{2+} present. White ppt is magnesium hydroxide/$Mg(OH)_2$ or calcium hydroxide/$Ca(OH)_2$.

Test and testing for	How to carry out the test	Typical observations	Deductions from observations (Detail of reactions)
Sodium hydroxide solution Testing for the ammonium ion (NH_4^+) (This test is also used to test for hydroxide ions. Add an ammonium compound to a suspected hydroxide and warm. Ammonia gas should be given off if hydroxide ions are present.)	Place a few cm³ of sodium hydroxide solution in a test tube and add a spatula measure of the sample; warm gently. Test any gas evolved using damp red litmus paper or damp universal indicator paper or test any gas evolved using a glass rod dipped in concentrated hydrochloric acid	pungent gas evolved damp indicator paper changes to blue white smoke with glass rod dipped in concentrated hydrochloric acid	ammonium ion/NH_4^+ present. Ammonia released from action of alkali (NaOH) on an ammonium compound. Ammonia gas is alkaline. White smoke is ammonium chloride from reaction of NH_3 (g) with HCl (g) from the concentrated HCl

TIP

The reactions of the solid halides with concentrated sulfuric acid as detailed on pages 233–236 and the displacement reactions of halide ions (page 232) could also be used for identification purposes.

REQUIRED PRACTICAL

Tests for ions

A student carried out a series of tests on a mixture of two ionic compounds labelled **A**. The compounds have a common cation.

1 Complete the table by writing an appropriate deduction for each observation.

Test	Observation	Deduction
Make a solution of **A** by dissolving a spatula of **A** in a test tube half full of water. Warm gently.	Colourless solution formed	
Add a few drops of 0.4 mol dm^{-3} sodium hydroxide to the test tube, then add a further 3 cm^3 of the sodium hydroxide solution to the test tube.	White ppt. White ppt is insoluble in excess sodium hydroxide	
Make a solution of **A** by dissolving half a spatula of **A** in a test tube half full of nitric acid solution. Warm gently. Transfer 1 cm^3 of the solution into each of two separate test tubes.	No effervescence	
Add a few drops of 0.05 mol dm^{-3} silver nitrate to the test tube, then add about 2 cm^3 of dilute ammonia to the same test tube.	White ppt. Ppt dissolves in dilute ammonia to produce a colourless solution	
Add a few drops of 0.1 mol dm^{-3} barium chloride solution to the second test tube.	White ppt	

2 Name the two compounds present in **A**.

Figure 10.14 A useful test for the presence of some Group 2 cations is to carry out a flame test, by dipping a nichrome wire in concentrated hydrochloric acid then into the compound to be tested and then into a blue Bunsen flame. A calcium ion flame test is brick red, strontium ion is red, and barium ion, as shown in this figure is green.

TEST YOURSELF 4

1 A white solid X is dissolved in water and acidified silver nitrate solution was added. A white precipitate was observed which was soluble in aqueous ammonia.
   a) Identify the white precipitate.
   b) Write the simplest ionic equation for the formation of the white precipitate.
   c) Identify the halide ion present in solid X.

2 Concentrated sulfuric acid is added to a solid sodium halide. The white solid changed to black, a yellow solid was formed and a gas was released, which smelled of rotten eggs.
   a) Identify the yellow solid formed.
   b) Identify the gas which smelled of rotten eggs.
   c) Identify the solid halide.

3 Explain why silver nitrate is acidified when testing for chloride ions in solution.

4 Identify a reagent which could be used to distinguish between solutions of silver chloride and silver bromide.

5 Sodium bromide reacts with concentrated sulfuric acid according to the equations:

Equation 1: $NaBr + H_2SO_4 \rightarrow NaHSO_4 + HBr$

Equation 2: $2HBr + H_2SO_4 \rightarrow Br_2 + SO_2 + 2H_2O$

Equation 2 represents a redox reaction.
   a) Describe what would be observed as HBr was formed?
   b) Explain, using oxidation numbers, why Equation 2 represents a redox reaction.

Practice questions

1 Which atom has the highest electronegativity?

   A F   **B** Cl

   C Br   **D** I   *(1)*

2 Which of the following is the compound in which chlorine has the highest oxidation state?

   A Cl_2O_7 **B** $KClO_3$

   C $MgCl_2$ **D** NaClO   *(1)*

3 Which one of the following solids would react with concentrated sulfuric acid to form sulfur?

   A potassium fluoride

   B potassium chloride

   C potassium bromide

   D potassium iodide   *(1)*

4 a) Explain why the melting points of the halogens increase down the group.   *(3)*

   b) i) What is meant by the term *electronegativity*?   *(1)*

     ii) State the trend in electronegativity of the halogens down the group.   *(1)*

     iii) Explain the trend in electronegativity of the halogens.   *(2)*

5 Chlorine reacts reversibly with cold water.

   a) Write an equation for the reversible reaction which occurs when chlorine reacts with water.   *(1)*

   b) Name the products of this reaction.   *(2)*

   c) Chlorine is used to treat water used in swimming pools. Explain why chlorine is used despite it being very toxic.   *(1)*

   d) In strong sunlight chlorine reacts with water to form oxygen gas. Write an equation for this reaction.   *(1)*

6 Chlorine reacts with cold, dilute, aqueous sodium hydroxide according to the equation given below.

$$2NaOH + Cl_2 \rightarrow NaCl + NaClO + H_2O$$

   a) Give the oxidation state of chlorine in NaClO.   *(1)*

   b) Give the IUPAC name of NaClO.   *(1)*

   c) Explain, using oxidation states, why this reaction is described as redox.   *(3)*

   d) Name the products formed when bromine reacts with cold, dilute, aqueous potassium hydroxide solution.   *(2)*

7 When chlorine is bubbled through an aqueous solution of potassium iodide, a redox reaction occurs.

$$2KI(aq) + Cl_2(g) \rightarrow 2KCl(aq) + I_2(aq)$$

   a) State what would be observed during this reaction?   *(1)*

   b) Write the simplest ionic equation for this reaction.   *(1)*

   c) Explain why this reaction is described as a redox reaction.   *(3)*

8 The presence of halide ions can be detected using silver nitrate solution and ammonia solution.

   a) Write the simplest ionic equation for the reaction between silver ions and iodide ions.   *(1)*

   b) Copy and complete the table below.

Halide	Colour of silver halide	Effect of adding dilute ammonia solution to the silver halide
chloride		
bromide		
iodide		

        (3)

9 The hydrogen halides can be formed by the reaction of concentrated sulfuric acid with the corresponding solid sodium halide.

   a) Write an equation for the reaction of concentrated sulfuric acid with sodium chloride at room temperature.   *(1)*

   b) Misty white fumes are released when sodium chloride reacts with concentrated sulfuric acid. Identify the product responsible for the misty white fumes.   *(1)*

   c) Identify two products, other than hydrogen iodide, which are formed when sodium iodide reacts with concentrated sulfuric acid.   *(2)*

10 Potassium bromide reacts with concentrated sulfuric acid.

$$2KBr + 3H_2SO_4 \rightarrow 2KHSO_4 + Br_2 + SO_2 + 2H_2O$$

a) Write the two half equations for the redox reaction. (2)

b) Identify the product which is seen as brown fumes. (1)

c) Sulfuric acid is described as an oxidising agent in this reaction. In terms of electrons, state the meaning of the term oxidising agent. (1)

d) Suggest why bromide ions are stronger reducing agents than chloride ions. (2)

11 a) The formulae below are for halogen containing compounds.

NaClO	NaI	NaCl
NaBr	NaF	NaClO$_3$
HBr	AgI	AgCl

i) Identify the compounds which are formed when chlorine reacts with cold, dilute, aqueous sodium hydroxide solution. (2)

ii) Identify the compound which is yellow. (1)

iii) Identify the compound which would appear as misty white fumes. (1)

iv) Identify the compound in solution which would react with silver nitrate solution to form a white precipitate. (1)

v) Identify the compound in which the halogen atom has an oxidation state of +1? (1)

vi) Identify the compound which reacts with concentrated sulfuric acid to give a rotten egg smell. (1)

b) Sodium fluoride reacts with concentrated sulfuric acid.

(i) Write an equation for the reaction of sodium fluoride with concentrated sulfuric acid. (1)

(ii) Explain why misty fumes are observed during this reaction. (1)

11 Introduction to organic chemistry

Organic chemistry was first defined in the early 1800s as a branch of modern science. Previously chemical compounds were classified into two main groups: organic and inorganic. According to prevailing scientific thought, organic compounds could only be synthesised from living organisms or their remains while inorganic compounds were synthesised from non-living matter. It was thought that a 'vital force' was needed for the formation of organic compounds. However, in 1828, a German chemist Friedrich Wöhler challenged this philosophy when he synthesised urea, H_2NCONH_2, from ammonium cyanate, NH_4OCN. He had succeeded in producing an organic compound, urea, from an inorganic one, ammonium cyanate. Today, chemists consider organic compounds to be those which contain carbon.

Carbon – a unique element

Carbon is truly an extraordinary element. There are approximately ten million compounds containing carbon and hydrogen whose formulae are known to chemists. This is far in excess of the total number of compounds of any of the other elements. Carbon can form this array of compounds because it has the unique ability to catenate, which means to form long chains. The carbon atoms covalently bond together to form a wide variety of chains and rings. Carbon forms four covalent bonds in all its compounds. Compounds that contain carbon and hydrogen only are called **hydrocarbons**.

Bonding in carbon compounds

Carbon can form four covalent bonds. The four bonds can be single bonds as in the compound methane. Alternatively, carbon can form or a mixture of single bonds and double bonds, as in the compound ethene. Carbon–carbon bonds and carbon–hydrogen bonds are relatively strong and non-polar. Chains and rings of carbon and hydrogen atoms form the skeleton of most organic compounds.

Surprisingly, considering the number of compounds classified as organic, the study of organic chemistry is very manageable. Organic compounds can be placed into groups where the physical properties follow a simple pattern and the chemical properties are similar. The first step in becoming a proficient organic chemist is to become familiar with the language of this branch of chemistry.

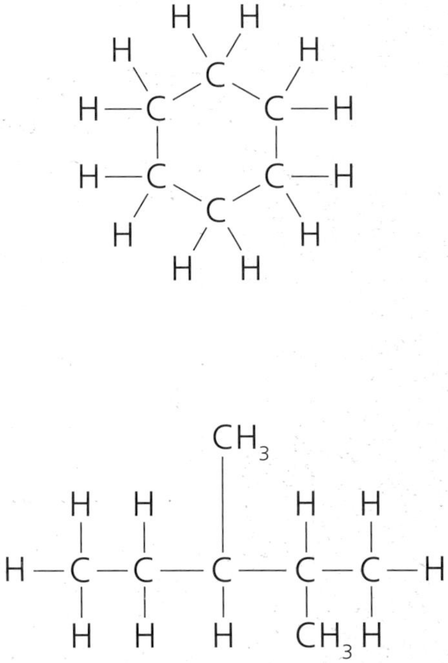

Figure 11.1 The origin of the name carbon comes from the Latin word 'carbo' for charcoal. This photo shows lump charcoal, a form of charcoal produced by slow burning lumps of hardwood. It is primarily used as a fuel, similar to coal.

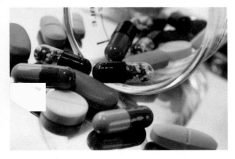

Figure 11.3 Most medicinal drugs are organic compounds. Discovering, making and testing new medicines is the business of the pharmaceutical industry. This is a very important industry in the UK. Pharmaceutical exports are worth an estimated £16 billion per year.

Figure 11.2 Examples of ring and branched hydrocarbons; cyclohexane and 2,3-dimethylpentane.

Functional groups and homologous series

Consider the organic compounds ethane, ethanol and propanol.

Ethane and ethanol compounds have two carbon atoms and single bonds. In ethane the carbon atoms are covalently bonded to each other and to three hydrogen atoms. In ethanol, one of the hydrogen atoms is replaced by a hydroxyl (—OH) group. Table 11.1 compares the physical and chemical properties of ethane, ethanol and propanol.

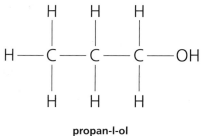

ethane

ethanol

propan-l-ol

Figure 11.4

Table 11.1 A comparison of the physical and chemical properties of ethane, ethanol and propanol.

Physical/chemical property	Ethane	Ethanol	Propanol
Melting point/°C	–183	–114	–126
Boiling point/°C	–89	78	98
State at room temperature	gas	liquid	liquid
Solubility in water	insoluble	soluble	soluble
Reaction with sodium	No reaction	$H_2(g)$ evolved	$H_2(g)$ evolved

The —OH group in ethanol has a pronounced effect on its chemistry. Ethanol has a boiling point that is almost 170 °C higher than the boiling point of ethane. Ethanol is completely soluble in water while ethane is insoluble. When a piece of sodium is placed in ethanol it will react producing bubbles of hydrogen gas but sodium does not react with ethane.

The chemical and physical properties of ethanol and propanol are similar as they both have an –OH group. The —OH group is an example of a functional group. A functional group is a group of atoms which are responsible for the characteristic reactions of a compound. Molecules that contain the same functional group all belong to the same chemical family, a homologous series. A homologous series is a group of compounds which have the same general formula. Successive members in a series differ from each other by a —CH_2 group. They have similar chemical reactions and show a gradation in physical properties. As the length of the carbon chain increases, the influence of the functional group on the compounds' properties decreases. Ethanol and propanol belong to the homologous series of alcohols. Ethane belongs to the homologous series of alkanes.

All members of a homologous series:

● have the same general formula
● show a gradation in physical properties
● have similar chemical properties
● have the same functional group
● differ from successive members by a —CH_2 group.

Viewing organic compounds as members of a homologous series and looking for patterns in their reactivity greatly simplifies their study. During this AS course you will study the properties of four homologous series of organic compounds: alkanes, halogenoalkanes, alkenes; and alcohols. You will also meet several more including aldehydes, ketones and carboxylic acids.

Formulae of organic compounds

Organic chemists use different ways of writing formulae to emphasise different aspects of a compound or to describe a chemical reaction. There are six types of formulae: molecular, empirical, displayed, structural, skeletal and general.

Molecular formulae

The **molecular formula** shows the actual number of atoms of each element in a compound as shown in the table below.

Table 11.2 Combination of atoms in ethane, propene, bromobutane and pentanol.

Compound	Molecular formula	Formed from
Propene	C_3H_6	three C atoms and six H atoms
Bromobutane	C_4H_9Br	four C atoms, nine H atoms and one Br atom
Pentanol	$C_5H_{10}O$	five C atoms, ten H atoms and one O atom

Displayed formulae

All atoms and **all** covalent bonds for a compound are shown in its displayed formula. Ionic parts of a molecule are shown using charges as shown in the displayed formula of sodium ethanoate below.

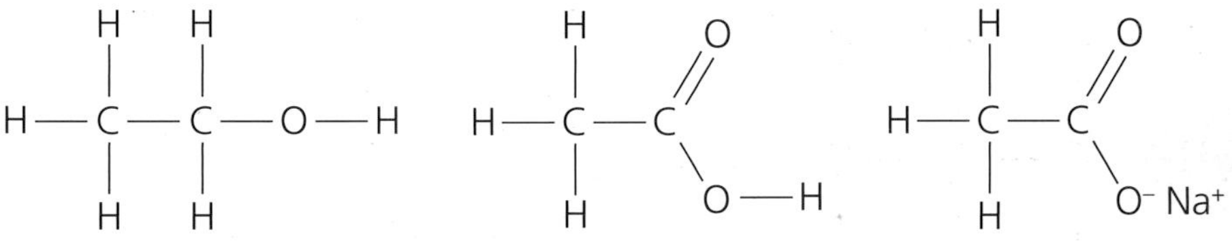

ethanol   ethanoic acid   sodium ethanoate

Figure 11.5

Structural formulae

The **structural formula** of a compound shows the arrangement of atoms, carbon by carbon with the attached hydrogens and functional groups, without showing the bonds. Each carbon is written separately followed by the atoms which are attached to it. When a group of atoms are attached to a carbon, brackets are used in the structural formula to indicate that the group is not part of the main carbon chain.

Table 11.3 The structural and displayed formulae for propan-1-ol, propan-2-ol and 2,3-dimethylpentane.

Name	Structural formula	Displayed formula
Propan-1-ol	$CH_3CH_2CH_2OH$	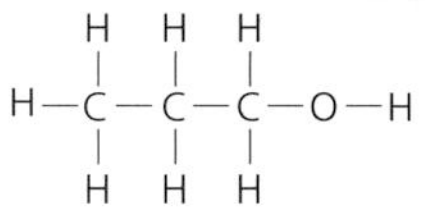
Propan-2-ol	$CH_3C(OH)HCH_3$	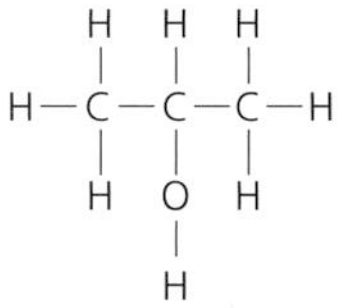
2,3-dimethylpentane	$CH_3CH_2CH(CH_3)CH(CH_3)_2$	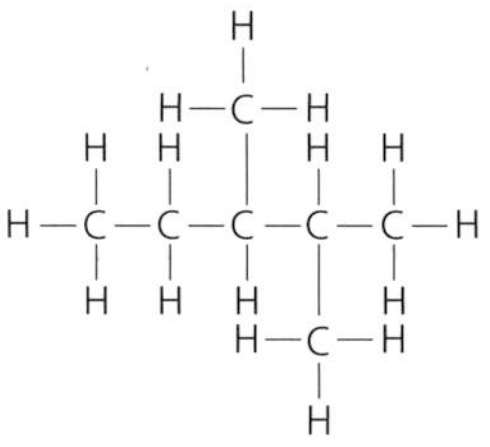

In some cases structural formula can be condensed as shown below. This is useful when there are long hydrocarbon chains.

Structural formula	Condensed structural formula
$CH_3CH_2CH_2CH_2CH_2CH_2CH_2CH_2CH_2COOH$	$CH_3(CH_2)_8COOH$

When the focus is on the reactions of the functional groups in a compound, the length and structure of the carbon chain is less important. In the case of decanoic acid, for example, the formula can then be written as $C_9H_{19}COOH$. Please note that this is neither a true molecular formula nor a true condensed structural formula but it can be more useful.

For compounds which contain double or triple bonds it is also useful to include information about the position of these bonds in the structure formulae, for example, $CH_3CHCHCH_3$ is the true structural formula of but-2-ene but in the format $CH_3CH=CHCH_3$ the position of the double bond is shown clearly.

Empirical formulae

An **empirical formula** is the simplest whole number ratio of the atoms of each element in a compound.

Table 11.4 A comparison of the molecular and empirical formulae for butane, butene, butanol and butanoic acid.

Name of compound	Molecular formula	Empirical formula
Butane	C_4H_{10}	C_2H_5
Butene	C_4H_8	CH_2
Butanol	$C_4H_{10}O$	$C_4H_{10}O$
Butanoic acid	$C_4H_8O_2$	C_2H_4O

TIP
Establishing the empirical formulae of a compound often requires a calculation. The empirical formula is worked out from the ratio of the number of moles of each atom present in a compound, which is calculated using the mass of each atom in the compound. Check out examples of this in Chapter 2.

TEST YOURSELF 2

1 A 10.0 g sample of hydrocarbon B (M_r = 58.0) contains 8.27 g of carbon.
   a) Calculate the empirical formula of the compound.
   b) Deduce the molecular formula of hydrocarbon B.
2 A 20.0 g sample of an organic compound contains hydrogen, 7.98 g of carbon and 10.68 g of oxygen.
   a) Why is this compound not classified as a hydrocarbon?
   b) Calculate the empirical formula of the compound.
   c) The mass of one mole of the compound is 60.00 g. What is its molecular formula?

Skeletal formulae

Skeletal formulae are an abbreviated diagrammatic description of a compound. They are bare 'stick-like' drawings. They do not show carbon or hydrogen atoms attached to the carbon chain but they do show other atoms, for example, oxygen, nitrogen and hydrogen atoms attached to atoms other than carbon. Each line represents a carbon—carbon bond as shown in Table 11.5.

Table 11.5 The displayed and skeletal formulae for ethane, propane and butane.

Name	Displayed formula	Skeletal formula
Ethane	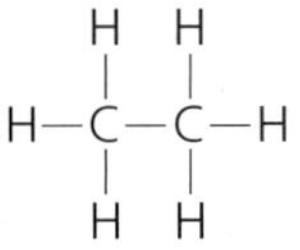 H H | | H—C—C—H | | H H	 /
Propane	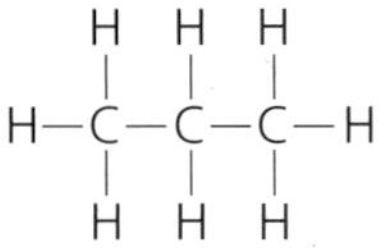 H H H | | | H—C—C—C—H | | | H H H	 ⌃
Butane	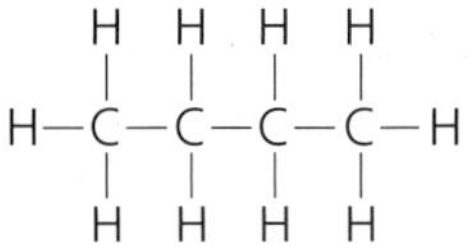 H H H H | | | | H—C—C—C—C—H | | | | H H H H	 ⌵

Note the zig-zag shape of the carbon chains. This makes it clear how many C—C bonds are in the chain. When you make a ball and stick model of an alkane (Figure 11.6), you will notice that the carbon chain in the molecule adopts this zig-zag shape due to the tetrahedral arrangement of the bonds around each carbon atom.

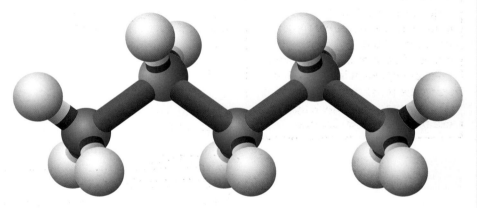

Figure 11.6 Model of pentane illustrating the zig-zag shape of the carbon chain. The blue spheres represent carbon atoms.

Compare the displayed and skeletal formulae for methanoic acid (Figure 11.7). You will note that it has been necessary to include the hydrogen atom attached to the oxygen atom as part of the formula.

The hydrogen atom must be placed at the end to show that there is only one carbon in the molecule. Note also that all of the bonds are shown in the skeletal formula. The OH group is represented as —O—H and not just as —OH. To illustrate this point, compare the skeletal formula of methanoic acid with that of ethanoic acid.

Carbon—carbon double bonds are also represented in skeletal formula as shown in Figure 11.8.

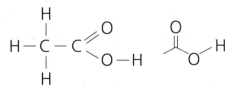

Figure 11.7 The displayed and skeletal formulae for methanoic acid and for ethanoic acid.

Figure 11.8 The skeletal formulae for hex-3-ene which has the structural formula $CH_3CH_2CHCHCH_2CH_3$.

EXAMPLE 1

Draw the following formulae for the compound which has the structural formula $CH_2CHCH(OH)CH_2CHCH_2$.

1 displayed formula
2 skeletal formula
3 molecular formula
4 empirical formula

Answers

1 Displayed formula
Remember the bond between the oxygen atom and the hydrogen atom.

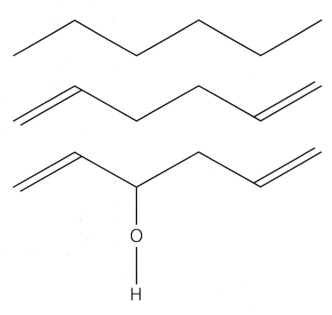

2 Skeletal formula
Draw the longest unbranched chain initially using single bonds only. There are six carbon atoms in the chain therefore there are five bonds.

Then add in the double bond positions.

Then add in any side groups.

3 Molecular formula
The molecular formula is $C_6H_{10}O$. It is not C_6H_9OH. This is a condensed structural formula showing the functional group is a —OH group.

4 Empirical formula
In this case the molecular and empirical formula are the same.

Table 11.6 Examples of general formulae for different homologous series.

Homologous series	General formula
Alkanes	C_nH_{2n+2}
Alkenes	C_nH_{2n}
Alcohols	$C_nH_{2n+1}OH$
Halogenoalkanes	$C_nH_{2n+1}X$, where X is a halogen

General formulae

A **general formula** is a type of empirical formula that represents the composition of any member of a homologous series. The general formula for the homologous series of alkenes is C_nH_{2n}. So for an alkene which has 10 carbon atoms then n = C = 10 and H = 2n = 20). The molecular formula of this alkene is therefore $C_{10}H_{20}$.

The general formula of the homologous series of alkanes is C_nH_{2n+2}. For an alkane which has 6 carbon atoms, the number of hydrogen atoms can be calculated $(2 \times 6 = 12) + 2 = 14$.

Therefore the molecular formula of this alkane is C_6H_{14}.

EXAMPLE 2

The general formula of the homologous series of alcohols is $C_nH_{2n+1}OH$.

What is the molecular formula of an alcohol which has seven carbon atoms?

Answer

n = 7

2n + 1 = 14 + 1 = 15

Therefore the formula = $C_7H_{15}OH$

TEST YOURSELF 3

1 The structural formulae of three molecules X, Y and Z are given below

$X = CH_3CH(OH)CH_2CH_3$

$Y = CH_3CH_2CH(CH_3)CH_3$

$Z = CH_3C(CH_3)CHCH_3$

For each of the molecules provide the following formulae.

a) displayed formula

b) skeletal formula

c) molecular formula

d) empirical formula

2 Use the general formulae provided in Table 11.6 to calculate the molecular formulae for the following compounds.

a) an alcohol with two carbon atoms

b) an alkene with 16 carbon atoms

c) an alkane with eight carbon atoms

Figure 11.9

INTERNATIONAL UNION OF PURE AND APPLIED CHEMISTRY

Table 11.7 Root names used to indicate the number of carbon atoms in the longest chain.

Root name	Number of carbon atoms
meth	1
eth	2
prop	3
but	4
pent	5
hex	6

Nomenclature – naming organic compounds

Chemists use a standard system for the naming of organic chemicals. The system has been developed by The International Union of Pure and Applied Chemistry, IUPAC.

The following steps are involved in the naming of an organic compound

Step 1   Count the number of carbons in the longest continuous chain and name it (Table 11.7).

This is followed by a syllable which describes the bonding in the chain, —ane or —ene. —ane means there are only single bonds between the carbon atoms, —ene means there is a carbon–carbon double bond present.

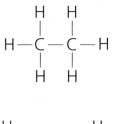

two carbon atoms   no double bonds present

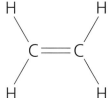

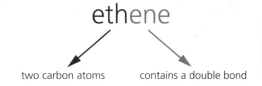

two carbon atoms   contains a double bond

Figure 11.10

Organic compounds do not always have straight chains of carbon atoms, they can be in a ring. These compounds are known as cyclic compounds. They have the prefix 'cyclo' in the name.

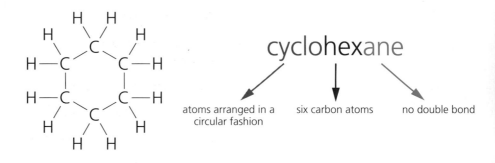

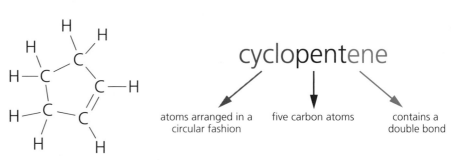

Figure 11.11

Step 2 Identify each side chain (or substituent) attached to the main carbon chain and name them.

The name of the side chain (or substituent) is added as a **prefix**. A list of common side chains and substituents with their applied prefix are provided in Table 11.8. The structural formulae for bromoethane and methylpropane illustrate this step.

Table 11.8 Prefixes used to name side groups or substituents attached to a carbon chain.

Prefix	Side chain group or substituent
methyl	$-CH_3$
ethyl	$-CH_2CH_3$
propyl	$-CH_2CH_2CH_3$
fluoro	$-F$
chloro	$-Cl$
bromo	$-Br$
iodo	$-I$

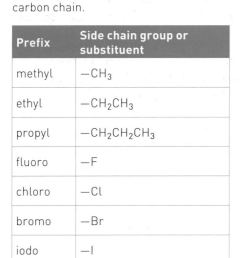

Figure 11.12

Step 3 Identify where each side chain or substituent group is attached and indicate its position along the chain by adding a number to the name.

Locant is the term used for the number which indicates the position of a side chain or substituent group within a molecule as illustrated in 2-bromobutane.

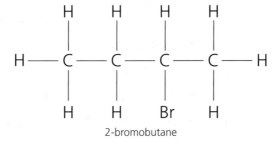

2-bromobutane

Figure 11.13

Variations to be aware of:

1 If there are two or more side chains or substituent groups, the following rules apply.

   a) Names are placed in alphabetical order.

   b) A separate number is needed for each side chain or group.

   c) Hyphens (-) are placed between numbers and letters separating them.

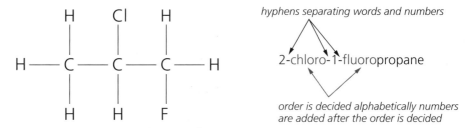

Figure 11.14

In the example in Figure 11.14, the '1' (the locant) within the name indicates that the fluorine atom is covalently bonded to the 1st carbon i.e. at the end of the chain. The '2' indicates that the chlorine is attached to the 2nd carbon in a chain of three.

The name 2-chloro-3-fluoropropane would be incorrect for this compound because the IUPAC standard is to use **the lowest number sequence possible**. Note that to name it 1-fluoro-2-chloro would also be incorrect as the names of the groups must be placed in alphabetical order.

2 If there are two or more of the **same** side chains or substituent groups present these compounds require an additional prefix. The prefix and number of chain or groups that each relates to are shown in Table 11.9.

Prefix	Number of the same side chains or groups present
di	two
tri	three
tetra	four
penta	five
hexa	six

Table 11.9 Prefixes and the number of groups or side chains they represent.

To correctly name a molecule with two or more of the same side chains or substituent groups, the following rules apply:

1 Apply the prefixes.

2 Give each group a number.

3 Separate the numbers with commas.

4 Prefixes do not affect the alphabetical order.

Review the structural formulae for 2,3-dimethylpentane and 1,1,2-trichloro-2,5-difluoro-3,4-dimethylhexane shown in Figure 11.15.

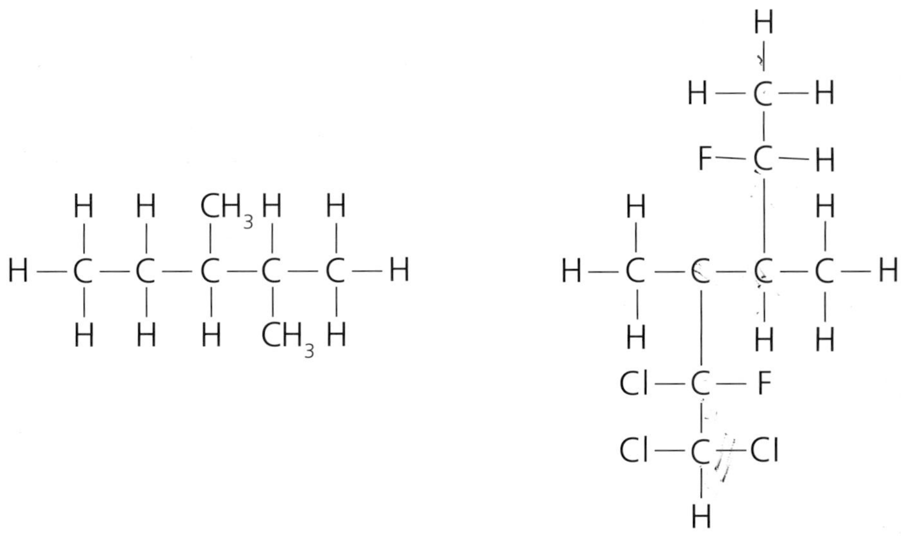

each methyl group requires a number

2,3-dimethylpentane

prefix indicates two methyl groups present

1,1,2-trichloro-2,5-difluoro-3,4-dimethylhexane

The longest chain has six carbons (marked in red).

The groups attached to the longest carbon chain are written in alphabetical order; chloro, fluoro and methyl;

The numbers used are the lowest combination possible.
The prefixes indicating the quantity of side groups eg di and tri, do not affect the alphabetical order.

Figure 11.15

Finally, the positions of double bonds are also indicated with numbers., as shown in Figure 11.16.

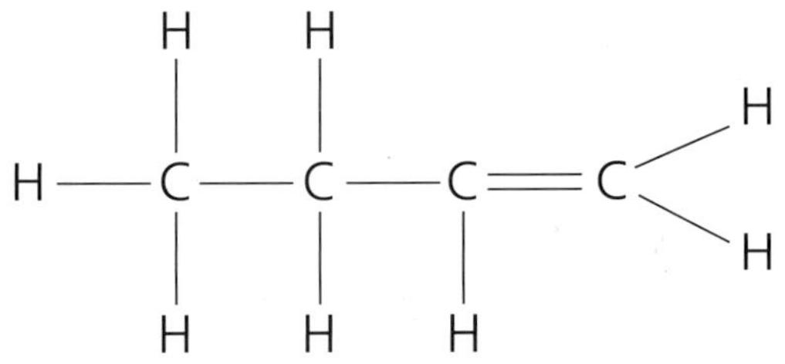

but-1-ene

double bond is between carbon 1 and carbon 2

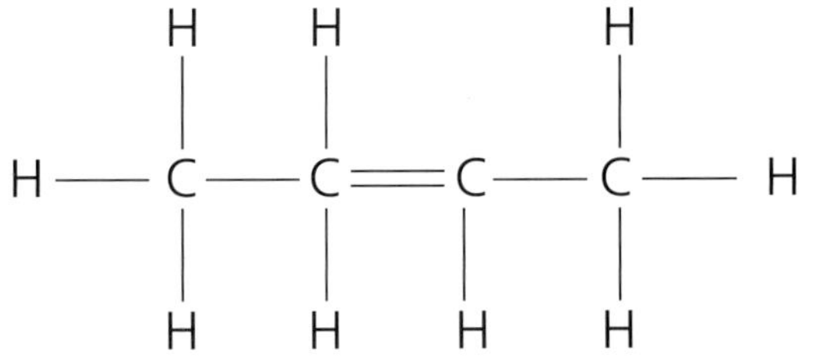

but-2-ene

double bond is between carbon 2 and carbon 3

Figure 11.16

Naming compounds with functional groups

There are a number of functional groups to be aware of. A selection are shown in Table 11.10.

Table 11.10 Common functional groups.

Homologous series	Functional group	Suffix	Example
carboxylic acid	(C with =O and OH)	-oic acid	propanoic acid
ester	(C with =O and O—)	-oate	ethyl ethanoate
acyl chloride	(C with =O and Cl)	-oyl chloride	butanoyl chloride
nitrile	—C≡N	-nitrile	propanenitrile
aldehyde	(C with =O and H)	-al	ethanal
ketone	(C with =O)	-one	propanone
alcohol	—OH	-ol	butanol
amine	—NH$_2$	-amine	ethylamine
alkene	C=C	-ene	propene
halogenoalkane	—C—X, X = halogen	named as a substituted hydrocarbon	1-bromobutane, chloroethane

Increasing priority when naming (arrow pointing up)

The functional group is usually named as part of the longest unbranched carbon chain by adding a suffix, a group of letters at the end of the name.

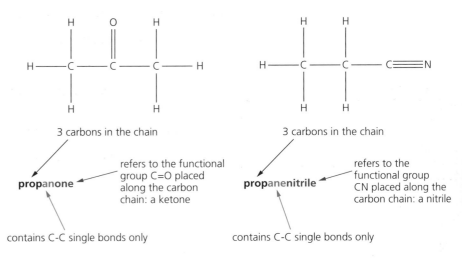

3 carbons in the chain

propanone

refers to the functional group C=O placed along the carbon chain: a ketone

contains C-C single bonds only

3 carbons in the chain

propanenitrile

refers to the functional group CN placed along the carbon chain: a nitrile

contains C-C single bonds only

Figure 11.17

Note propanone is **not** named propaneone. The letter 'e' is removed from the name of the chain when there are two vowels together.

Propanenitrile, is **not** propannitrile. The 'e' in 'ane' is not removed as it is not next to a vowel.

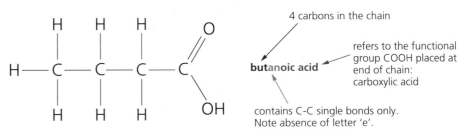

4 carbons in the chain

butanoic acid

refers to the functional group COOH placed at end of chain: carboxylic acid

contains C-C single bonds only. Note absence of letter 'e'.

Figure 11.19

When a compound contains **two** different functional groups, one is named as part of the unbranched chain and the other as a substituent. The group with the highest nomenclature priority is named as part of the chain as shown in Figure 11.20.

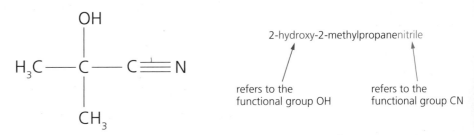

2-hydroxy-2-methylpropanenitrile

refers to the functional group OH

refers to the functional group CN

Figure 11.20

You will not be expected to remember priority at this level of study. It is included in Table 11.10 for information only.

You do not have to learn all the names of the different functional groups included in the course at this stage. You will become familiar with them as you study the chemical reactions and the physical properties of different homologous series.

Figure 11.18 Butanoic acid is responsible for the smell of rancid butter and also gives vomit its unpleasant smell. Perhaps not surprisingly, this acid is used in stink bombs.

In summary, the sequence for naming organic compounds is as follows:

1 Name the longest unbranched carbon chain.

2 C—C or C=C in chain?

3 Add suffix for functional groups if they are present. Decide if the letter 'e' should be removed or not.

4 Add prefix for substituent or side groups.

5 Add numbers to indicate the position of functional groups and side groups as appropriate.

TIP

It can be useful to draw the displayed formula to find the longest unbranched chain.

EXAMPLE 3

Name the organic compound which has the structural formula $CH_3CH_2CH(CH_3)CH_2Br$.

Answer

Step 1   Name the longest unbranched carbon chain (in red). There are four carbon atoms in the longest chain and only single bonds, therefore the 'ane' suffix is required: butane.

Step 2   Add a prefix for any side groups (circled). There are two side groups. They have to be placed in alphabetical order, therefore bromo comes before methyl. The name so far is bromomethylbutane.

Step 3   Add locants to indicate the position of the side groups. The lowest numbering system begins on the right of the structure 1-bromo-2-methyl, rather than 4-bromo-3-methyl. This compound is 1-bromo-2-methylbutane.

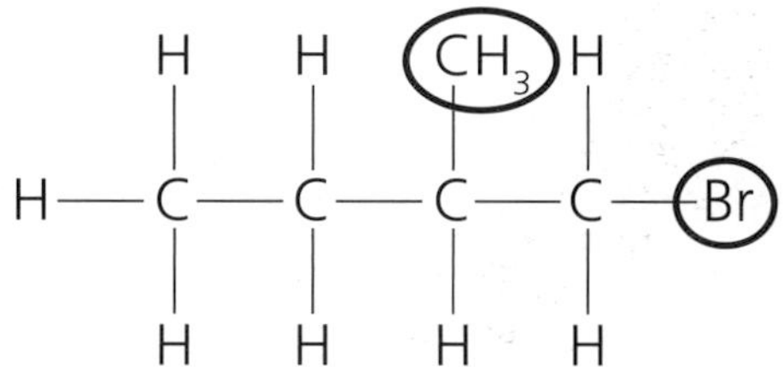

Figure 11.21

Figure 11.22

EXAMPLE 4

What is the IUPAC name for the following compound? The formula used is not a true displayed formula, but it will make the identification of the longest unbranched chain a little easier.

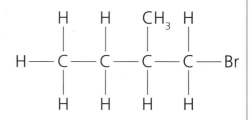

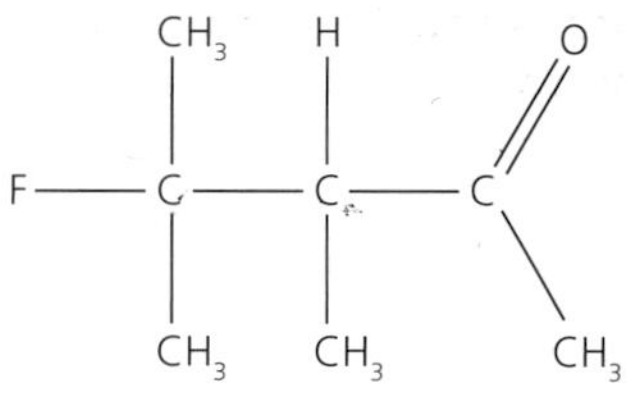

Figure 11.23

Answer

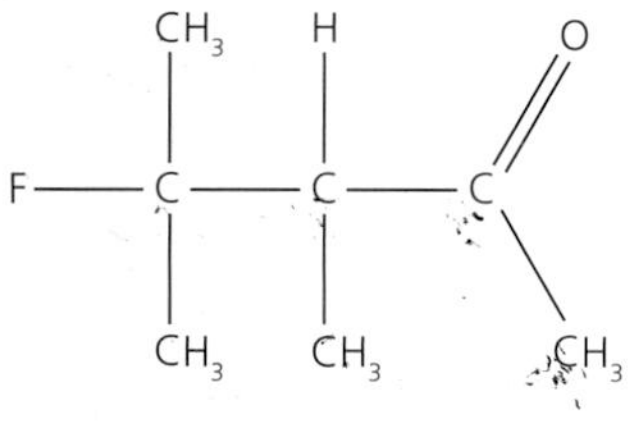

The longest unbranched chain has 5 carbon atoms: 'pent'

There are only single C—C bonds in the chain: pentane

Figure 11.24

257

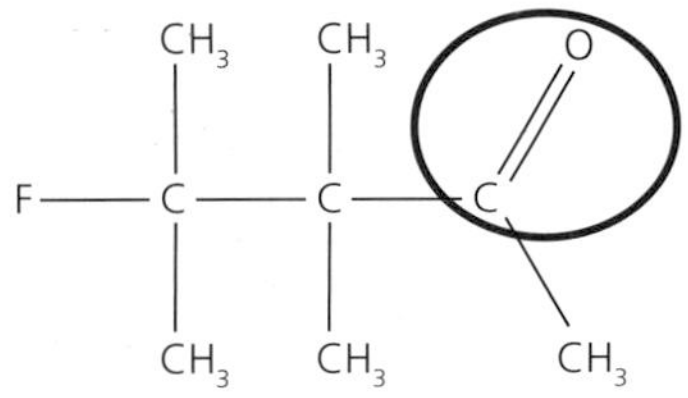

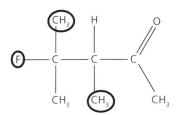

Figure 11.25

This molecule has one functional group; this is a ketone. A suffix is added to name the functional group; which is 'one' (pronounced own). The letter 'e' from 'ane' is dropped. The name so far is pentanone.

Figure 11.26

There are three side groups: two methyl groups and one fluoro group. These are named in alphabetical order, i.e. **f**luoro di**m**ethyl.

So, the name at this point is fluoro dimethyl pentanone.

Finally add the numbers to indicate the position of the side group. In this case a functional group is also present. It is numbered as part of the chain. This functional group will be numbered as the lowest possible number.

The compound is **4-fluoro-3,4-dimethylpentan-2-one**

Note the numbers are separated from each other by a comma and from words by a hyphen. There are no spaces in the name.

EXAMPLE 5

Name the compound which is represented by this skeletal formula.

Answer

Cycloalkanes and cycloalkenes can be treated as the longest unbranched chain and the groups connected to it are named as side groups.

This compound is, therefore, ethylcyclohexane.

Figure 11.27

TEST YOURSELF 4

1 a) What is the IUPAC name for each of the organic compounds listed?
  i) $CH_3CH_2CH_2Br$
  ii) $CH_3CHClCH_3$
  iii) CH_2CHCH_2I
  iv) $CH_3CHCHCH_3$
  v) $CH(CH_3)_2CH_2CH_2OH$
  vi) $CH(CH_3)_2CH(OH)CH_2CH_2CH_3$
  b) Explain why compound (iv), is the only hydrocarbon in the list.
2 Draw the displayed formula for the following compounds.
  a) propene
  b) butan-2-ol
  c) 2-chloro-1,1-difluoropropene
  d) pentan-2-ol
  e) hexa-2,4-diene
  f) but-2-en-1-ol
  g) 1,2,2-triiodocyclohexane

TIP
In part g) there are two vowels placed together in this name. Note that the only vowel which can be removed from compound names is the 'e' from 'ane' or 'ene'. In part e), a diene has 2 double bonds.

Figure 11.29 Limonene takes its name from the word lemon, as lemon rind, and the rind of other citrus fruits, contain considerable amounts of this compound, which contributes to their characteristic odour.

ACTIVITY

Preparation of limonene

Limonene is an unsaturated hydrocarbon with the structure shown below.

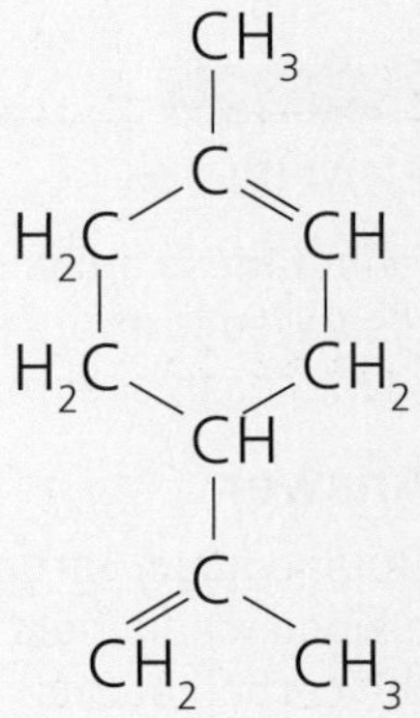

Limonene is present in orange peel and can be extracted using the following procedure.

- Blend together the peel of three oranges and $70\,cm^3$ of water. Distil the resultant solution until $30\,cm^3$ of distillate are obtained.
- Transfer the distillate into a separating funnel and shake with 1,1,1-trichloroethane.
- 1,1,1-trichloroethane has a density of $1.3\,g\,cm^{-3}$. It is run off into a beaker containing solid anhydrous sodium sulfate.
- The anhydrous sodium sulfate is then removed and the 1,1,1-trichlorethane is distilled off to leave limonene.

Figure 11.28

1 What is the molecular formula of limonene?
2 What is the empirical formula of limonene?
3 Draw the displayed formula of 1,1,1-trichlorethane.
4 Explain why limonene dissolves in 1,1,1-trichloroethane but is insoluble in water.
5 In the separating funnel, two layers form. State, and explain, in which layer the 1,1,1-trichloroethane is present.
6 The peel of three oranges had a mass of 120.0 g. Calculate the percentage by mass of limonene in the orange peel if $1.2\,cm^3$ of limonene (density $= 0.8\,g\,cm^{-3}$) was obtained.
7 Suggest why all of the limonene present in the orange peel was not obtained.

Isomerism

Compounds that have the same molecular formula but a different structural formula are known as **structural isomers**.

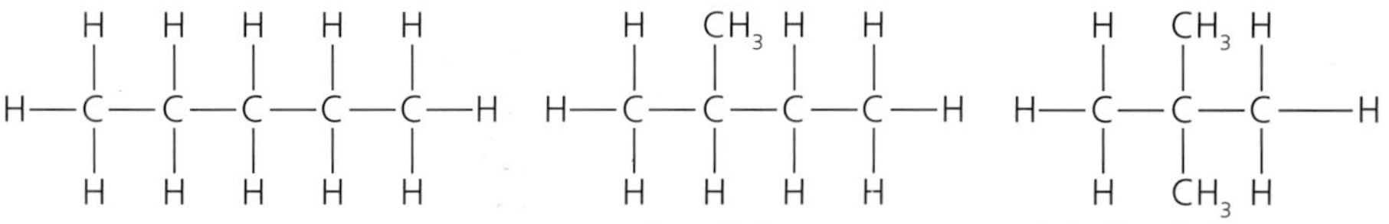

pentane   2-methylbutane   2,2-dimethylpropane

Figure 11.30

C_5H_{12} can have three possible structural formulas; it has three isomers, as shown in Figure 11.30.

There are three types of structural isomers: chain, positional and functional group isomers.

Chain isomerism

Chain isomerism occurs when there is more than one way of arranging carbon atoms in the longest chain. C_5H_{12} exists as three chain isomers as illustrated in Figure 11.30.

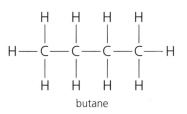

butane

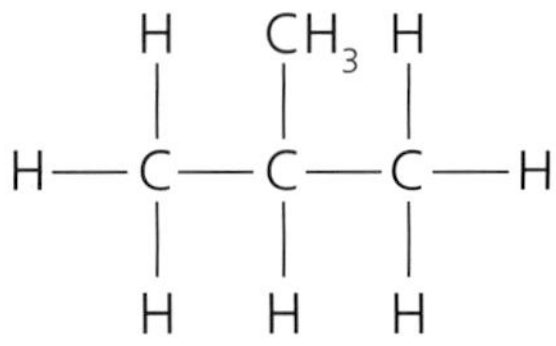

2-methylpropane

Figure 11.31 There are two structural isomers of C_4H_{10}.

Chain isomers have similar chemical properties but slightly different physical properties. The more branched the isomer, the weaker the van der Waals' forces between molecules and the lower the boiling point.

EXAMPLE 6

Isobutane is a chain isomer of butane. Draw the displayed formula of isobutane and state the IUPAC name for this isomer.

Answer

Butane has 4 carbon atoms and 10 hydrogen atoms in a single chain held together by single bonds between the carbon atoms.

Isobutane is a chain isomer of this molecule. It will therefore have the same number of carbon atoms and hydrogen atoms but one of the carbon atoms will form a branch from the main chain.

The compound shown in Figure 11.33 has the same number of carbon atoms and hydrogen atoms as butane but only three carbon atoms are in the chain. The fourth carbon atom forms a branch in the chain. The IUPAC name for this chain isomer is 2-methylpropane.

Figure 11.32 Isobutane and butane are examples of chain isomers. Butane is used as a fuel in gas hair straighteners and curlers.

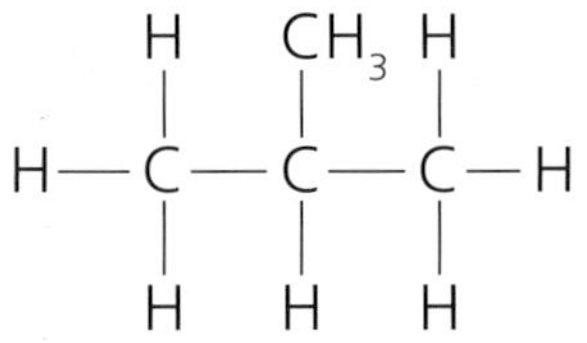

Figure 11.33

TIP

2-methylpropane may be called methylpropane as there is only one carbon (2) where the methyl group can be positioned.

Positional isomerism

Positional isomers have the same carbon chain and the same functional group but it is attached at different points along the carbon chain.

C_3H_8O exists as two isomers: the alcohols, butan-1-ol and butan-2-ol. These compounds differ only in the position of the —OH group (Figure 11.34).

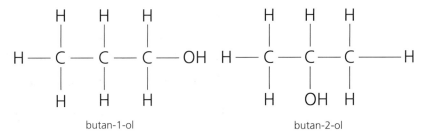

butan-1-ol   butan-2-ol

Figure 11.34

Functional group isomerism

Functional group isomers are compounds with the same molecular formula that have different functional groups. A compound with the molecular formula C_3H_6O could either belong to the homologous series of aldehydes or ketones as illustrated by propanal and propanone in Figure 11.35.

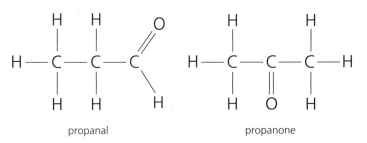

propanal   propanone

Figure 11.35

Cyclohexane, a cycloalkane, is a functional group isomer of hexene, an alkene. Both compounds have the molecular formula C_6H_{12} but only hexene has a carbon—carbon double bond.

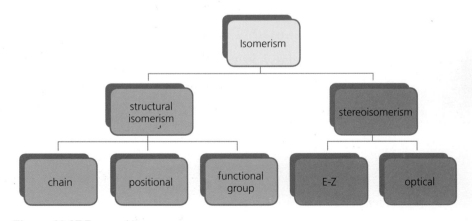

Figure 11.36

Esters and alkanoic acids can be functional group isomers. For example, propanoic acid and methyl ethanoate are isomers of $C_3H_6O_2$.

TIP

It is useful to remember that cycloalkanes are functional group isomers of alkenes. It can also help to draw the displayed formula, or the skeletal formula, of a compound before you begin to look for the isomer.

TEST YOURSELF 5

1 C_6H_{14} has five isomers.
   a) What is meant by the term isomer?
   b) Draw and name all isomers of C_6H_{14}.
   c) What type of structural isomerism is shown by the five isomers?
2 What type of structural isomerism is shown in the following pairs of compounds.
   a) Pentene and cyclopentane
   b) $CH_3CH(OH)CH_2CH_3$ and $CH_2(OH)CH_2CH_2CH_3$
   c) $CH_3CH(OH)CH_2CH_3$ and $CH_3C(CH_3)(OH)CH_3$

Stereoisomerism

Stereoisomers are molecules which have the same molecular and structural formula but a different arrangement of atoms in 3D space.

There are two types of stereoisomerism. E-Z isomerism which you will study in Chapter 14 and optical isomerism which you will study in Year 2. Figure 11.37 summarises the key types of both structural isomerisation and stereoisomerism.

```
                          Isomerism
                    ┌──────────┴──────────┐
              structural              stereoisomerism
              isomerism
        ┌─────────┼─────────┐         ┌──────┴──────┐
      chain   positional  functional  E-Z        optical
                          group
```

Figure 11.37 Types of isomers.

Other general terms used in organic chemistry

Aliphatic compounds are hydrocarbon molecules which have single, double or triple bonds. The carbon chain can be straight, branched or cyclic. Chemists use the letter R as short-hand to indicate the aliphatic group, e.g. an alcohol molecule (ROH) or a carboxylic acid molecule (RCOOH).

Aromatic compounds are organic molecules which contain a delocalised ring of electrons. The simplest aromatic compound is benzene. The chemistry of benzene is covered in Year 2.

Practice questions

1 Which one of the following is the IUPAC name for the hydrocarbon shown in Figure 11.38 below? *(1)*

A 2-ethyl-3-methylbutane

B 2,3-dimethylpentane

C 2-methyl-3-ethylbutane

D 2,3-dimethylheptane

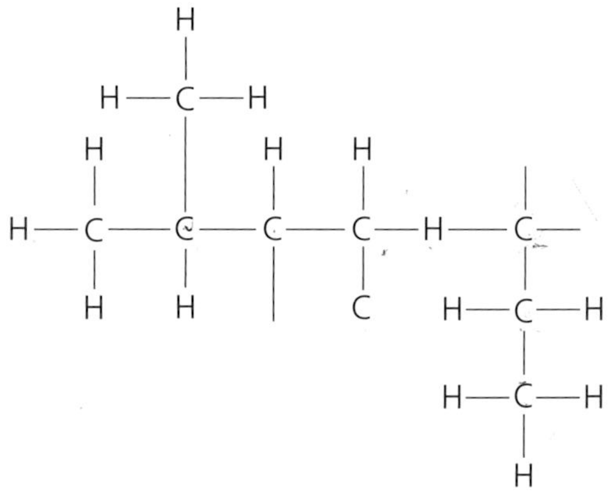

Figure 11.38

2 Which of the following is the IUPAC name for the molecule shown in Figure 11.39 below? *(1)*

A 4-methylpent-3-ene-2-ol

B 2-methyl-4-hydroxy-pent-2-ene

C 4-hydroxy-2-methylpent-2-ene

D 4-methylpentan-2-ol

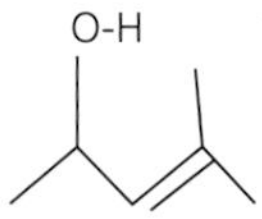

Figure 11.39

3 Ethanoic acid belongs to the homologous series of carboxylic acids as it has a COOH functional group.

   a) Explain the meaning of:

   i) homologous series *(1)*

   ii) functional group *(1)*

   iii) structural formula. *(1)*

   b) Write the empirical formula and the displayed formula for ethanoic acid. *(2)*

4 Refer to Table 11.6 on page 250:

   a) Decide in which homologous series each of the following compounds belong:

   i) $CH_3CH(OH)CH_3$ *(1)*

   ii) CH_3CH_2CHO *(1)*

   iii) $CH_3CH_2COCH_3$ *(1)*

   iv) $CH_3CH=CHCH_2CH_3$ *(1)*

   v) $CH(CH_3)_2CH_2COOH$ *(1)*

   vi) $CH_3CH_2NH_2$ *(1)*

   vii) CH_3CH_2CN. *(1)*

   b) Name each of the compounds (i–vii) in part (a). *(7)*

5 The nine molecules (A–I) are displayed below using three different types of molecular formulae.

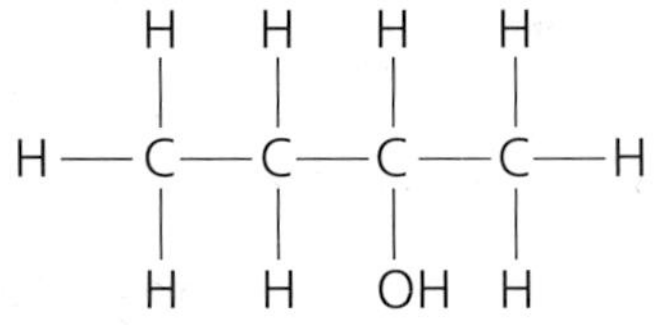

A

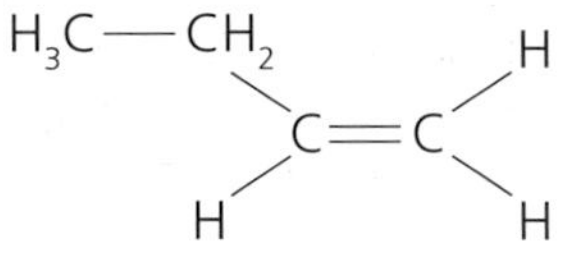

```
    H   H   H   H
    |   |   |   |
H — C — C — C — C — H
    |   |   |   |
    H   H   OH  H
```
A

```
H₃C — CH₂      H
          \   /
           C=C
          /   \
         H     H
```
B

```
    H   H   CH₃ H
    |   |   |   |
H — C — C — C — C — H
    |   |   |   |
    H   H   H   H
```
C

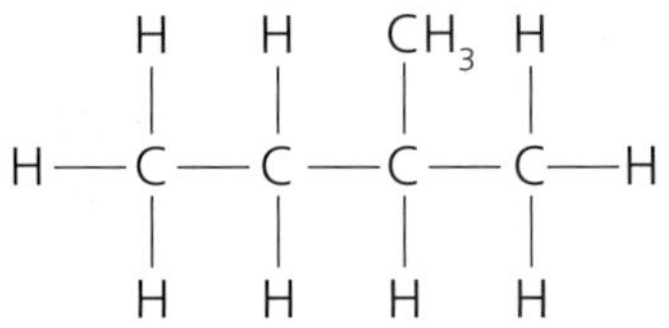

```
H₃C        H
    \      /
     C = C
    /      \
   H    D   CH₃
```
D

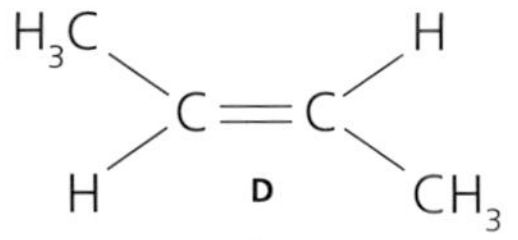

CH₃C(CH₃)(OH)CH₃

E

CH₃CH₂CHCH₂

F

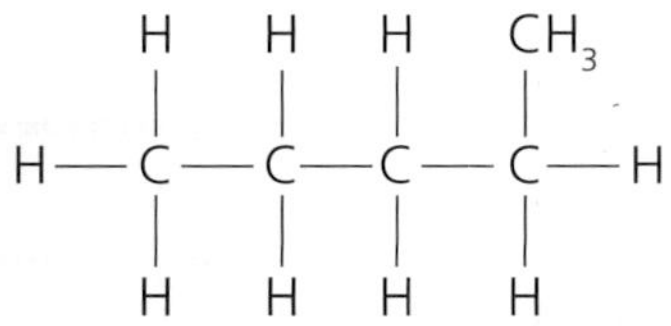

```
    H   H   H   CH₃
    |   |   |   |
H — C — C — C — C — H
    |   |   |   |
    H   H   H   H
```
G

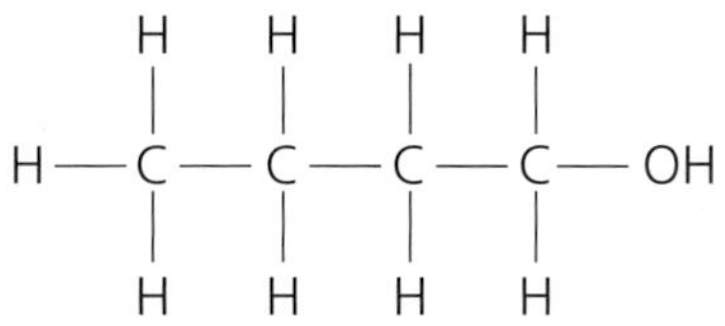

```
    H   H   H   H
    |   |   |   |
H — C — C — C — C — OH
    |   |   |   |
    H   H   H   H
```
H

```
    H   H   CH₃
    |   |   |
H — C — C — C — H
    |   |   |
    H   CH₃ H   H
```
I 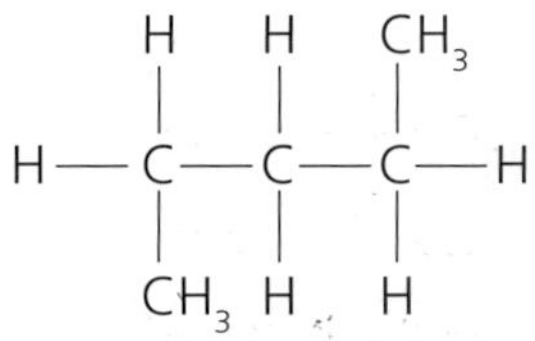

a) Using the letters A–I, identify:

  i) three compounds that are isomers   *(1)*

  ii) two compounds that are identical   *(1)*

  iii) two compounds that are chain isomers   *(1)*

  iv) two compounds that are positional isomers.   *(1)*

b) Name each substance (A–I)   *(3)*

12 Alkanes

TEST YOURSELF ON PRIOR KNOWLEDGE 1

1 Name and explain the origin of the intermolecular forces between methane molecules.
2 The skeletal formula and the displayed formula for a straight chain isomer of a molecule with four carbon atoms is shown below.
   a) Name the molecule.
   b) Draw the displayed formula and skeletal formula for all of the isomers of this molecule.

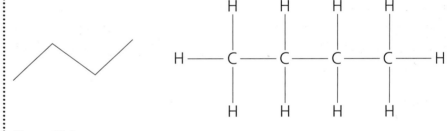

Figure 12.1

Alkanes are a homologous series of saturated hydrocarbons. The alkanes are classed as **hydrocarbons** as they contain atoms of hydrogen and carbon *only*. Each successive member differs from the next by a $-CH_2$ group. This is true of any homologous series of compounds. Each carbon atom in an alkane has four single bonds. The molecule is said to be **saturated** as each carbon atom has the maximum number of single bonds. A saturated hydrocarbon is one which contains only single bonds between the carbon atoms.

A **hydrocarbon** is a compound which contains hydrogen and carbon atoms only.

A **saturated hydrocarbon** is one which contains only single bonds between the carbon atoms.

Figure 12.2 One cow can release 500 dm³ of methane a day! Cows and sheep are thought to be responsible for one fifth of global methane production.

The name, molecular formula and displayed formula for the first four members of the alkane homologous series are given in Table 12.1.

The general formula of the alkanes is:

$$C_nH_{2n+2}$$

where n is the number of carbon atoms.

Methane is the shortest chain alkane with just one carbon atom. It is produced naturally as a product of some forms of anaerobic respiration which occur in the gut of ruminant mammals, such as cows and sheep, and in compost heaps, rice fields and landfill waste tips.

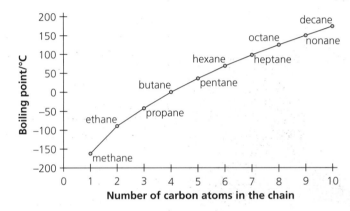

Figure 12.3 The boiling points of straight chain alkanes increase with increasing relative molecular mass. Shorter chain alkanes are gases at room temperature. C_5–C_{17} are liquids and the longer chain alkanes are waxy solids.

Table 12.1 Successive members of the homologous series of the alkanes.

Name	Molecular formula	Displayed formula (simplest straight chain structure)	Structural formula
Methane	CH_4	H—C—H (with H above and below)	CH_4
Ethane	C_2H_6	H—C—C—H (with H above and below each C)	CH_3CH_3
Propane	C_3H_8	H—C—C—C—H (with H above and below each C)	$CH_3CH_2CH_3$
Butane	C_4H_{10}	H—C—C—C—C—H (with H above and below each C)	$CH_3CH_2CH_2CH_3$

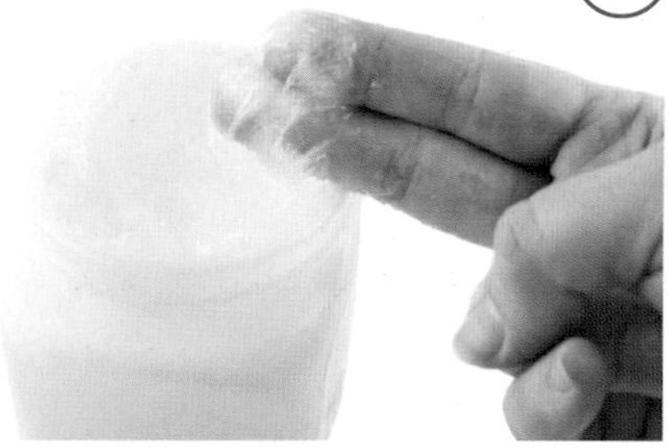

Figure 12.4 Paraffin wax, such as Vaseline, is a mixture of solid alkanes, carbon chain length ≥ C_{25}. It can be applied to cuts and grazes as a physical barrier against entry of bacteria or to lips and hands to prevent chapping.

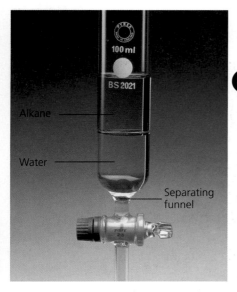

Figure 12.5 This channel swimmer is covered in a mixture of lanolin and paraffin wax, which is a mixture of long chain alkanes. The waxy solid prevents friction and chaffing.

Figure 12.7 Mixtures of alkanes and water can be separated in a separating funnel as they are immiscible.

The physical properties of the alkanes

Alkane molecules are non-polar. The difference in electronegativity between the carbon atom and the hydrogen atom is extremely small. This means there is only one type of intermolecular force between alkane molecules, van der Waals' forces. The strength of van der Waals' forces between molecules increases as the relative molecular mass of the molecule increases and as a result the melting point and boiling point increase.

The graph in Figure 12.3 is drawn using the boiling points of the straight chain isomers only. The branched chain isomers have lower melting and boiling points than the straight chain molecules with the same number of Waals carbon atoms. Molecules with branched chains have a smaller surface area in contact with each other. This decreases the strength of the van der Waals forces between the molecules. Branched molecules do not fit together as closely as straight chain molecules and this too decreases the strength of van der Waals forces. In Figure 12.6, the boiling points of a straight chain and a branched chain isomer are compared.

molecular formula C_4H_{10}

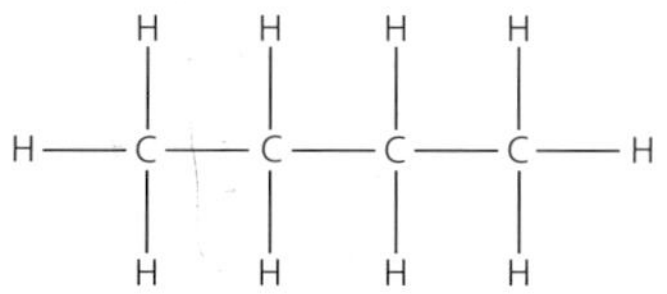

butane – a straight chain isomer of C_4H_{10}
boiling point 0°C

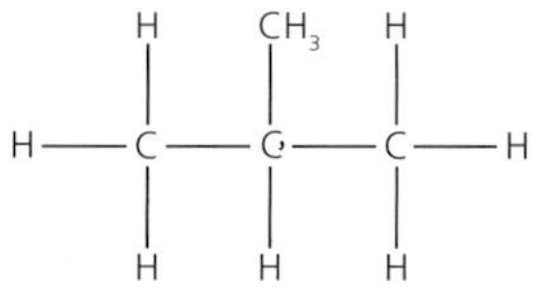

2-methylpropane – a branched isomer of C_4H_{10}
boiling point −12°C

Figure 12.6 Chain branching affects physical properties of a molecule such as boiling point.

Alkanes are insoluble in water but will dissolve in other non-polar liquids such as hexane and cyclopentane. Mixtures of alkanes are separated by fractional distillation while liquid alkane and water mixtures can be separated in the laboratory using a separating funnel (Figure 12.7).

TEST YOURSELF 2

1 An alkane molecule is known to contain 14 carbon atoms.
  a) What is the molecular formula of the alkane?
  b) Why is this compound described as a hydrocarbon?
2 a) Draw the displayed formula for the following molecules.
    i) pentane
    ii) 2,2-dimethylpropane
    iii) butane
  b) Explain why pentane has a higher boiling point than
    i) butane
    ii) 2,2-dimethylpropane
3 Alkanes are a homologous series of hydrocarbons. Explain, using examples, what is meant by a homologous series.

Figure 12.8 Crude oil.

Fractional distillation is the continual evaporation and condensation of a mixture causing the components to separate because of a difference in their boiling points.

A fraction is a group of compounds that have similar boiling points and are removed at the same level of a fractionating column.

Source of organic chemicals

Crude oil (Figure 12.8) is a dark yellow to black, sticky, viscous liquid which is found deposited in rock formations below the surface of the earth. It is a rich and varied mixture of over 150 carbon-based compounds and is the world's main source of organic chemicals. It is formed from the remains of plants and animals, which over many millions of years became covered in mud, silt and sand. High pressures and temperatures change the mud, sand and silt to rock and the animal and plant remains complete their slow transformation to oil. The compounds in crude oil are separated by fractional distillation.

Crude oil (**petroleum**) is a mixture containing mainly alkane hydrocarbons that can be separated by fractional distillation

Fractional distillation of crude oil

Fractional distillation is the continual evaporation and condensation of a mixture causing the components to separate because of a difference in boiling point. Fractional distillation does not separate the crude oil mixture into individual compounds; instead several less complex mixtures are formed. These distillates are known as fractions. A fraction is a group of compounds which have similar boiling points and are removed at the same level of a fractionating column. Crude oil is heated in a furnace until a portion changes state into a vapour. The liquid-vapour mixture is passed up a fractionating tower (Figure 12.9), which is cooler at the top than the bottom. When a substance reaches a layer which is cool enough, it condenses and is piped off. The shorter chain molecules are collected at the top of the tower, where it is cooler, as they have lower boiling points.

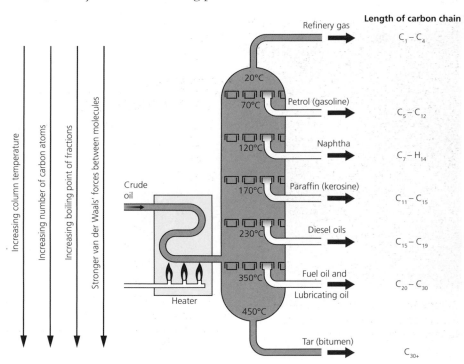

Figure 12.9 Fractional distillation of crude oil.

267

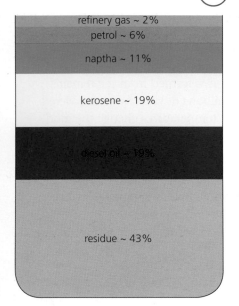

refinery gas ~ 2%

petrol ~ 6%

naptha ~ 11%

kerosene ~ 19%

diesel oil ~ 19%

residue ~ 43%

Figure 12.10 Approximate percentage composition of a sample of North Sea crude oil

....

Cracking is a process where long chain hydrocarbon molecules are broken into shorter chain molecules which are in high demand.

Industrial cracking of hydrocarbons

Figure 12.9 shows some of the fractions which are separated from crude oil. Some of the fractions are much more useful than others. Generally it is the fractions with the lower boiling point that are more useful. However, these fractions are the least abundant ones in a crude oil sample, as shown in Figure 12.10. The refinery gas, petrol and naphtha fractions are in huge demand for fuels and by the chemical industry but these fractions comprise only 19% of the sample illustrated. To meet the demand, the long chain molecules in the less useful fractions are broken down into shorter chain molecules, in a process known as **cracking**. Cracking is a process where long chain hydrocarbon molecules are broken into shorter chain molecules which are in high demand because they are more useful.

Thermal cracking

During thermal cracking, the long chain alkanes are heated to a very high temperature, typically between 1000 and 1200 K, at extremely high pressures, up to 70 atm, for a very short space of time, 1 s. These conditions produce cracking products which contain shorter chain alkanes and are rich in alkenes. The high temperature and pressures employed in thermal cracking could decompose the molecule completely and produce carbon and hydrogen. To prevent this, the time during which these conditions are applied is very short. This balanced symbol equation describes the thermal cracking of a molecule of pentadecane – an alkane with 15 carbon atoms.

$$C_{15}H_{32} \rightarrow C_7H_{16} + 4C_2H_4$$

The heptane, C_7H_{16}, produced is a major component of paints and is used in the manufacture of pharmaceuticals. The ethene is used in the manufacture of polythene.

Catalytic cracking

During catalytic cracking, the long chain alkane is heated under pressure in the presence of a zeolite catalyst. The temperature used is approximately 800–1000 K and the pressure used between 1–2 atm. A zeolite is an acidic mineral which has a honeycomb structure (Figure 12.11) made from aluminium oxide and silicon dioxide. The honeycomb structure gives it a large surface area which increases the rate of reaction. Modern catalytic cracking units are operated continuously for up to three years at a time producing a product that consists mainly of branched alkanes, cycloalkanes and aromatic compounds. This method of cracking is used to produce fuels for road vehicles.

While the temperatures employed in the catalytic cracking process are lower than those used in the process of thermal cracking, they are still high. High temperatures are needed as the carbon-carbon bonds in the alkane molecules, which are cracked, are extremely strong and therefore difficult to break.

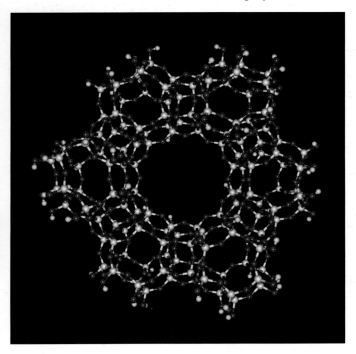

Figure 12.11 The honeycomb structure of a zeolite, which is used as a catalyst for hydrocarbon cracking.

Table 12.2 A comparison of the conditions and products of different cracking processes.

	Cracking process	
	Thermal	**Catalytic**
Temperature	1000–1200 K (high)	800–1000 K (high)
Pressure	70 atm (high)	1–2 atm (low)
Time	1 s	2–4 s
Catalyst	none	aluminosilicate zeolite
Products	high percentage of alkenes some short chain alkanes	mainly aromatic hydrocarbons and motor fuels

TEST YOURSELF 3

1 a) Explain what is meant by the following chemical terms:
   i) fractional distillation
   ii) cracking
b) Why is the cracking of hydrocarbons necessary?
c) i) State the conditions necessary for thermal cracking.
   ii) Why are the conditions applied for a short period of time?
   iii) Name the types of compounds present in the product mixture.
d) Complete the equation to show the thermal cracking of decane, $C_{10}H_{22}$.
   $$C_{10}H_{22} \rightarrow C_6H_x + \underline{} C_yH_z$$

The combustion of the alkanes

Alkanes are used as fuels as they combust readily releasing vast amounts of heat energy.

Complete combustion

Alkanes burn in a plentiful supply of oxygen to produce carbon dioxide and water only. Methane is a major component of natural gas, the gas found with oil and coal deposits.

$$\text{methane} + \text{oxygen} \rightarrow \text{carbon dioxide} + \text{water}$$
$$CH_4(g) + 2O_2(g) \rightarrow CO_2(g) + 2H_2O(g)$$

Butane is a major component of liquid petroleum gas produced by the fractional distillation of crude oil.

$$\text{butane} + \text{oxygen} \rightarrow \text{carbon dioxide} + \text{water}$$
$$C_4H_{10}(g) + 6\tfrac{1}{2}O_2(g) \rightarrow 4CO_2(g) + 5H_2O(g)$$

Figure 12.12 Carbon monoxide detector

Table 12.3 The symptoms of carbon monoxide poisoning.

mild exposure	slight headache, nausea, vomiting, unusual tiredness
medium exposure	severe throbbing headache, drowsiness, disorientation, confusion, fast heart rate
extreme exposure	unconsciousness, convulsions, cardio-respiratory failure, death

TIP

If you are asked to write an equation for reaction of propane in a limited supply of air to produce a **solid** and water only, then the equation to use is

$$C_3H_8 + 2O_2 \rightarrow 3C + 4H_2O$$

Incomplete combustion

In a limited supply of air, alkanes will burn to form water and carbon monoxide.

methane + (limited) oxygen → carbon monoxide + water

$$CH_4(g) + 1\tfrac{1}{2}O_2(g) \rightarrow CO(g) + 2H_2O(g)$$

butane + (limited) oxygen → carbon monoxide + water

$$C_4H_{10}(g) + 4\tfrac{1}{2}O_2(g) \rightarrow 4CO(g) + 5H_2O(g)$$

Carbon monoxide is a toxic, colourless and odourless gas which is often referred to as the 'silent killer'. Carbon monoxide reacts with the haemoglobin in red blood cells preventing them from carrying the oxygen to all of the cells in your body. It is absorbed 200 times faster than oxygen and is so difficult to eliminate it is classified as a cumulative poison (Table 12.3).

All gas fuelled appliances must be regularly maintained to prevent the formation of carbon monoxide due to a lack of oxygen present for combustion. Carbon monoxide detectors can warn of dangerous levels of the gas (Figure 12.12).

If the supply of oxygen is further limited solid carbon particles (soot) are formed.

methane + (very limited) oxygen → carbon + water

$$CH_4(g) + O_2(g) \rightarrow C(s) + 2H_2O(g)$$

butane + (very limited) oxygen → carbon + water

$$C_4H_{10}(g) + 2\tfrac{1}{2}O_2(g) \rightarrow 4C(s) + 5H_2O(g)$$

Combustion in an internal combustion engine

There are more than 27 million cars on the roads in the UK; the vast majority have an alkane-fuelled internal combustion engine (Figure 12.13). A small amount of fuel mixed with a large excess of air is drawn into a combustion chamber. The mixture is compressed and ignited with an extreme temperature spark. The mixture burns explosively forcing movement of the engine parts. The products of combustion exit via the exhaust. This process of intake, compression, combustion and exhaust takes place hundreds of times per

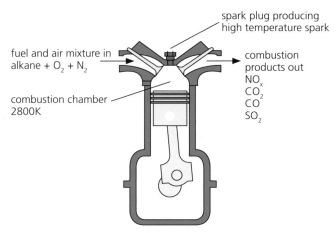

Figure 12.13 The internal combustion engine.

minute. The primary reaction facilitating the change from chemical energy to kinetic energy is the combustion of the alkane fuel in oxygen. However, the pressure and extreme temperature of the combustion chamber cause unwanted side reactions. The normally unreactive nitrogen, approximately 78% of the air intake, combines with oxygen producing a series of nitrogen oxides, mainly NO and NO_2, (NO_x).

$$\text{nitrogen} + \text{oxygen} \rightarrow \text{nitrogen(II) oxide}$$
$$N_2(g) + O_2(g) \rightarrow 2NO(g)$$

$$\text{nitrogen} + \text{oxygen} \rightarrow \text{nitrogen(IV) oxide}$$
$$N_2(g) + 2O_2(g) \rightarrow 2NO_2(g)$$

Sulfur dioxide can also be present in the exhaust mixture. The sulfur originates from impurities in crude oil which end up in the fuel.

$$\text{sulfur} + \text{oxygen} \rightarrow \text{sulfur dioxide}$$
$$S + O_2(g) \rightarrow SO_2(g)$$

For example, when 1,1-thiobisethane, a sulfur containing compound in crude oil burns, the equation is:

$$\text{1,1-thiobisethane} + \text{oxygen} \rightarrow \text{carbon dioxide} + \text{water} + \text{sulfur dioxide}$$
$$C_4H_{10}S_2 + 8\tfrac{1}{2}O_2 \rightarrow 4CO_2 + 5H_2O + 2SO_2$$

Consequently, a chemical cocktail of combustion products are released into the environment by our use of alkane-based fuels in the internal combustion engine. This mixture contains a number of pollutants including NO_x, CO, carbon particles, SO_2 and unburnt hydrocarbons (Table 12.4).

Table 12.4 Identity and source of combustion products from burning alkane-based fuels.

Compounds present in the combustion mixture	Origin	Conditions	Pollution caused
$CO_2(g)$	C in fuel compound	in excess O_2	Global warming
$CO(g)$	C in fuel compound	in limited O_2	Toxic gas:
$C(s)$	C in fuel compound	in very limited O_2	Particles exacerbate asthma
$NO_x(g)$	N_2 in the air	at extremely high temperatures	Acid rain and photochemical smog
$SO_2(g)$	S from fuel impurities		Acid rain
hydrocarbons	fuel compounds which remain unburned		React with NO_x to form ground level ozone which causes respiratory problems
H_2O	H in fuel compound		

Environmental consequences of burning alkane-based fuels

Global warming

The Earth is surrounded by an atmosphere of several layers of gas. Infrared radiation from the sun passes through these layers to reach the Earth and warms it up. Infrared radiation from the Earth travels back through the atmosphere where some of it is prevented from escaping into space by atmospheric gases, such as carbon dioxide, water and methane. The effect of trapping the energy from the sun is known as The Greenhouse Effect and the gases which cause the phenomenon are known as greenhouse gases. This is an important natural process as without it the average temperature of the atmosphere at the surface of the Earth would be 60 °C lower and the Earth would not be able to sustain life.

Table 12.5 The greenhouse gases.

The greenhouse gases	
water vapour	occurs naturally in the atmosphere
carbon dioxide	produced during respiration and as a product of combustion
methane	produced as a product of digestion by cows and sheep, and by other natural processes

TIP

Note that alkane fuels, e.g. butane, contribute to the greenhouse effect because when they combust, they produce carbon dioxide and water which are *both* greenhouse gases.

Global warming is the term given to the increasing average temperature of the atmosphere at the surface of the Earth.

Global warming is the term given to the increasing average temperature of the atmosphere at the surface of the Earth. It is caused by changing the balance of the concentration of greenhouse gases. Recent human activity has rapidly increased the concentration of carbon dioxide in the atmosphere, due to increased burning of fossil fuels. As a result, more heat from the sun is trapped and so the Earth's temperature increases.

Acid rain

Rain water is a weak acid, pH ~5.5, due to the naturally occurring carbon dioxide present in the atmosphere. Acid rain is rain water which is more acidic than this, with a pH lower than 5.5. The nitrogen oxides produced by high temperature combustion of fossil fuels contribute to acid rain but the main contributor is the sulfur dioxide gas produced when the impurities in fossil fuels are burned. The sulfur dioxide reacts with water in the air to produce sulfurous acid (sulfuric(IV) acid), H_2SO_3, which is oxidised in the air to form sulfuric acid (sulfuric(VI) acid), H_2SO_4. The overall equation can be written as:

$$SO_2(g) + H_2O(g) + \tfrac{1}{2}O_2(g) \rightarrow H_2SO_4(aq)$$

Acid rain destroys trees and vegetation, corrodes buildings and kills fish in lakes. Acid rain may fall far from the source of the polluting gases and as a consequence, spreads the effects of acid rain to many other areas. This makes acid rain a global environmental issue.

Photochemical smog

Photochemical smog is caused by pungent, toxic gases and minute solid particles suspended in the air close to the surface of the Earth. Smog is formed when nitrogen oxides (NO_x), sulfur dioxide and unburnt hydrocarbon fuels react with sunlight (Figure 12.15). It also includes carbon atoms produced when hydrocarbon fuels are burned in a very limited supply of air. Smog can form in almost any climate in industrialised cities and causes a variety of health problems from relatively minor red irritated eyes and nasal congestion to severe lung diseases such as emphysema.

Figure 12.14 Forest destruction by acid rain.

Figure 12.15 Photochemical smog lying over the Hollywood Hills in Los Angeles, USA. The brown colouration of the clouds is due to the presence of nitrogen oxides in the air.

Removal of air pollutants

Chemists continue to improve and devise methods to either remove pollutants or prevent them from entering the atmosphere.

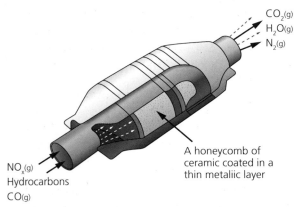

Figure 12.16 Catalytic converter.

Catalytic convertors

A catalytic convertor reduces the amount of carbon monoxide, nitrogen oxides and unburnt hydrocarbons released into the air from an internal combustion engine, by converting them into less toxic gases. It is fitted to the exhaust system of the engine. At its centre is a honeycomb of ceramic material covered in a thin layer of platinum and rhodium. The platinum and rhodium are the catalysts. The thin layer coating the honeycomb structure provides a large surface area for the reaction, increasing the rate of conversion and also ensuring that as little as possible of the expensive catalytic metals are required (Figure 12.16).

As the gases pass over the catalyst, they react with each other producing less polluting products:

nitrogen(II) oxide + carbon monoxide → carbon dioxide + nitrogen

$$2NO(g) + 2CO(g) \rightarrow 2CO_2(g) + N_2(g)$$

octane + nitrogen(II) → carbon + nitrogen + water
(unburnt   oxide   dioxide
hydrocarbon)

$$C_8H_{18}(g) + 25NO \rightarrow 8CO_2(g) + 12\tfrac{1}{2}N_2(g) + 9H_2O(g)$$

Removing sulfur

To decrease the amount of sulfur dioxide released into the atmosphere, most of the sulfur containing compounds are removed from petrol and diesel before use. However, removal of sulfur before combustion is not economically viable for fuels used in power stations. The sulfur dioxide is removed from the combustion emissions instead in a process known as flue gas desulfurisation. The gases are passed through a wet semi-solid mixture, a slurry, containing calcium oxide or calcium carbonate. Calcium oxide and calcium carbonate are bases. They **neutralise** the acidic sulfur dioxide to form calcium sulphite (calcium sulfate(IV)). This product has little commercial value but rather than dump it in waste pits, it is oxidised to calcium sulfate (calcium sulfate(VI)) which is commercially useful as a construction material.

calcium oxide + sulfur dioxide → calcium sulfite

$$CaO(s) + SO_2(g) \rightarrow CaSO_3(s)$$

calcium carbonate + sulfur dioxide → calcium sulfite + carbon dioxide

$$CaCO_3(s) + SO_2(g) \rightarrow CaSO_3(s) + CO_2$$

calcium sulfite +   oxygen   → calcium sulfate
            (from an oxidising agent)   (gypsum)

$$CaSO_3(s) + [O] \rightarrow CaSO_4(s)$$

Image labels: CO₂(g) H₂O(g) N₂(g); NOₓ(g) Hydrocarbons CO(g); A honeycomb of ceramic coated in a thin metaliic layer

EXAMPLE 1

Exhaust emissions from a petrol engine were found to contain traces of a molecule 'T' with the molecular formula C_7H_{16}.

1 Explain how this compound came to be present in the exhaust fumes.
2 Suggest other compounds which may be present in the fumes.
3 Write a balanced equation to show how 'T' is removed from the exhaust emissions, after passing through a catalytic converter.

Answers

The formula of 'T' conveys a lot of information. It is an alkene with seven carbon atoms. It belongs to the petrol fraction from the fractional distillation of crude oil. It is therefore used as a fuel in petrol engines. Some of this fuel will pass through the engine unburnt.

1 C_7H_{14} is unburnt hydrocarbon which is used as fuel for the petrol engine.
2 Oxides of nitrogen can be present (NO, NO_2) due to the combustion of nitrogen by the high temperature spark in the combustion engine. Sulfur oxides are present as the fuel may contain some sulfur-containing compounds.
3 You need to recall that the unburnt hydrocarbons are removed by reacting with nitrogen(II) oxide to produce nitrogen, carbon dioxide and water. Write the formulae in an equation form.

$$C_7H_{16} + NO \rightarrow N_2 + CO_2 + H_2O$$

To balance the equation, balance the C atoms and the H atoms followed by the O atoms before finally balancing the N atoms.

$$C_7H_{16} + 22NO \rightarrow 11N_2 + 7CO_2 + 8H_2O$$

TEST YOURSELF 5

1 a) What is the function of a catalytic converter?
  b) Write balanced symbol equations to show how the following gases are removed from vehicle exhaust fumes.
   i) CO
   ii) NO_2
   iii) NO
   iv) C_5H_{12}
  c) Why is the catalyst constructed in a honeycomb shape?
  d) State two ways in which the cost of the catalytic converter is kept to a minimum when the metals used are very expensive.

ACTIVITY

Burning petrol in car engines

In a car engine, petrol vapour and air are compressed and ignited. The compression can cause pre-ignition which results in an undesirable knocking sound from the engine. Branched chain alkanes, such as that shown in Figure 12.17, are less likely to cause pre-ignition.

Figure 12.17

1 Name the alkane in Figure 12.17.
2 State the molecular formula of this alkane.
3 Draw and name a straight chain alkane isomer of this molecule.

Figure 12.18 A car engine – an internal combustion engine – which is powered by petrol, a mixture of hydrocarbons.

275

4 Write a balanced symbol equation for the complete combustion of this alkane.

5 Explain why incomplete combustion of this alkane may produce smoke.

6 State one other environmental problem caused by the burning of petrol in a car engine.

7 The boiling point of this branched alkane is 109 °C. Suggest the boiling point of the straight chain isomer named in question 3. Explain your answer.

8 Carbon monoxide and oxides of nitrogen NO_x are formed when burning hydrocarbon fuels in car engines. Environmental damage due to these gases can be reduced if cars are fitted with catalytic converters.

What are carbon monoxide and nitrogen oxides changed into in a catalytic converter?

9 Explain in terms of energy, how a catalyst enables a reaction, such as this, to proceed more quickly.

Removal of oxides of nitrogen from diesel engine exhaust gases is problematic. One solution to aid the removal of nitrogen oxides is to inject ammonia into the exhaust gases before they enter the catalytic converter.

$$4NH_3 + 4NO + O_2 \rightarrow 4N_2 + 6H_2O$$

10 What volume of oxygen gas is needed to react completely with $1.6\,dm^3$ of nitrogen(II) oxide at 20 °C and 1 atmosphere pressure.

TIP

The chlorination of alkanes is discussed in Chapter 13, Halogenoalkanes under Synthesis and reactions of the halogenoalkanes.

Environmental chemistry is a lively and complicated area of research. Chemists must consider a myriad of factors when evaluating the impact of an action on our world. It is interesting to note that the town of Norilsk in Russia produces part of the world's supply of the catalytic metals for use in catalytic converters but is one of the ten most polluted towns on earth. The pollution has arisen from the many metallic production processes carried out in this Russian town.

Practice questions

1 How many branched chain isomers are there with the molecular formula C_6H_{14}? *(1)*

A 4   **B** 5
C 7   **D** 8

2 Hexane is an alkane.

a) State the general formula of the alkanes. *(1)*

b) Give the molecular formula for hexane. *(1)*

c) Alkane A can be cracked to form one molecule of hexane and two molecules of propene, C_3H_6. What is the molecular formula of A? *(1)*

3 Cracking is a process used in the petrochemical industry.

a) Explain what is meant by the term cracking. *(1)*

b) One type of cracking produces a high percentage of alkenes.

i) Name this type of cracking. *(1)*

ii) State the conditions used in this type of cracking. *(2)*

c) Give one economic reason why cracking is necessary. *(1)*

4 Crude oil is a complex mixture mainly consisting of hydrocarbons. It is fractionally distilled to produce less complex mixtures, known as fractions.

a) Explain why the shorter chain hydrocarbons are collected at the top of the fractionating column. *(2)*

b) The compounds from the heavier fractions can be decomposed into shorter chain molecules using a zeolite catalyst.

　i) Give two additional conditions necessary for this process. *(2)*

　ii) Describe the structure of a zeolite and suggest why this is an important feature of this type of cracking process. *(2)*

5 Octane is used as a fuel for internal combustion engines.

a) Write an equation for the incomplete combustion of octane to produce carbon monoxide and carbon in a ratio of 1:1. *(1)*

b) Why does the presence of carbon monoxide in vehicle emissions cause concern? *(1)*

c) Catalytic converters are used to remove carbon monoxide from vehicle emissions.

　i) Write an equation for the reaction between carbon monoxide and nitrogen(II) oxide, NO, that occurs in a catalytic converter. *(1)*

　ii) Name a metal which can be used as a catalyst in a catalytic converter. *(1)*

　iii) Suggest a reason, other than that of cost, why the catalytic metal is thinly coated onto a ceramic honeycomb. *(1)*

6 Butane, pentane and hexane are members of the homologous series of alkanes.

a) State two characteristics of a homologous series. *(2)*

b) The boiling points of the straight chain isomers of butane, pentane and hexane are given in the table below.

Alkane	butane	pentane	hexane
Boiling point /°C	−0.5	36	69

　i) Explain the trend in the boiling points. *(2)*

　ii) Name a process which can be used to separate these alkanes. *(1)*

c) Alkane fuels derived from crude oil may contain trace impurities which burn to give toxic gases. A toxic gas, T, is removed by passing it through calcium oxide slurry.

　i) Name the toxic gas. *(1)*

　ii) Name the type of reaction which happens between the toxic gas, T, and the calcium oxide. *(1)*

d) Butane is formed when hydrocarbon X is subjected to a cracking process. One molecule of X produces a molecule of pent-1-ene, two molecules of ethene and one molecule of butane.

　i) Identify hydrocarbon X. *(1)*

　ii) Deduce the type of cracking used. *(1)*

13 Halogenoalkanes

278

PRIOR KNOWLEDGE

- IUPAC naming system for organic compounds
- different types of formulae
- nature of intermolecular forces, especially van der Waals' forces
- polar and non-polar molecules
- structural isomerism
- definition of base
- bond enthalpy
- knowledge of the chemistry of the alkanes.

TEST YOURSELF ON PRIOR KNOWLEDGE 1

1 Alkanes are a homologous series of saturated hydrocarbons.
   a) Explain what is meant by the following chemical terms:
      i) saturated
      ii) hydrocarbon
      iii) homologous series.
2 a) Write the molecular formula of an alkane with 7 carbon atoms.
   b) Give the structural formula and skeletal formula of two chain isomers of this molecule.
3 Write a balanced symbol equation for the combustion of hexane.

The halogenoalkanes are a homologous series of saturated carbon compounds containing one or more halogen atoms (Figure 13.1). Halogenoalkanes have a wide variety of commercial uses. They are used as refrigerants, propellants, solvents, flame retardants, anaesthetics and pharmaceuticals. In recent decades many halogenoalkanes have been proven to cause pollution, which has led to the depletion of the ozone layer (Page 283).

3-bromo-1-fluorobutane

1,2-dichloropropane

Figure 13.1

Halogenoalkanes are named as a substituted alkane with the position of the halogen atom indicated by numbers where necessary (Figure 13.1).

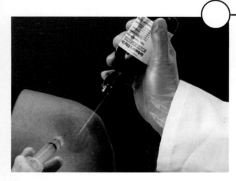

Figure 13.2 The halogenoalkane chloroethane acts as a mild local analgesic (pain killer) when sprayed onto the surface of the skin. It evaporates very quickly and cools the skin. It can be used to temporarily relieve minor sports injuries.

TIP

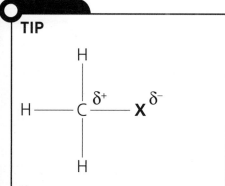

Figure 13.3

The carbon–halogen bond in a halogenoalkane molecule is polar, thus dipole-dipole interactions are present between molecules.

Physical properties of the halogenoalkanes

Halogenoalkanes contain the functional group C–X, where X is a halogen atom, F, Cl, Br or I. The general formula of the homologous series is $C_nH_{2n+1}X$.

The nature of the C–X bond in the halogenoalkanes

The C–X bond in halogenoalkanes is polar because the halogen atom is more electronegative than the C atom. The electronegativity of the halogen atom decreases as Group 7 is descended and thus the C–X bond becomes less polar (see Table 13.1).

Table 13.1 Electronegativity values for carbon and halogens.

Element	Electronegativity	Polarity of the C–X bond
fluorine	4.0	Extremely polar but not ionic
chlorine	3.0	
bromine	2.8	*decreases*
iodine	2.5	Almost non-polar

Boiling point

There are two separate types of intermolecular forces between halogenoalkane molecules; van der Waals' forces and permanent dipole–dipole interactions.

Consider first the pattern in boiling points of the chloroalkanes as the carbon chain length increases. The intermolecular force of attraction due to van der Waals' forces increases as the relative molecular mass increases, therefore the boiling point of the chloroalkanes increases as the chain length increases. This pattern is repeated for the bromoalkanes and the iodoalkanes.

Now consider the trend in boiling point when the carbon chain length is kept the same and the halogen atom is changed. Although permanent dipole–dipole interactions are greater the more polar the carbon–halogen bond, the changing van der Waals' forces have a greater effect on the boiling point of the molecule. As the relative molecular mass of the halogen increases the boiling point increases, thus the boiling point of an iodoalkane is greater than the boiling point of a bromoalkane, which is greater than the boiling point of a chloroalkane, for compounds with identical carbon chains (Figure 13.4).

Figure 13.4 The bar chart shows the boiling points of some halogenoalkanes. Three of these have boiling points below room temperature and are gases at room temperature.

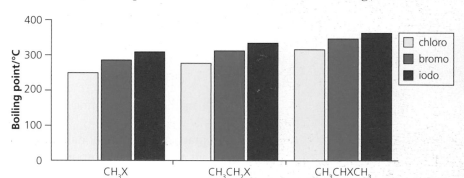

Solubility in water

The halogenoalkanes, despite the polar nature of the carbon-halogen bond, are insoluble or only very slightly soluble in water. They are soluble in organic solvents and due to their ability to mix with other hydrocarbons are used extensively as dry cleaning fluids and degreasing agents.

TEST YOURSELF 2

1 Draw the displayed formula for the following halogenoalkanes:
  i) 1,1,1-trichloroethane
  ii) 2,3-dibromobutane
  iii) 2-chloro-1-fluoropropane.

2 Draw the displayed formula for bromoethane illustrating the polarity of the carbon-bromine bond.

3 Explain why the carbon-bromine bond is polar and compare it to the carbon-chlorine bond.

4 Name the following halogenoalkanes.
  i) $CH_2BrCH_2CH_2Br$
  ii) $CH_3CHICHClCH_2CH_3$
  iii) $C_3H_7CCl_2F$
  iv) $CH_3CBr(CH_3)CH_2CH_2Br$.

5 Which of each pair has the greater boiling point?
  i) C_2H_5Br and C_2H_5F
  ii) C_2H_5Br and CH_3Br
  iii) $CH_3CH_2CH_2CH_2Br$ and $CH_3CH(CH_3)CH_2Br$

Synthesis and reactions of the halogenoalkanes

Synthesis of the chloroalkanes via the photochemical chlorination of the alkanes

Chlorine will react with methane in the presence of sunlight forming a mixture of chloroalkanes and fumes of hydrogen chloride gas.

$$CH_4 \quad + \quad Cl_2 \quad \xrightarrow{uv} \quad CH_3Cl \quad + \quad HCl$$
methane + chlorine → chloromethane + hydrogen chloride

or

$$CH_4 \quad + \quad 2Cl_2 \quad \xrightarrow{uv} \quad CH_2Cl_2 \quad + \quad 2HCl$$
methane + chlorine → dichloromethane + hydrogen chloride

or

$$CH_4 \quad + \quad 3Cl_2 \quad \xrightarrow{uv} \quad CHCl_3 \quad + \quad 3HCl$$
methane + chlorine → trichloromethane + hydrogen chloride

or

$$CH_4 \quad + \quad 4Cl_2 \quad \xrightarrow{uv} \quad CCl_4 \quad + \quad 4HCl$$
methane + chlorine → tetrachloromethane + hydrogen chloride

In practice a mixture of halogenoalkanes are produced with some longer chain alkanes. The components of the liquid mixture can be separated using fractional distillation.

The presence of the longer chain alkanes in the mixture helped chemists to formulate a mechanism for the reaction. A **mechanism** is a detailed step-by-step sequence illustrating how an overall chemical reaction occurs.

Consider the reaction between methane and chlorine to form chloromethane.

$$CH_4 \ + \ Cl_2 \ \rightarrow \ CH_3Cl \ + \ HCl$$

methane + chlorine → chloromethane + hydrogen chloride

One of the H atoms in the methane molecule has been removed and replaced by a chlorine atom. This type of reaction is called a **substitution** reaction. A substitution reaction is one in which an atom or group of atoms is replaced by another atom or group of atoms. This particular substitution is a **free-radical substitution** reaction.

Step 1   Initiation

The molecules of chlorine absorb energy supplied by the UV light and the chlorine–chlorine covalent bond breaks symmetrically. Each atom from the bond leaves with one electron from the shared pair of electrons. Two chlorine atoms are formed. Each atom has an unpaired electron in the outer shell. They are called **free radicals** and are extremely reactive.

$$Cl_2 \rightarrow Cl\bullet + Cl\bullet$$

Step 2   Propagation

The highly reactive chlorine free radicals, Cl•, react with the methane molecules forming hydrogen chloride gas and leaving a methyl free radical, •CH₃. In turn the methyl free radical reacts with a second chlorine molecule forming chloromethane, CH_3Cl, and another chlorine free radical. This continuing process of producing free radicals is known as a **chain reaction**.

$$Cl\bullet + CH_4 \rightarrow HCl + \bullet CH_3$$

$$\bullet CH_3 + Cl_2 \rightarrow CH_3Cl + Cl\bullet$$

Step 3   Termination

In order for the reaction to stop two free radicals must collide and react to form a molecule. There are several possible combinations of free radicals.

$$Cl\bullet + Cl\bullet \rightarrow Cl_2$$

Two chlorine radicals form a chlorine molecule. However the UV light would break the chlorine molecule down again, so this combination would not terminate (or end) the reaction.

$$Cl\bullet + \bullet CH_3 \rightarrow CH_3Cl$$

One chlorine radical and one methyl radical react to form a molecule of chloromethane.

$$\bullet CH_3 + \bullet CH_3 \rightarrow C_2H_6$$

Ethane is formed by the reaction between two methyl radicals. This combination explains the existence of longer chain alkanes in the reaction mixture.

In summary, halogenoalkanes are the organic product of the photochemical reaction of the halogens with alkanes in UV light. They are produced via a free radical substitution mechanism in a chain reaction.

A **substitution** reaction is one in which an atom or group of atoms is replaced by another atom or group of atoms.

A **free radical** is an atom or group with an unpaired electron.

TIP

A dot • is used to represent the unpaired electron in a radical. The dot can be on either side of the species, e.g. Cl• or •Cl. **Homolytic fission** is the breaking of a covalent bond to produce two free radicals. Each atom in the bond leaves with one electron from the shared pair.

e.g. $CH_3Br \rightarrow \bullet CH_3 + Br\bullet$

EXAMPLE 1

1,2-dichloroethane reacts with chlorine in UV light to produce a mixture of further substituted halogenoalkanes.

1 Write two equations showing the propagation of this chain reaction to produce 1,1,2-trichloroethane.
2 Traces of 1,2,3,4-tetrachlorobutane are found in the reaction mixture. Write an equation to show how this product is formed.
3 Write a balanced symbol equation to illustrate the overall reaction between 1,1,2-trichloroethane with chlorine in UV light to form hexachloroethane.

Answers

1 The stem of the question gives a reaction condition and two reactants one of which is 1,2-dichloroethane. It is useful to write the structural formula, CH_2ClCH_2Cl.

You should recognise this as a free radical substitution reaction, which has three distinct steps in the mechanism of which propagation is the second. You should recall that the propagation steps for the chain reaction of methane and chlorine are:

$CH_4 + Cl\bullet \rightarrow HCl + \bullet CH_3$

$\bullet CH_3 + Cl_2 \rightarrow CH_3Cl + Cl\bullet$

1,2-dichloroethane will react by a similar mechanism.

$CH_2ClCH_2Cl + Cl\bullet \rightarrow CH_2ClCHCl\bullet + HCl$

$CH_2ClCHCl\bullet + Cl_2 \rightarrow CH_2ClCHCl_2 + Cl\bullet$

It is useful to understand what is happening. In a first propagation step the Cl• reacts with the organic molecule thus extracting a H atom to form HCl and leaving the organic molecule with an unpaired electron where the hydrogen atom was removed.

In the second propagation step this free radical which is extremely reactive reacts with a molecule of chlorine breaking the Cl—Cl covalent bond, to replace the lost hydrogen and forming another Cl•.

2 The third step in this free radical mechanism is a termination step. This involves two free radicals joining together to form one molecule. The two free radicals are Cl• and $CH_2ClCHCl\bullet$. The structural formula of 1,2,3,4-tetrachlorobutane is $CH_2ClCHClCHClCH_2Cl$ and so it is formed when two $CH_2ClCHCl\bullet$ radicals react. The equation is:

$CH_2ClCHCl\bullet + CH_2ClCHCl\bullet \rightarrow CH_2ClCHClCHClCH_2Cl$

or it can be represented using displayed formula as in Figure 13.5.

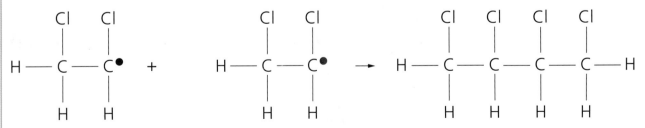

Figure 13.5

3
$CHCl_2CH_2Cl + 3Cl_2 \rightarrow CCl_3CCl_3 + 3HCl$

> **TIP**
> Check your equations. Propagation steps always have a molecule and a free radical as reactants and a different molecule and free radical as products. The equation does not represent a propagation step if there are two molecules on one side.

Free radical substitution reactions in the ozone layer

Ozone, O_3, is an allotrope of oxygen. The highest levels of ozone are in the stratosphere, a region between approximately 10 km and 50 km above the surface of the earth, known as the ozone layer. In this region, while the concentrations of ozone are only 2–10 ppm, the ozone performs an essential function. It enhances the absorption of harmful ultraviolet light by nitrogen and oxygen, preventing a proportion of these harmful rays from reaching the surface of the Earth. Ultraviolet light that reaches the surface of the earth can cause sunburn but is also necessary for the production of vitamin D in humans. In the absence of ozone, ultraviolet light of wavelengths 200 nm to 300 nm can also reach the Earth. These wavelengths cause skin cancer, cataracts in eyes, damage to plant tissue and reduce the plankton population in the oceans.

Over the past three to four decades, scientists have observed a steady decrease in the ozone present in the stratosphere and have concluded that this decrease is caused by photochemical chain reactions by halogen free radicals. The source of these halogen free radicals are the halogenoalkanes, which are extensively used as solvents, propellants, flame retardants and anaesthetics.

Chlorine free radicals cause the greatest destruction due to the widespread use of chlorofluoroalkanes (CFCs) which were valued both for their lack of toxicity and non-flammability. Unfortunately, because CFCs are stable in the lower atmosphere they do not degrade. They diffuse into the stratosphere where UV light causes homolysis of the carbon–chlorine bond and produces a chlorine free radical, $Cl^\bullet$.

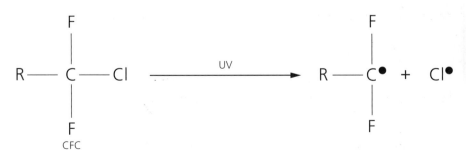

Figure 13.6

The chlorine free radical reacts with ozone, O_3, decomposing it to form oxygen O_2.

$$Cl^\bullet + O_3 \rightarrow ClO^\bullet + O_2$$

The chlorine free radical is reformed by reacting with more ozone molecules.

$$ClO^\bullet + O_3 \rightarrow 2O_2 + Cl^\bullet$$

In this sense the chlorine free radical is a catalyst for the decomposition of ozone and so contributes to a hole in the ozone layer. It is estimated that one molecule of chlorine can decompose 100 000 molecules of ozone.

Overall:

$$2O_3 \rightarrow 3O_2$$

There was a realisation by governments and chemists that solutions to the problem of ozone depletion had to be found. Almost 200 countries have pledged to phase out the production of ozone depleting agents (The Montreal Protocol) and chemists have developed and synthesised alternative chlorine-free compounds with low toxicity, which do not deplete the ozone layer, such as hydrofluorocarbons (HFCs), for example, trifluoromethane, CHF_3.

The reactions of the halogenoalkanes

Nucleophilic substitution reactions

The carbon–halogen bond is polar due to the difference in electronegativity between the carbon atom and the halogen atom. The δ^+ carbon atom in the carbon–halogen bond in a halogenoalkane is susceptible to nucleophilic attack. A **nucleophile** is an electron pair donor.

Halogenoalkanes react with nucleophiles such as:

hydroxide ions, OH^-,   cyanide ions, CN^-,   and ammonia, NH_3

$$:\overline{O}H \qquad\qquad :\overline{C}N \qquad\qquad :NH_3$$

The first two nucleophiles are negatively charged while the ammonia is neutral.

In general the negatively charged nucleophiles react via the following mechanism, where Nu represents the nucleophile.

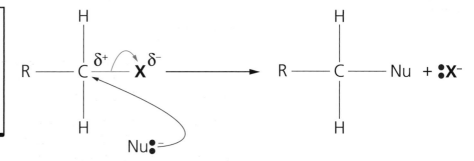

Figure 13.8

Chemists use the '**curly arrow**' (actual technical term!) in diagrams representing organic reaction mechanisms to illustrate the movement of a pair of electrons (Figure 13.8). The pair of electrons can be a lone pair or a bonded

---

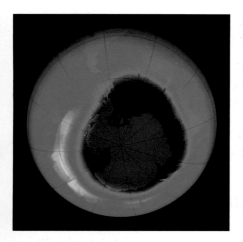

Figure 13.7 Satellite image of the ozone hole (purple) over Antarctica (centre) on 25th September 2010. Ozone layer thicknesses are colour coded from purple (lowest) through blue, cyan and green to yellow (highest). The ozone hole was at its annual maximum, covering an area of over 22 million square kilometres.

A **nucleophile** is an electron pair donor. 'Phile' originates from the ancient Greek word philia meaning love. Nucleophile, loosely translated, means loves nucleus or positive centre.

TIP
Make sure, for negative nucleophiles that you show both the lone pair of electrons, $:$, and the position of the negative charge on the ion.

TIP
Note the position of the lone pair of electrons, $:$, in the molecule or ion. It is at this point that the bond will be made with the δ^+ carbon atom of the organic molecule.

pair. A curly arrow must begin at a lone pair or in the centre of a bond and must end at an atom or in the centre of a bond.

In this case the lone pair of electrons on the nucleophile is attracted to the $\delta+$ charge on the carbon atom attached to the halogen atom (X). The nucleophile makes a bond with the carbon atom. This is followed by both electrons in the C–X bond moving to the X atom. The curly arrow showing the movement of this pair of electrons begins on the C–X bond and ends on the halogen atom. It leaves the molecule as a halide ion. The *leaving group* is shown with a lone pair of electrons.

Nucleophilic substitution by a hydroxide ion

The reaction between aqueous sodium hydroxide (or potassium hydroxide) and a halogenoalkane can take place at room temperature but it is slow and is often **refluxed** gently to improve the rate and the yield (Figure 13.9). The halogenoalkane is dissolved in a little ethanol as it is insoluble in water. The aqueous solution of the hydroxide ions and the ethanolic solution of the halogenoalkane are miscible. The organic product is an alcohol.

$$R–CH_2X + NaOH \rightarrow R–CH_2OH + NaX$$

For example, in the reaction of bromoethane with an aqueous solution of hydroxide ions.

- dissolve bromoethane in a small volume of ethanol
- add an **aqueous** solution of sodium hydroxide
- reflux gently

$$CH_3CH_2Br + NaOH \rightarrow CH_3CH_2OH + NaBr$$
bromoethane + sodium hydroxide → ethanol + sodium bromide

The **mechanism** is:

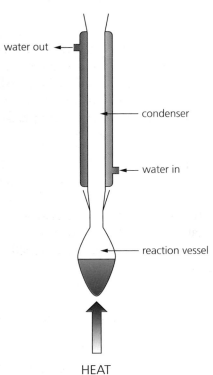

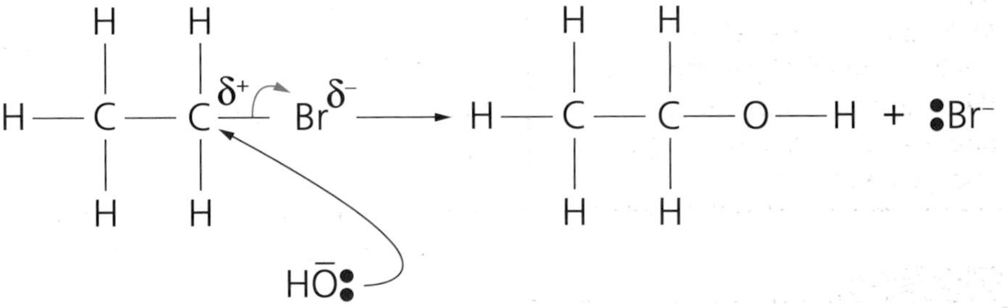

Figure 13.10

This reaction can also be classified as a hydrolysis reaction. **Hydrolysis** is the breaking of chemical bonds with water. The term also refers to the process of breaking bonds with hydroxide ions.

Rate of hydrolysis of the halogenoalkanes
The rate of the nucleophilic substitution/hydrolysis of the halogenoalkanes depends on the ease of breaking the carbon–halogen bond. The stronger the carbon–halogen bond the more difficult it will be to break and the slower the reaction.

Reflux is the continuous boiling and condensing of a reaction mixture.

water out ←

condenser

water in ←

reaction vessel

HEAT

Figure 13.9 A reflux apparatus. A condenser is placed in a vertical position on the reaction vessel. Any vapourised reactants will condense and return to the reaction flask.

Table 13.2 The average standard bond enthalpy of the carbon-halogen bonds.

Bond	C–X bond enthalpy/kJ mol⁻¹
C–F	484
C–Cl	338
C–Br	276
C–I	238

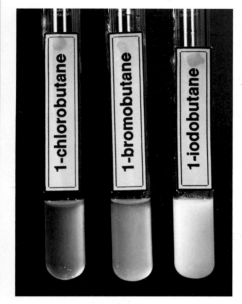

Figure 13.11 The halogenoalkanes 1-chlorobutane, 1-bromobutane and 1-iodobutane have been hydrolysed with ethanolic silver nitrate (AgNO₃) giving a precipitate of a silver halide. In terms of rate of reaction, 1-iodobutane reacts fastest, then 1-bromobutane and finally 1-chlorobutane. Can you explain why?

Table 13.2 gives the average standard bond enthalpy or bond strength for the carbon–halogen bonds. The strongest bond is the C—F bond. Bond strength decreases as Group 7 is descended. Therefore, the C—I bond is the most reactive. The C—F bond is so strong that fluoroalkanes are extremely unreactive. (This is one reason why hydrofluorocarbons have been developed as alternatives to CFCs – see page 284).

However, the most stable ion is the fluoride ion, F⁻. The least stable is the iodide, I⁻, ion. Combining these two factors, the order of reactivity of the halogenoalkanes is:

$$RI > RBr > RCl$$

Nucleophilic substitution by a cyanide ion

The cyanide ion contains a carbon atom bonded via a triple covalent bond to a nitrogen atom (C≡N).

An aqueous solution of potassium cyanide, KCN, is mixed with an ethanolic solution of the halogenoalkane and refluxed gently. The organic product is a nitrile.

$$R–CH_2X + KCN \rightarrow R–CH_2CN + KX$$

For example, in the reaction of bromoethane with an aqueous solution of cyanide ions:

- dissolve bromoethane in a small volume of ethanol
- add an aqueous solution of potassium cyanide
- reflux gently

$$CH_3CH_2Br \quad + \quad KCN \quad \rightarrow \quad CH_3CH_2CN \quad + \quad KBr$$

bromoethane + potassium → propanenitrile + potassium bromide
cyanide

Mechanism

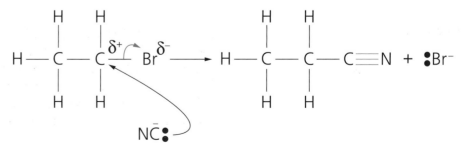

Figure 13.12

The initial organic molecule has two carbons, the organic product has three. This reaction increases the length of the carbon chain by one. This can be useful in organic synthesis when the available starting material contains one less carbon atom than is required in the product.

Take note of the name of the product. The CN group when present in inorganic compounds is named cyanide; KCN is potassium cyanide. When present in organic compounds, the CN group is named nitrile; CH_3CH_2CN is propanenitrile.

Nucleophilic substitution by an ammonia molecule

Ammonia is a nucleophile because it can donate a pair of electrons. It is a neutral nucleophile. The mechanism is a little more complex than the mechanism by which the negative nucleophiles react. A mixture of a concentrated solution of ammonia and the halogenoalkane dissolved in a little ethanol is placed in a sealed container under pressure. The subsequent reaction produces a primary amine as the organic product.

$$R–CH_2X + 2NH_3 \rightarrow R–CH_2NH_2 + NH_4X$$

For example, in the reaction of bromoethane with a concentrated solution of ammonia:

- dissolve bromoethane in a small volume of ethanol
- add a concentrated solution of ammonia in excess
- in a sealed container under pressure

$$CH_3CH_2Br + 2NH_3 \rightarrow CH_3CH_2NH_2 + NH_4Br$$

bromoethane + ammonia → ethylamine + ammonium bromide

> **TIP**
> If you need to outline a mechanism, then simply draw Figure 13.13

Mechanism

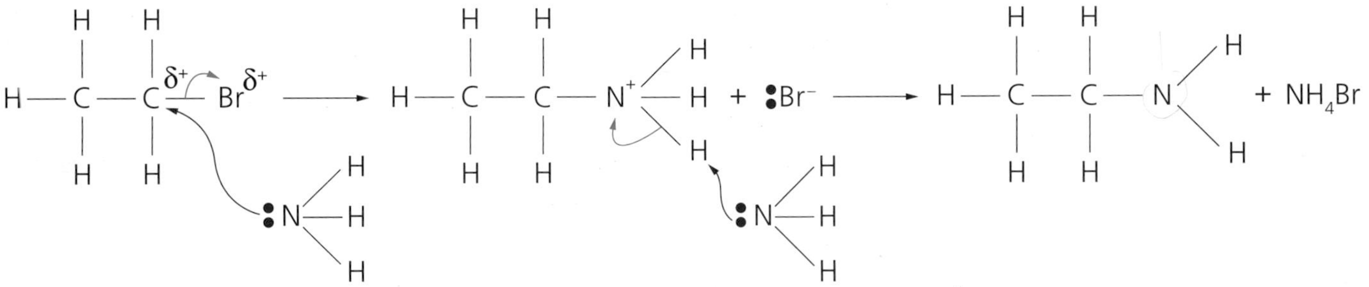

Figure 13.13

The lone pair of electrons on the nitrogen atom of the nucleophile is attracted towards the slightly positive carbon atom of the halogenoalkane. The electrons in the carbon–halogen bond move to the halogen atom and it leaves as a halide ion. The organic species now contains a nitrogen atom, which has four bonds. This nitrogen is positively charged. A second ammonia molecule is attracted to the positive organic intermediate and removes a hydrogen ion.

> **TIP**
> $CH_3CH_2CH_2CN$ is **but**an**e**nitrile. The carbon in the -CN group is counted as part of the carbon chain. Remember the presence of the letter **e**. It is not butannitrile.

TIP

The balanced symbol equation for the overall reaction between ammonia and a halogenoalkane contains two ammonia molecules. The first acts as a nucleophile and substitutes the halogen. The second acts as a base and accepts a proton, a hydrogen ion, from the reactive intermediate.

Ethylamine is classified as a **primary amine** as it has one R group attached to the N atom. In practice the product mixture contains diethylamine, a **secondary amine** and triethylamine, a **tertiary amine**.

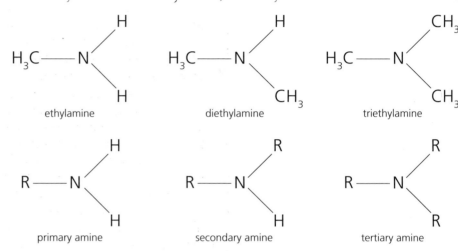

Figure 13.14 Primary, secondary and tertiary amines.

The yield of primary amine is encouraged by the use of a concentrated solution of ammonia. This prevents further substitution of the N atom. Further substitution is possible as the product, the primary amine, is also a nucleophile. The mechanism below illustrates how the further substituted amines are formed.

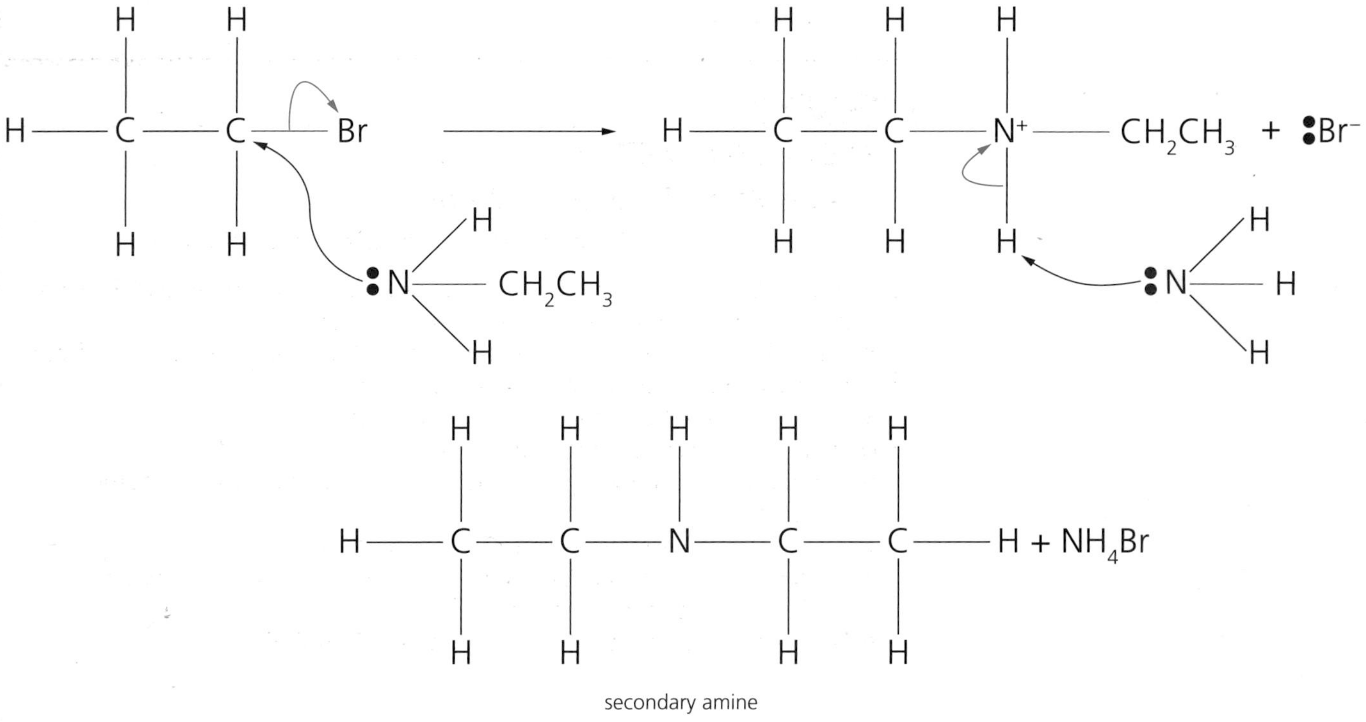

Figure 13.15

TEST YOURSELF 4

1 Chemists use *curly arrows* to show *mechanisms* of chemical reactions such as those of *nucleophiles* or *bases* with halogenoalkanes.
Explain what is meant by the following terms:
  a) mechanism
  b) curly arrows
  c) nucleophile
  d) base.

2 Butan-2-ol can be formed by the reaction of 2-bromobutane with hydroxide ions.
  a) Explain using a reaction mechanism how butan-2-ol can be formed from 2-bromobutane.
  b) Name the reagent and state the reaction conditions.
  c) Name the type of reaction.
  d) Write a balanced symbol equation for the overall reaction.

3 2-bromobutane is a liquid which is insoluble in water. Adding it to aqueous potassium cyanide produces two immiscible layers.
  a) How can the halogenoalkane be encouraged to mix with the aqueous layer so that a reaction can take place?
  b) Outline the reaction mechanism for aqueous potassium cyanide reacting with 2-bromobutane.
  c) Name the organic product.
  d) Write a balanced symbol equation for the overall reaction.

4 Ammonia reacts with 2-chloropropane to form an amine.
  a) Write an overall equation for the reaction to show the formation of a primary amine.
  b) Name the organic product.
  c) Explain the function of the second molecule of ammonia.
  d) State the reaction conditions.
  e) Why are di- and tri-substituted amines are found in the product mixture?

REQUIRED PRACTICAL

Preparation of 1-bromobutane

The following method can be used to prepare 1-bromobutane.

- Place $30\,cm^3$ of water, $40\,g$ of powdered sodium bromide and $21.8\,g$ of butan-1-ol in a $250\,cm^3$ round bottomed flask.
- Allow $25\,cm^3$ of concentrated sulfuric acid to drop into this flask from a tap funnel, cooling the flask in an ice bath.
- Reflux the reaction mixture for 45 minutes.
- Distil off the crude 1-bromobutane.
- Wash the distillate with water before washing with concentrated sulfuric acid.
- Finally wash with sodium carbonate solution.
- Remove the 1-bromobutane layer and add a spatula of anhydrous sodium sulfate, swirl and filter.
- Distil and collect the 1-bromobutane at 99–102 °C.

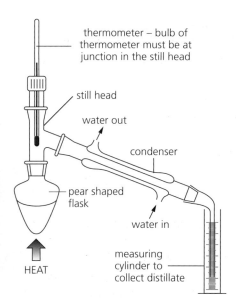

Figure 13.16 Distillation apparatus.

thermometer – bulb of thermometer must be at junction in the still head

still head

water out

condenser

pear shaped flask

water in

HEAT

measuring cylinder to collect distillate

1 Write the equation for the formation of 1-bromobutane from butan-1-ol.

2 a) Write an equation for the formation of hydrogen bromide from sodium bromide and sulfuric acid.

   b) Explain how bromine and sulfur dioxide are formed during the preparation.

TIP
Refer to the reactions of solid halides and concentrated sulfuric acid in Chapter 12.

3 Suggest why the concentrated sulfuric acid is added slowly and not all at once.

4 What is added to the distillation flask to ensure even boiling?

5 State three changes made to the apparatus set-up to change from reflux to distillation.

6 a) Explain how the process of washing with sodium carbonate solution is carried out.

   b) Suggest what is removed by the sodium carbonate solution.

7 Suggest what is removed by adding anhydrous sodium sulfate.

8 If 3.55 g of 1-bromobutane were obtained, calculate the percentage yield.

9 Referring to practical and theoretical considerations, suggest why the percentage yield is not 100%.

TIP
Potassium hydroxide is more soluble in ethanol than sodium hydroxide is and it is used in preference to sodium hydroxide as a source of hydroxide ions.

Elimination reactions of the halogenoalkanes

An aqueous solution of potassium hydroxide contains hydroxide ions. The aqueous hydroxide ions will react as nucleophiles with halogenoalkanes, forming an alcohol in a nucleophilic substitution reaction. However, when dissolved in **ethanol** the hydroxide ions act as a **base** and accept a proton (a hydrogen ion) to form water. The consequence is that the halogenoalkane molecule loses a hydrogen atom and a halogen atom. The organic product is an alkene.

$$R–CH_2CH_2X + KOH \rightarrow R–CH=CH_2 + H_2O + KX$$

Note that the hydrogen and bromine atoms which are removed from the halogenoalkane do not form hydrogen bromide. Study of the **mechanism** illustrates why water is formed by the hydrogen leaving.

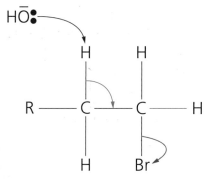

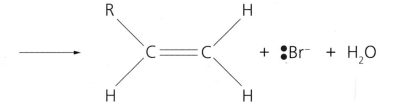

Figure 13.17

The hydrogen atom is removed by the hydroxide ions to form water. A carbon–carbon double bond is formed and the bromine leaves as a bromide ion. The hydrogen removed by the hydroxide ion is attached to a carbon adjacent to the carbon attached to the bromine.

The conditions of an organic reaction are extremely important. 2-chloropropane reacts with an **ethanolic** solution of potassium hydroxide to form propene but in an **aqueous** solution of potassium hydroxide, propan-2-ol is the main organic product.

This type of reaction is called an elimination reaction. An **elimination reaction** is one in which a small molecule is removed from the organic compound. For example, in the reaction of bromoethane with an ethanolic solution of hydroxide ions:

- dissolve bromoethane in a small volume of ethanol
- add an **ethanolic** solution of potassium hydroxide

$$KOH_{(in\ ethanol)} + CH_3CH_2Br \rightarrow H_2C{=}CH_2 + H_2O + KBr$$

potassium hydroxide + bromoethane → ethene   + water + potassium bromide

The formation of isomers in the elimination product mixture
When 2-chloropentane is refluxed in hot ethanolic potassium hydroxide two isomeric alkenes, pent-1-ene and pent-2-ene, are formed. In Figure 13.18, the ** indicates the hydrogen atoms which may bond to the OH$^-$ ion. These hydrogens are attached to carbon atoms adjacent to the C—Br bond. Depending on which H bonds to the OH$^-$, two different structures may form.

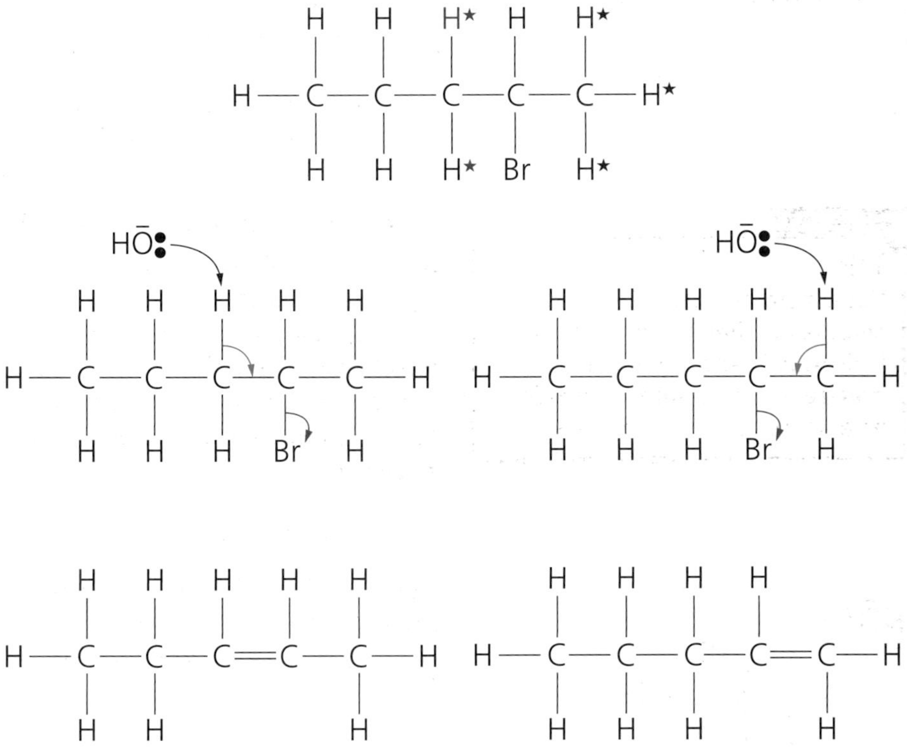

Figure 13.18

pent-2-ene   pent-1-ene

Pent-2-ene exists as two stereoisomers. The nature of stereoisomers is discussed in Chapter 14.

ACTIVITY

Preparation of methylpropene gas

Methylpropene gas can be prepared by heating 2-chloro-2-methylpropane in the apparatus shown in Figure 13.19. The mineral wool is soaked in 2-chloro-2-methylpropane and ethanolic potassium hydroxide.

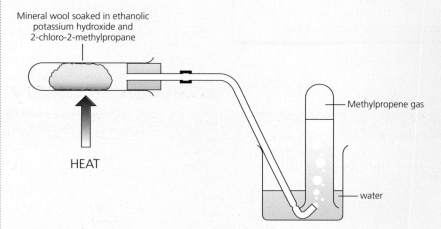

Figure 13.19 Laboratory preparation of methylpropene.

1 Draw the displayed formula of 2-chloro-2-methylpropane.
2 Explain why 2-chloro-2-methylpropane is insoluble in water.
3 **a)** Write an equation for the reaction of 2-chloro-2-methylpropane with ethanolic potassium hydroxide.
  b) State the name of this type of reaction.
4 A potential danger in heating reagents and collecting the gaseous product over water is that after heating, water rises back up into the delivery tube. Explain why this happens and suggest how this potential danger could be removed.
5 Calculate the volume of gaseous methylpropene formed at a temperature of 298 K and a pressure of 100 kPa, from 2 cm³ of liquid 2-chloro-2-methylpropane.
  (Density of 2-chloro-2-methylpropane: 0.8 g cm⁻³)
6 2-chloro-2-methylpropane will react with aqueous sodium hydroxide to form an alcohol.
  a) Write a balanced symbol equation for this reaction.
  b) State the name for this type of reaction.

Concurrent elimination and nucleophilic substitution reactions

In practice, both elimination and substitution reactions occur simultaneously. By manipulating the conditions the chemist can ensure a high yield of the nucleophilic substitution product or a high yield of the elimination product. However, the conditions imposed on the reaction mixture are not the only factors which determine the main product. The position of the functional group can also have an effect on the nature of the product. Elimination is more likely for a tertiary halogenoalkane and substitution is more likely for a primary halogenoalkane (Figure 13.20).

Figure 13.20 Classification of halogenoalkanes.

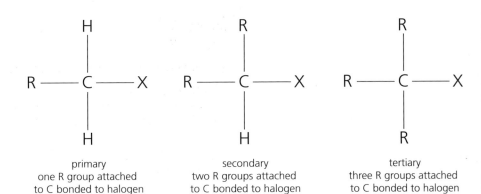

primary
one R group attached
to C bonded to halogen

secondary
two R groups attached
to C bonded to halogen

tertiary
three R groups attached
to C bonded to halogen

Reaction summary

Reagent	Conditions	Mechanism	Product	
potassium hydroxide	aqueous reagent; dissolve halogenoalkane in a little ethanol; reflux	nucleophilic substitution	alcohol	
potassium cyanide	aqueous reagent; dissolve halogenoalkane in a little ethanol; reflux	nucleophilic substitution	nitrile	carbon chain length is increased by 1
concentrated ammonia	excess ammonia; in a sealed tube; under pressure	nucleophilic substitution	amine	excess ammonia encourages a high yield of primary amine; discourages production of further substituted amines
potassium hydroxide	dissolve reagent in ethanol; dissolve reactants in ethanol; no water present	elimination	alkene	product may be a mixture of positional isomers

Practice questions

1 Methane reacts with chlorine to produce chloromethane and a gas, **G**.

a) State the conditions necessary for the reaction. *(1)*

b) Name gas, **G**, produced in the reaction. *(1)*

c) Using a series of equations describe the mechanism for this reaction. *(4)*

d) Name the type of mechanism. *(1)*

2 Bromoethane is susceptible to nucleophiles such as hydroxide ions and cyanide ions.

a) What is a nucleophile? *(1)*

b) Write a balanced symbol equation for the reaction of bromoethane with an aqueous solution of:

   i)   potassium hydroxide *(1)*

   ii) potassium cyanide. *(1)*

c) Draw the displayed formula for the organic product produced by the reaction in 2 b) ii) above. *(1)*

d) Explain why the reaction in 2 b) ii) is especially useful in organic synthesis. *(1)*

3 The following five compounds can be synthesised from compounds belonging to the homologous series of halogenoalkanes. In each case give the reagents and conditions necessary for their production and the name of the starting halogenoalkane.

a) pent-2-ene *(2)*

b) propanenitrile *(2)*

c) propan-2-ol *(2)*

d) methylamine *(2)*

e) diethylamine. *(2)*

4 The reaction of bromine with ethane in uv light is similar to that of chlorine with methane.

a) Write equations for the steps in the mechanism for the bromination of ethane to form a bromoethane. (4)

b) Name two other halogenoalkanes which may be present in the product mixture. (2)

Butane may also be present in the product mixture.

c) Write an equation for the termination reaction showing how butane may be formed. (1)

5 Write a balanced symbol equation to show the overall reaction of chlorine with propane to form 1,1,2,3,3-pentachloropropene. (1)

6 When hydroxide ions react with 1-chloropropane, two competing reactions take place.

a) Name and give the displayed formula of the two products that form when potassium hydroxide reacts with 1-chloropropane. (4)

b) Give the conditions necessary for favouring each product. (2)

7 Tetrachloromethane can be formed from trichloromethane in a free radical substitution reaction.

a) Write equations for:

i) initiation

ii) propagation

iii) termination steps (3)

b) Write an equation for the overall reaction to form tetrachloromethane from dichloromethane. (1)

c) Tetrachloromethane is a colourless liquid with a sweet smell. It was used extensively as a dry cleaning fluid.

i) Explain, with the aid of equations showing intermediates that form in the ozone layer, why the use of tetrachloromethane is severely restricted in over 200 countries. (3)

ii) Give two other uses of tetrachloromethane. (2)

d) Chemists have developed compounds such as 1,1,1,2-tetrafluoroethane that can be employed as alternatives to tetrachloromethane.

i) Draw the displayed formula for 1,1,1,2-tetrafluoroethane. (1)

ii) Give a reason why the use of 1,1,1,2-tetrafluoroethane is more environmentally acceptable than tetrachloromethane. (1)

8 Nucleophiles react with bromoethane in substitution reactions as shown in the scheme below.

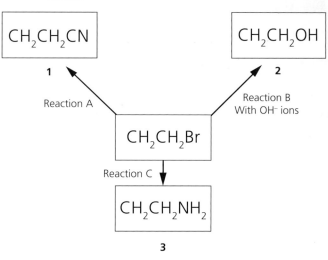

Figure 13.22

a) What is meant by the term nucleophile? (1)

b) Name the organic compounds labelled **1**, **2** and **3** in the scheme above. (3)

c) Name the nucleophile which reacts with the bromoethane in

i) Reaction A and

ii) Reaction C. (2)

d) Outline a mechanism for

i) Reaction B

ii) Reaction C. (4)

e) Give reagents and reaction conditions for reaction B. (1)

f) Explain why an excess of ammonia is needed for reaction C to produce a high yield of compound 3. (1)

g) When hydroxide ions react with bromoethane, ethene may be formed rather than compound 2.

    i) Give the reagents and conditions required for the formation of ethene from bromoethane. *(2)*

    ii) Name the type of reaction. *(1)*

9 The reaction shown below

$$C_2H_5Br + KOH \rightarrow KBr + H_2O + C_2H_4$$

is an example of:

A dehydration

B elimination

C free radical substitution

D nucleophilic substitution *(1)*

10 Which one of the following is a propagation step in the chlorination of methane?

A $Cl_2 \rightarrow 2Cl\bullet$

B $CH_4 + Cl\bullet \rightarrow CH_3Cl + H\bullet$

C $CH_4 \rightarrow CH_3\bullet + H\bullet$

D $CH_4 + Cl\bullet \rightarrow CH_3\bullet + HCl$ *(1)*

14

Alkenes

TEST YOURSELF ON PRIOR KNOWLEDGE 1

1 But-1-ene can be formed from the elimination reaction of a halogenoalkane, H, with hydroxide ions.
   a) Draw the displayed formula for the halogenoalkane, H.
   b) State the conditions for the formation of but-1-ene from H.
   c) Outline a mechanism for this reaction.
   d) Name one other organic product which may be produced.

Figure 14.1 This apparatus generates a steady regulated supply of ethene to encourage a controlled ripening of exotic fruits grown in hot sunny environments and destined for our supermarket shelves.

Figure 14.2 Home grown tomatoes can be picked when mature but green and placed in a box for 24 hours with a ripe banana. The ethene released by the banana will speed up the ripening process of the tomato to produce firm red fruit ready to eat.

Ethene (C_2H_4), a gas at room temperature and pressure, is the first member of the homologous series of alkenes. This simple molecule containing just two carbon atoms and four hydrogen atoms is one of the most versatile organic compounds in use today. World-wide production of ethene is in excess of 100 million tonnes, much more than any other organic compound. It is used in the production of plastics such as polythene, PVC and polystyrene. It is used to manufacture raw materials for other industrial processes e.g. ethanal and ethanol.

Alkenes are unsaturated hydrocarbons. An **unsaturated** compound contains at least one carbon–carbon double bond, C=C. They have a general formula C_nH_{2n}.

The presence of a double bond is indicated by the use of the letters 'ene' in the name of the chain. (Chapter 11). A number to show the position of the double bond may be necessary.

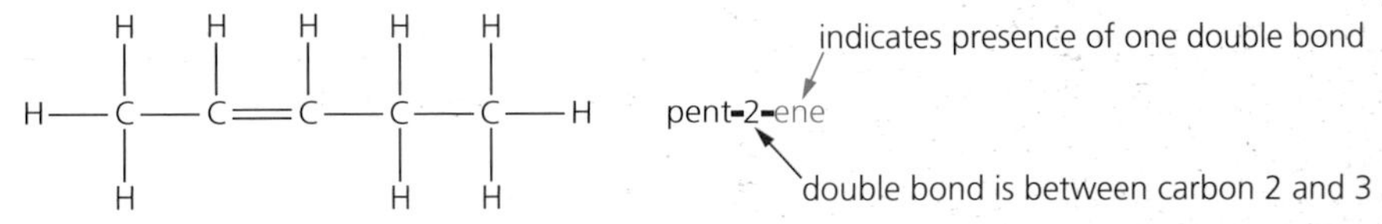

Figure 14.3

Alkenes have a carbon carbon double bond. It is a centre of **high electron density** and as a result the alkenes are more reactive than alkanes.

E-Z isomerism

Isomers are molecules which have the same molecular formula but a different structural formula (Chapter 11). But-1-ene and but-2-ene (Figure 14.4) are examples of structural isomers. The double bond is in a different position in each molecule. They are positional isomers.

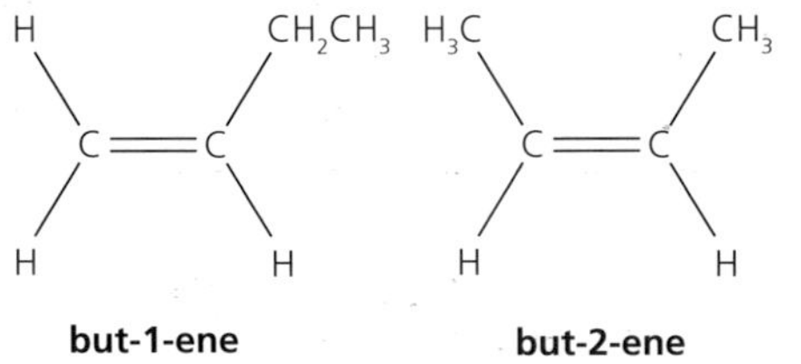

but-1-ene   **but-2-ene**

Figure 14.4

However, there are two possible structures, A and B, for but-2-ene.

To form molecule B from molecule A (Figure 14.5), part of the C=C must be broken. Bond breaking requires energy. This energy is not available at room temperature and pressure. We say there is **restricted rotation** around the planar double bond. Clearly molecules A and B are different; this is one example of a type of isomerism known as **stereoisomerism.** Stereoisomers are molecules which have the same molecular formula and structural formula but a different three-dimensional arrangement of their atoms in space. When the difference is due to the positions around the carbon–carbon double bond the isomers are said to be **E–Z** isomers. They are named either E or Z isomers according to the orientation of the groups attached to the carbon atoms of the double bond. The prefix Z indicates the groups are on the same side of the double bond on different carbons. The prefix E indicates the groups are on opposite sides of the double bond on different carbons.

> **Stereoisomers** are molecules which have the same molecular formula and structural formula but a different three-dimensional arrangement of their atoms in space.

Structure A   **Structure B**

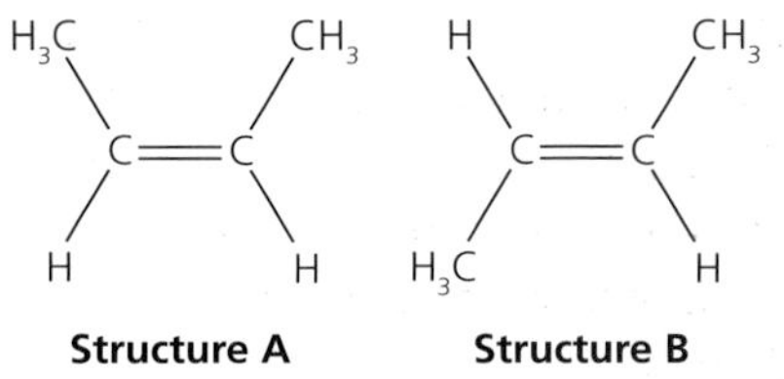

Figure 14.5

297

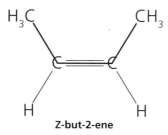

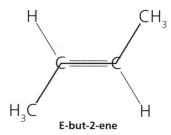

Z-but-2-ene

Z from the German word zusammen meaning altogether or along with

E-but-2-ene

E from the German word entgegen meaning to be opposite

Figure 14.6

More complicated molecules are designated as E or Z isomers by ranking the groups bonded on either side of the carbon–carbon double bond. The groups are ranked using a set of rules known as Cahn-Ingold-Prelog (CIP) priority rules. The higher the atomic number of the element bonded to the carbons in the C=C the higher the ranking. When the higher priority substituents are on the same side of the plane of the C=C the isomer is a Z isomer. When the higher priority substituents are on opposite sides of the plane of the C=C, the isomer is an E isomer.

To determine the rank order of the groups attached to the C=C, first consider the atomic number of the atoms attached to the carbons in the bond. If these atoms are the same consider the atomic number of the next atom.

Consider 1-bromo-2-chloro-1-fluoroethene. There are four different groups attached to the carbon atoms of the C=C. Consider each carbon of the C=C in turn and rank the groups (Figure 14.7).

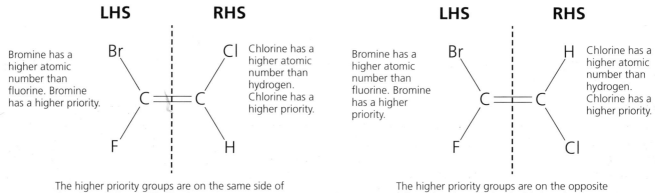

LHS **RHS**

Bromine has a higher atomic number than fluorine. Bromine has a higher priority.

Chlorine has a higher atomic number than hydrogen. Chlorine has a higher priority.

The higher priority groups are on the same side of the plane of the double bond.

Z-1-bromo-2-chloro-1-fluoroethene

LHS **RHS**

Bromine has a higher atomic number than fluorine. Bromine has a higher priority.

Chlorine has a higher atomic number than hydrogen. Chlorine has a higher priority.

The higher priority groups are on the opposite side of the plane of the double bond.

E-1-bromo-2-chloro-1-fluoroethene

Figure 14.7

Now consider the E–Z isomers of 2-hydroxy-3-methylpent-2-en-1-ol (Figure 14.8). To classify isomers as E or Z, decide which side of the plane of the double bond the higher priority groups are attached. Priority is decided by the atomic number of the atom attached directly to the C of the C=C.

Figure 14.8

Figure 14.9

LHS   **RHS**

The top group has a C atom bonded directly to the C=C, atomic number 6. The bottom group also has a carbon directly bonded to the C=C. This is the same priority. Now look at all the atoms bonded directly to these carbons. The top carbon has 3 H atoms. This is a total atomic number of 3. The bottom carbon has 2 H atoms and a C atom. The total atomic number is 8. Therefore the bottom group has the higher priority.

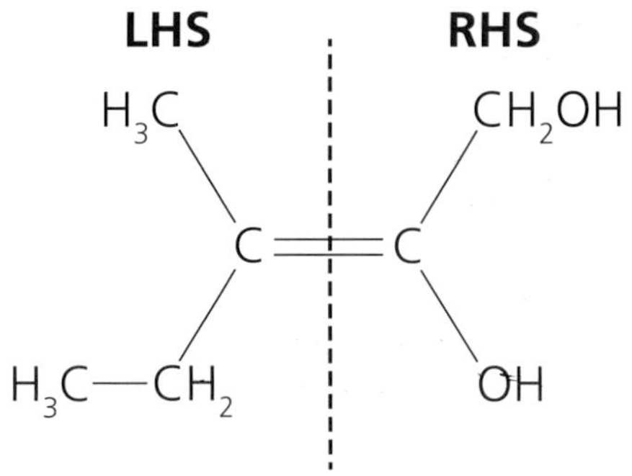

The top group has a C atom bonded directly to the C=C, atomic number 6. The bottom group has an oxygen atom bonded directly to the C=C, atomic number 8. The OH group has the higher priority.

The higher priority groups are on the same side of the plane of the double bond.

Z-2-hydroxy-3-methylpent-2-en-1-ol

LHS   **RHS**

As before, the atoms directly attached to the C atom of the C=C are the same and the priority is decided by the atoms next along the chain.

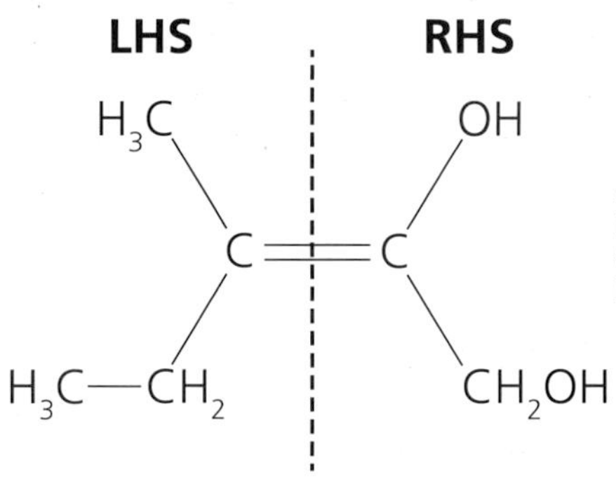

As before the OH group has the higher priority as it has the atom with the higher atomic numer bonded to the C atom in the C=C.

The higher priority groups are on the opposite side of the plane of the double bond.

E-2-hydroxy-3-methylpent-2-en-1-ol

Figure 14.9

EXAMPLE 1

An isomer of $C_7H_{12}O_2$ is shown in Figure 14.10. Classify the isomer as either an E or a Z isomer.

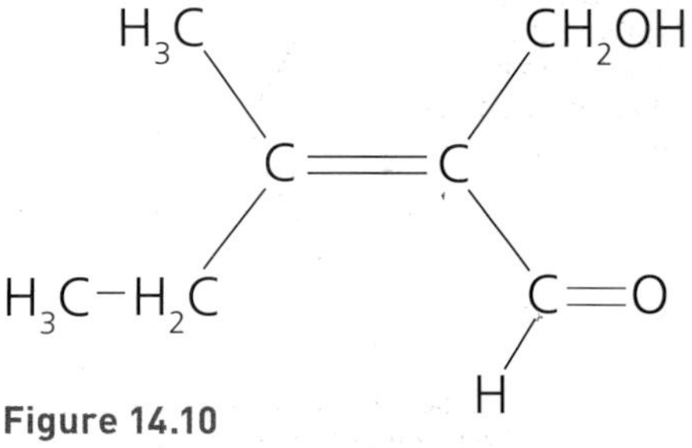

Figure 14.10

Answer

Using the CIP rules prioritise the groups attached to the carbon on the left-hand side then the right-hand side of the double bond. The atom attached with the higher atomic number is the higher priority atom. For identical atoms rank the next group of atoms directly attached. For atoms which are attached by a double bond the atomic number of the atom is counted twice.

LHS   **RHS**

The top group has a C atom bonded directly to the C=C, atomic number 6. The bottom group also has a C directly bonded to the C=C. This is the same priority. Now look at all the atoms bonded directly to these carbons. The top carbon has 3 H atoms. This is a total atomic number of 3. The bottom carbon has 2 H atoms and a C atom. The total atomic number is 8. Therefore the bottom group has the higher priority.

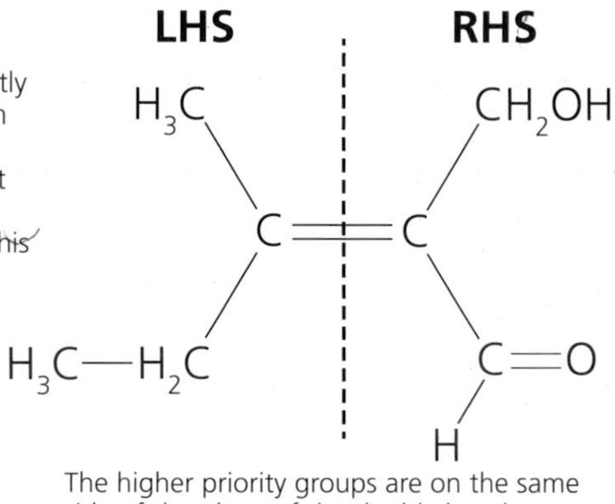

The top group and the bottom group both have C atoms bonded directly to the C=C, atomic number 6. Now consider the atoms bonded directly to this carbon. The top group has two H atoms and 1 O atom directly bonded giving a total atomic number of 10. The bottom group has one O atom and one H atom directly attached but the O is attached by a double bond. The O must be added twice. This gives a total atomic number 17. The CHO has a higher priority than the CH_2OH group.

The higher priority groups are on the same side of the plane of the double bond.

This isomer is the Z isomer

Figure 14.11

4 Write a balanced symbol equation for the dehydration of ethanol (C_2H_5OH).

5 Why was the gas allowed to escape for a few minutes before collecting the product?

6 At the end of the experiment the delivery tube is disconnected before turning off the heat. Why?

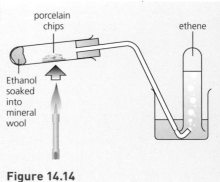

Figure 14.14

Addition polymers

Alkenes react with other alkene molecules to form polymers which are commonly used as plastics. The reaction is called addition polymerisation and the polymers referred to as addition polymers.

The polymers are named based on the alkene from which they are formed. Ethene forms poly(ethene), propene forms poly(propene). The brackets indicate that the structure does not contain the actual alkene but is the polymer formed from the alkene. Some polymers also have common names without brackets, for example poly(ethene) is commonly called polythene.

Poly(ethene)

Polythene is formed when ethene molecules add on to each other to form a chain of carbon atoms. The unit which breaks its double bond is called the **monomer** and the long chain formed is called the polymer.

A **polymer** is a long chain of repeating monomer units joined together.

This may be represented by the equation:

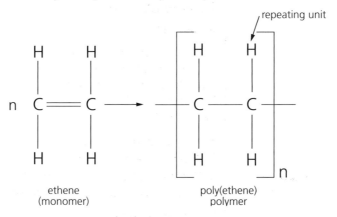

Figure 14.15

The monomer in this addition polymer is ethene and the n on both sides of the equation represent a large number of ethene molecules which have to add to each other.

The polymer which is poly(ethene) shows that the double bond has been broken and the unit is repeated in both directions for a large number of units (hence the n in the polymer structure as well). The structure inside the brackets is referred to as the repeating unit.

TIP

The equation may be simplified to:

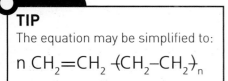

Poly(propene)

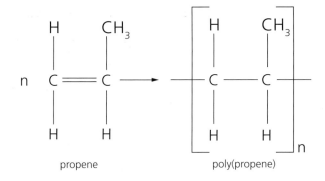

Figure 14.16

Poly(chloroethene)

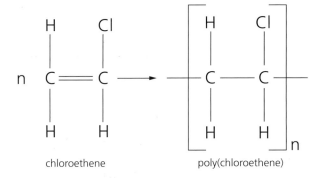

chloroethene   poly(chloroethene)

Figure 14.17

Poly(chloroethene) is commonly called PVC or polyvinylchloride as the monomer chloroethene was commonly called vinyl chloride.

Poly(phenylethene)

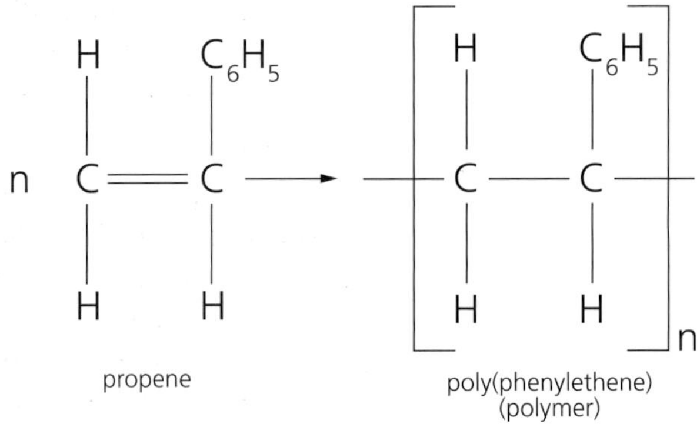

propene   poly(phenylethene)
(polymer)

Figure 14.18

Phenylethene is commonly called styrene so the polymer is often called polystyrene.

You may be asked to draw a section of the polymer chain comprising two or three repeating units. For example a section of poly(propene) showing two repeating units would look like this:

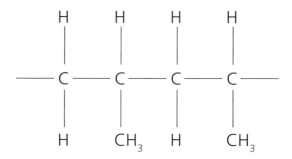

Figure 14.19

Or a section of poly(but-1-ene) showing three repeating units would look like this:

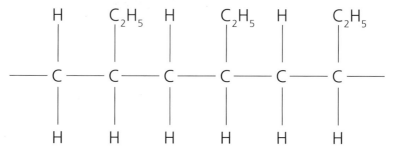

Figure 14.20

Production and uses of polymers

Addition polymers are unreactive. This is because, despite the name ending -ene, they are unreactive alkane molecules, without a reactive carbon-carbon double bond. This results in their widespread use as inert materials. They are very useful as insulators, as packaging and in making containers.

Two forms of poly(ethene) exist, high density (often abbreviated to HDPE) and low density (LDPE). HD poly(ethene) is less flexible and is used for kitchenware whereas LD poly(ethene) is more flexible and is used for plastic bags and some plastic bottles. A catalyst is used to produce HDPE from ethene.

Poly(chloroethene) or PVC is a rigid plastic used for window and door frames and drainpipes. However it may also be used for clothing and electrical wire insulation. The difference is in the presence of a plasticiser. uPVC is unplasticised PVC and is rigid and inflexible and used for window and door frames and drainpipes. If a plasticiser molecule is added (usually different phthalates), PVC becomes much more flexible and can be used for clothing like wellington boots or raincoats.

Polymer (recycling code)	Uses
Poly(ethene) (PE)	plastic bags, bottles, film wrapping, kitchenware
Poly(propene) (PP)	ropes, thinsulate clothing, carpets, crates, furniture
poly(chloroethene) (PVC)	wellington boots, raincoats, drainpipes, window frames, door frames, electrical wire insulation
Poly(phenylethene) (PS)	expanded polystyrene is used for insulation in houses and packing, unexpanded polystyrene is used for toys and containers

Strong covalent bonds join atoms to each other in individual polymer molecules. Weak van der Waals' intermolecular forces occur between polymer molecules. A polymer will melt when the intermolecular forces are overcome. The stronger the forces, the more energy is needed to break them, and the higher the material's melting point. Since the hydrocarbon chains are often very long, the van der Waals' forces between the chains are often very strong and the polymers have relatively high melting and boiling points. The chain length is variable, most polymers contain chains of a variety of different lengths. Thus the van der Waals' forces are of variable strength and these polymers tend to melt gradually over a range of temperatures rather than sharply at a fixed temperature.

Polymers which have very few branches are very compact and the chains can pack together very efficiently. Since the chains are closely packed, the van der Waals' forces between the chains are strong and these polymers tend to be stronger and harder as well, for example high density poly(ethene).

Polymers which are highly branched cannot pack together as well, for example low density poly(ethene). Since the chains are not closely packed, the van der Waals' forces between the chains are weaker and these polymers tend to be weaker and softer.

Repeating units

You may often be asked to circle or draw the repeating unit in a polymer or to draw the monomer from which the polymer is formed.

The structure below shows part of a polymer chain.

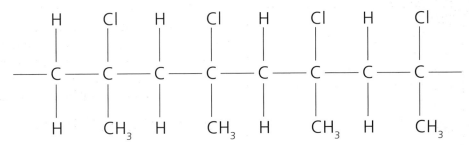

Figure 14.21

It is important to be able to identify the repeating unit from the polymer. There are two possibilities circled on the structure below.

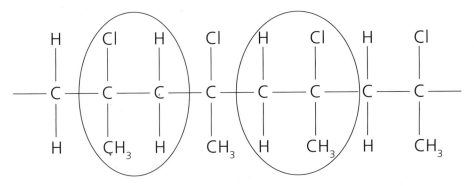

Figure 14.22

TIP
When asked to draw a repeating unit, you should not put brackets around it and do not put an n after it to indicate the repeat. For example for poly(ethene), the repeating unit is:

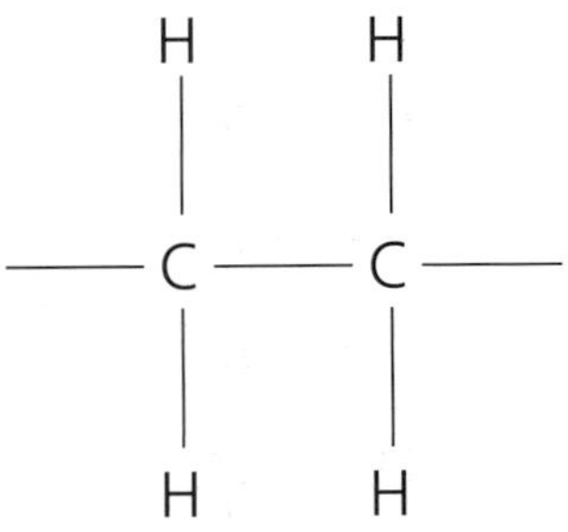

Figure 14.24

TEST YOURSELF 3

1 Name the following polymers:

(a) $\leftarrow\!CHCl{-}CH_2\!\rightarrow_n$

Figure 14.25

(b) $\leftarrow\!CF_2{-}CF_2\!\rightarrow_n$

Figure 14.26

2 State two uses of poly(chloroethene).

3 Show the repeating unit for the following polymers:

(a) poly(ethene)

(b) poly(propene)

(c) poly(1,1-dichloroethene)

Both of these repeating units would come from the same monomer:

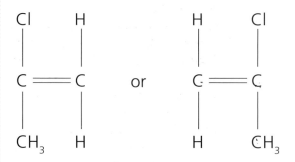

Figure 14.23

These monomers are both the same. The IUPAC name of the monomer is 2-chloropropene.

Drawing the repeating unit is also a common question.

All bonds need to be shown in a repeating unit including the bonds to show where the structure would connect to repeating units on either side.

Reactions of the alkenes

An **electrophile** is a substance which can accept a pair of electrons.

The carbon–carbon double bond in alkenes is an area of high electron density and makes members of this homologous series susceptible to attack from a class of substances known as electrophiles. **Electrophiles** are substances which can accept a pair of electrons. Polar molecules such as hydrogen bromide, HBr, are electrophiles. The δ^+ side of the polar molecule is attracted towards the area of high electron density between the carbon atoms of the double bond. Contrast this with the class of substances known as nucleophiles (Chapter 13). Neutral molecules such as bromine, Br_2, can also react with the C=C in alkenes. The high electron density concentrated at the C=C polarises the neutral bromine molecule and it becomes an electrophile.

TIP

A polar molecule is one which has a permanent separation of charge due to a difference in electronegativity between atoms in a covalent bond.

$H^{\delta+}{-}Br^{\delta-}$

Figure 14.27

Electrophilic addition reactions of the alkenes

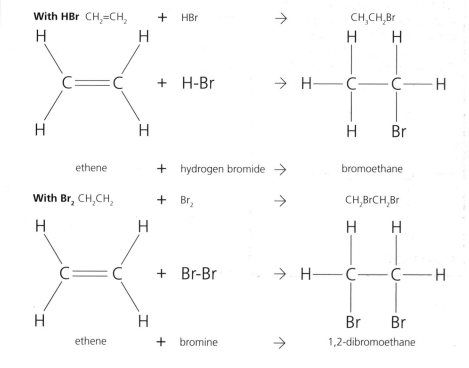

Figure 14.28

An **addition reaction** is a reaction in which two molecules join together to make one larger molecule.

The reactions described by the balanced chemical equations above are known as **addition** reactions as two molecules react together to form one molecule only. In the reaction between ethene and bromine the isomer 1,1-dibromoethane is not formed. The absence of this isomer is explained by considering the mechanism of the reaction between electrophiles and alkenes.

Electrophilic addition mechanisms

The mechanism of the reaction of alkenes with polar electrophiles $A^{\delta+}-B^{\delta-}$

Electrons move from an area of high electron density between the two carbon atoms to the $\delta+$ side of the electrophile. Electrons in the A—B bond move to the $B^{\delta-}$.

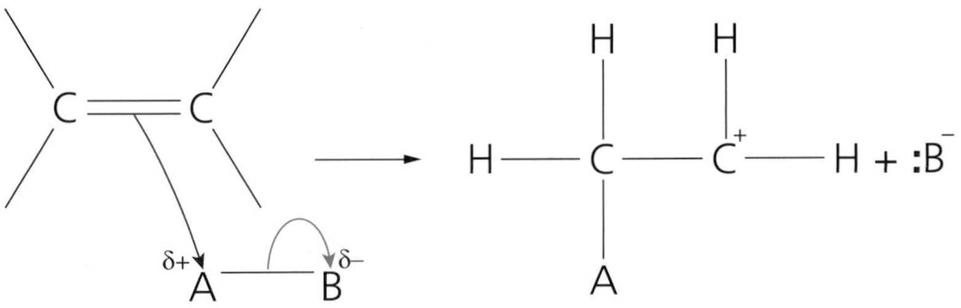

Figure 14.29

A species which contains a positive charge on a carbon atom is known as a **carbocation**.

One carbon is bonded to A, the other has a positive charge. A species which contains a positive charge on a carbon atom is known as a **carbocation**. B now has a negative charge and is shown with a lone pair of electrons.

The lone pair of electrons on the anion is attracted to the positive charge on the carbon atom and one molecule is formed. The double bond has been broken and A–B has been **added across this double bond**. A and B are now bonded to different carbons.

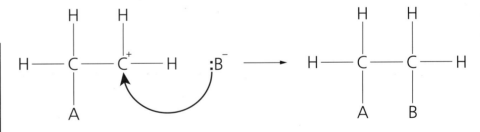

Figure 14.30

TIP
A curly arrow shows the movement of electrons and must begin at an area of high electron density such as a lone pair of electrons or from the centre of a bond.

Heterolytic fission occurs when a covalent bond breaks and both electrons in the bond move to one of the atoms. Two oppositely charged ions are formed.

The bond in the electrophile has broken unevenly. Both of the electrons in the shared pair have gone to one of the atoms. This is known as **heterolytic fission**. Contrast this with the homolytic fission of the chlorine bond in UV light in the free radical substitution of the alkanes to produce halogenoalkanes (Chapter 13). The carbon atoms in the double bond of the reactant are now bonded to one more atom each. The A—B bond has undergone heterolytic fission and each atom has bonded to different carbons. The product, unlike the organic reactant, is saturated.

The electrophilic addition of hydrogen bromide to an alkene

Hydrogen bromide is a polar electrophile. The addition product is a halogenoalkane. This reaction will take place at an appreciable rate even at temperatures below room temperature. For example, the reaction of ethene and hydrogen bromide:

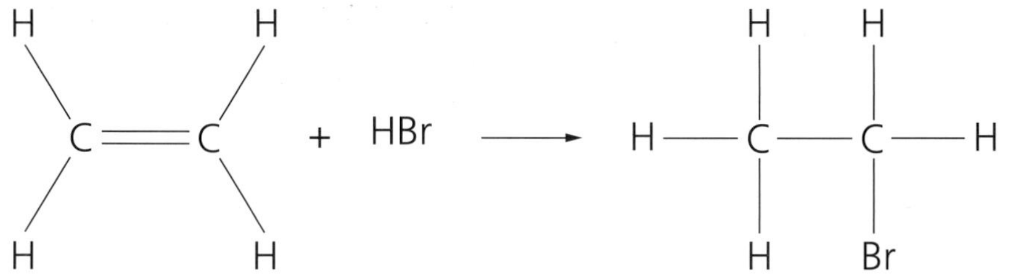

Figure 14.31

Another example is the reaction of but-2-ene with hydrogen bromide at room temperature.

but-2-ene + hydrogen bromide → 2-bromobutane

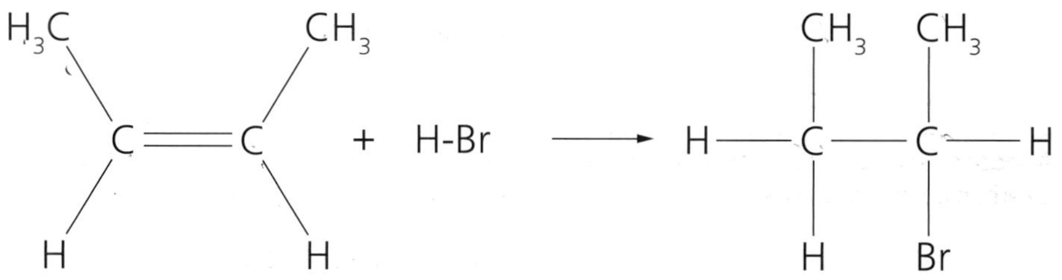

Figure 14.32

The **mechanism** for the reaction is:

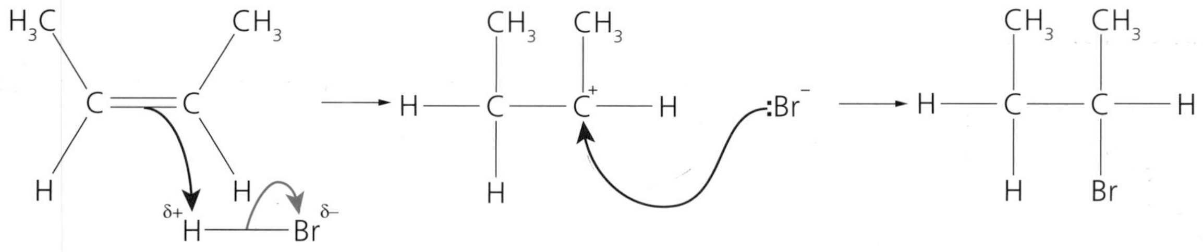

Figure 14.33

The electrophilic addition of concentrated sulfuric acid to an alkene

Concentrated sulfuric acid, H_2SO_4 is a polar electrophile. The addition product is an alkyl hydrogensulfate. The reaction will happen even at temperatures below room temperature.

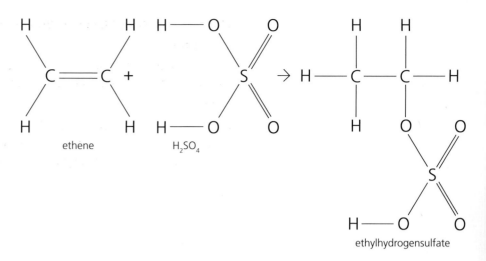

ethene H_2SO_4 ethylhydrogensulfate

Figure 14.34

The reaction of but-2-ene with sulfuric acid is shown below.

but-2-ene   **+**   sulfuric acid $\rightarrow$ 2-butylhydrogensulfate

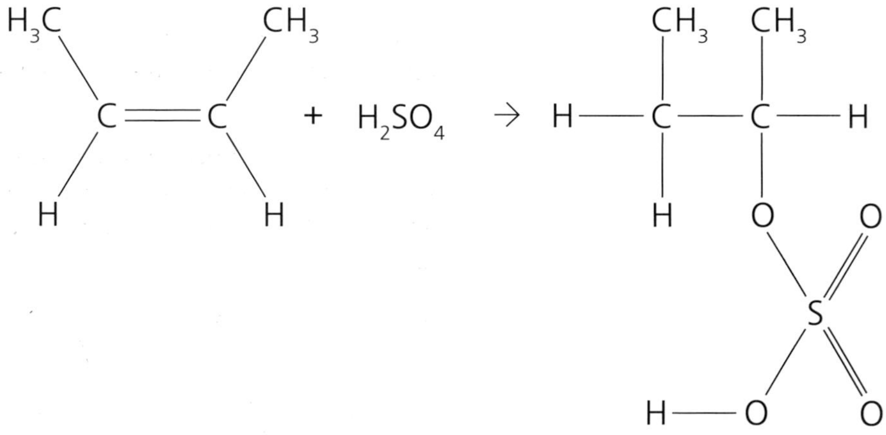

$$CH_3CH=CHCH_3 \quad + \quad H_2SO_4 \quad \rightarrow \quad CH_3CH_2CH(HSO_4)CH_3$$

Figure 14.35

The mechanism for the reaction is:

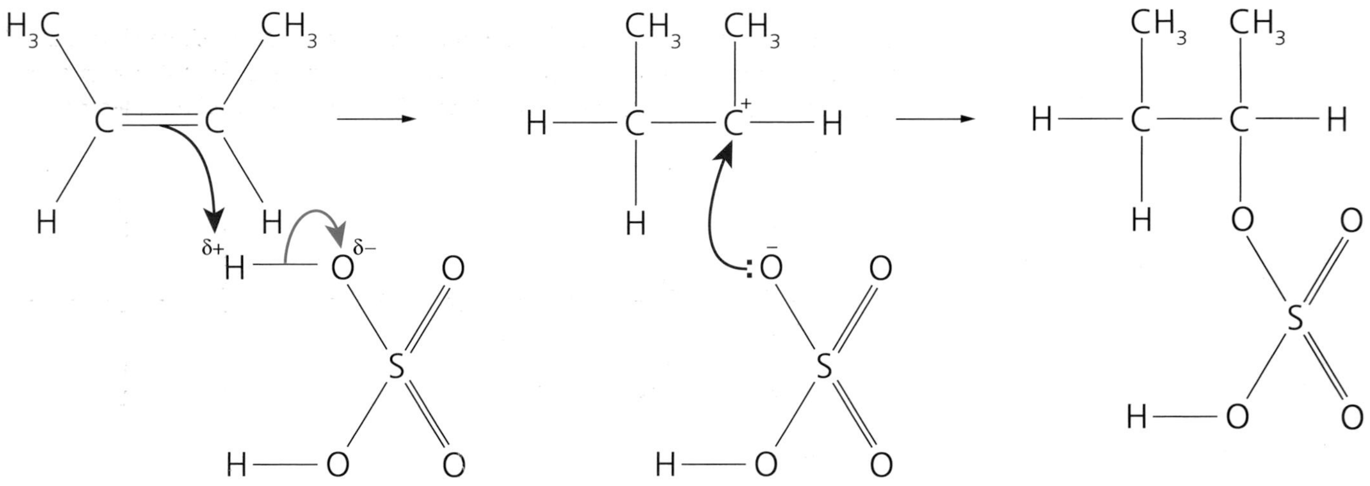

Figure 14.36

The mechanism of the reaction of alkenes with neutral electrophiles, A–A

This mechanism is similar to that of the electrophilic addition of polar nucleophiles. However, there is one extra initial step, polarisation of the A–A molecule.

The A-A molecule is not polar. When the A-A molecule moves close to the double bond which is an area of high electron density, it becomes polarised, the electrons in the C=C repel the electrons in the neutral molecule.

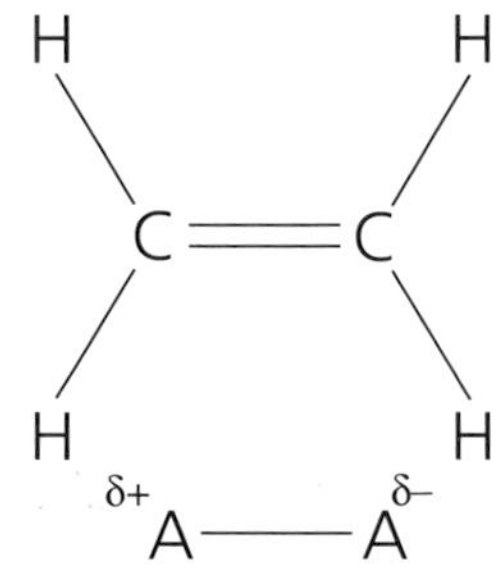

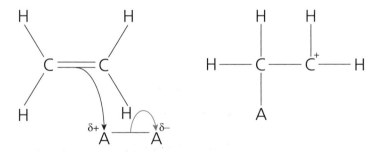

Figure 14.37

Electrons move from an area of high electron density between the two carbon atoms to the δ^+ side of the electrophile. Electrons in the A-A bond move to the A^{δ^+}

One carbon is bonded to A and the other has a positive charge. A species which contains a positive charge on a carbon atom is known as a carbocation. The other A now has a negative charge and is shown with a lone pair of electrons.

Figure 14.38

The lone pair of electrons on the anion is attracted to the positive charge on the carbon atom and one molecule is formed. The molecule, A-A, has broken and added across the double bond. A and A are now bonded to different carbons.

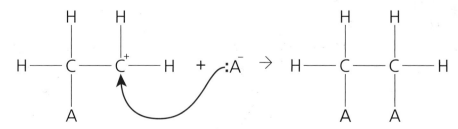

Figure 14.39

The electrophilic addition of bromine to an alkene

Bromine is a neutral electrophile. The addition product is a halogenoalkane. Bromine gas can be mixed with or bubbled through an alkene. The reaction will happen in the absence of light and at temperatures below room temperature. The reactant mixture is coloured due to the presence of bromine; the product mixture is colourless. Bromine will react with any molecule which has one or more double bonds.

The equation below shows the reaction of but-2-ene with bromine at room temperature.

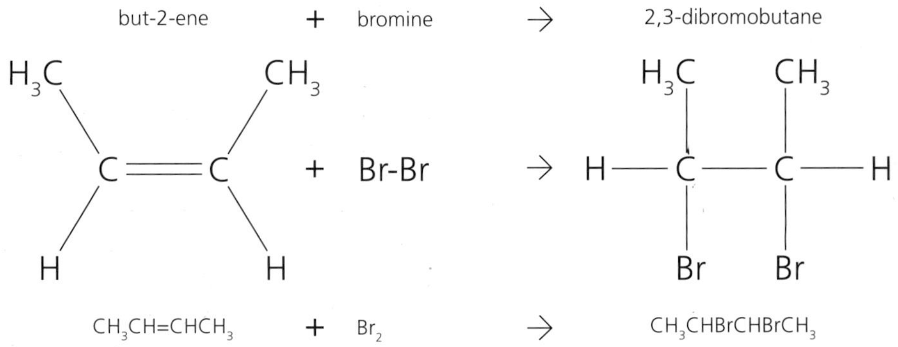

but-2-ene + bromine → 2,3-dibromobutane

$$CH_3CH=CHCH_3 + Br_2 \rightarrow CH_3CHBrCHBrCH_3$$

Figure 14.40

The **mechanism** for the reaction is:

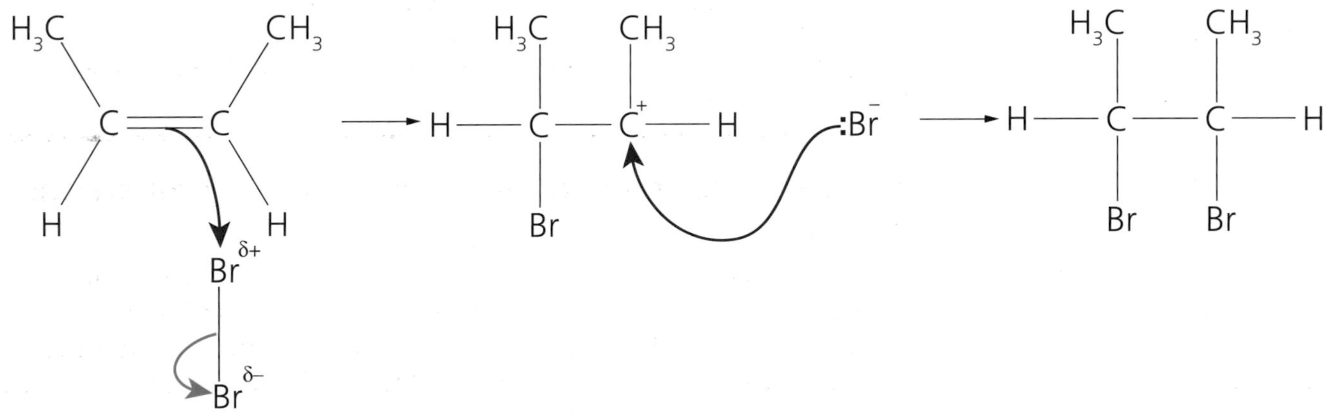

Figure 14.41

A variation of this reaction is used as a **test for unsaturation** or a test for the presence of a carbon-carbon double bond, C=C. Bromine is added to water to produce bromine water which is orange in colour. It will react with an alkene to form a colourless product.

$$Br_{2(as\ bromine\ water)} + alkene \rightarrow bromoalkane$$

orange solution   colourless solution

To carry out the test, bromine water is mixed with a sample of the organic substance. When the colour changes from orange to colourless without the release of a gas, the presence of a carbon–carbon double bond is confirmed (Figure 14.42).

Figure 14.42 Orange bromine water is decolourised by an alkene but remains unchanged in the presence of an alkane.

TEST YOURSELF 4

1 a) Explain what is meant by the following chemical terms.
    i) electrophile
    ii) nucleophile
    iii) heterolytic fission
    iv) homolytic fission.
  b) Write balanced symbol equations showing:
    i) the heterolytic fission of a bromine molecule
    ii) the homolytic fission of a bromine molecule.
2 a) Outline the mechanism for the reaction of ethene with hydrogen bromide.
  b) Write a balanced symbol equation for the overall reaction.
  c) Name the organic product.
  d) State the name of the type of mechanism.
  e) What is the name given to the type of organic intermediate produced in this reaction?
3 a) Compare and contrast the reaction of bromine with ethane to the reaction of bromine with ethene. Include in your answer,
    i) the name of the mechanisms,
    ii) type of fission of Br—Br bond,
    iii) names of the by-products, and
    iv) the conditions required for each reaction to take place.
  b) Write balanced symbol equations for:
    i) the reaction of bromine with methane, and
    ii) the reaction of bromine with pent-2-ene.
  c) Describe a chemical test for the presence of a C=C in an organic compound.

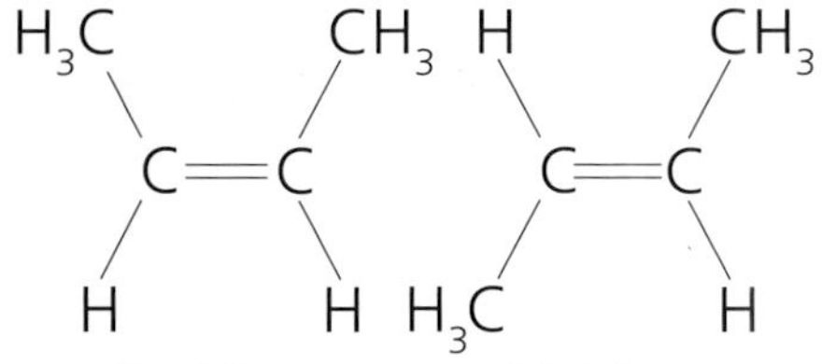

Electrophilic addition reactions to an unsymmetrical alkene

But-1-ene and but-2-ene are isomers. But-2-ene exists as two stereoisomers, Z-but-2-ene and E-but-2-ene (Figure 14.43). These stereoisomers are described as symmetrical alkenes as the groups on each carbon in the double bond are the same. Each carbon of the double bond has a hydrogen atom and a methyl, $-CH_3$, group attached.

But-1-ene, however, is an unsymmetrical alkene (Figure 14.44). Each carbon of the double bond is bonded to a different set of groups. One carbon atom is bonded to two hydrogen atoms while the other is bonded to an ethyl group, $-CH_2CH_3$, and a hydrogen atom. This difference does not affect the stages of the mechanism of electrophilic addition but it does lead to the possibility of two different products.

Z-but-2-ene   E-but-2-ene

Figure 14.43

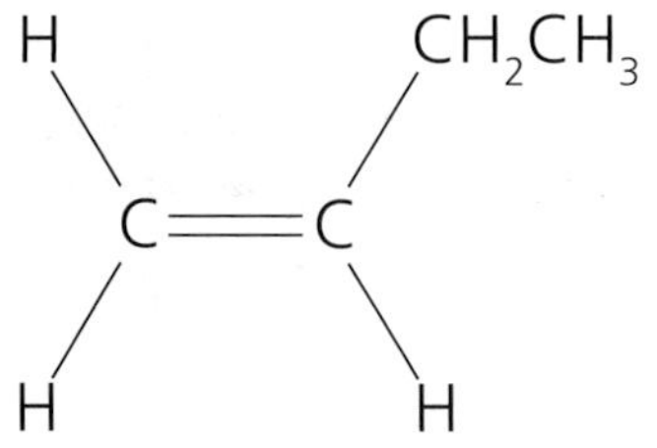

but-1-ene

Figure 14.44

TIP
Electrophilic addition of polar electrophiles to an unsymmetrical alkene produces two different products, one of which may be present in a greater proportion (the major product) than the other (the minor product).

311

Consider the reaction between but-1-ene and hydrogen bromide.

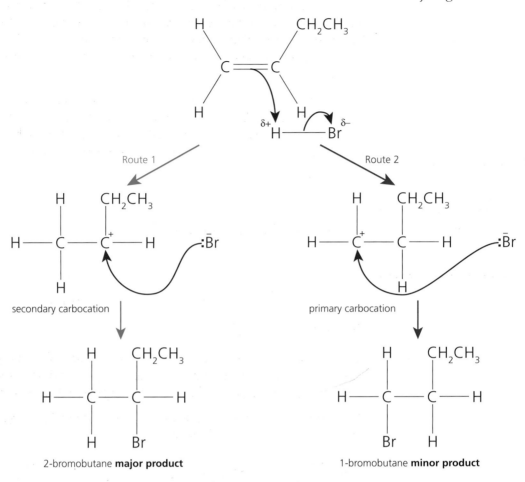

Figure 14.45 Two products form when hydrogen bromide reacts with but-1-ene.

There is the possibility of two different carbocations forming depending on which carbon the hydrogen from the HBr adds to. In Route 1 (Figure 14.45), H from HBr adds to the right-hand carbon producing a positive charge on the carbon on the right, which leads to the formation of 2-bromobutane. In Route 2 (Figure 14.45), H from HBr adds to the left-hand carbon producing a positive charge on the left-hand carbon thus forming 1-bromoethane as the addition product.

In practice both compounds are formed but approximately 90% of the mixture is 2-bromobutane; it is said to be the **major product**. This is due to the relative stability of the different carbocations which may form as the reaction progresses.

Carbocations are classified as, primary, secondary and tertiary depending on the number of alkyl groups, –R, attached (Figure 14.46).

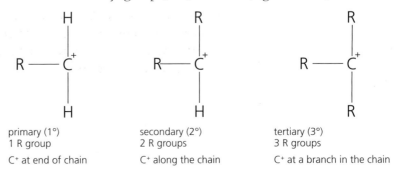

Figure 14.46 Carbocations.

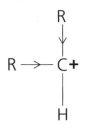

Figure 14.47

Alkyl groups, such as methyl (–CH$_3$) or ethyl (–CH$_2$CH$_3$) groups, have an **electron donating effect** relative to a hydrogen atom. They 'push' electrons away from the group, towards the opposite end of the bond. This is sometimes referred to as a positive inductive effect. The electron donating effect is usually shown by an arrow on the R–C bond as shown in Figure 14.47.

Electron donating groups decrease the size of the positive charge on the carbon and the carbocation becomes more stable. The greater the number of R groups the more stable the ion is. Tertiary carbocations are more stable than secondary carbocations, which are more stable than primary carbocations.

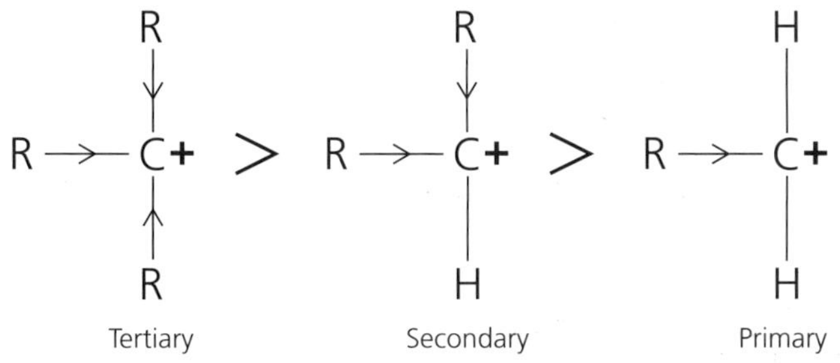

R ⟶ C+ > R ⟶ C+ > R ⟶ C+

Tertiary   Secondary   Primary

Figure 14.48

The relative stability of the carbocation intermediate effects the product formed. More stable carbocations remain for a longer time in the reaction mixture and have a greater chance of reacting with an anion to form a product.

Consider again the two possible carbocations formed in the reaction between but-1-ene and hydrogen bromide.

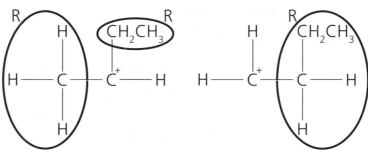

This carbocation has two R groups attached to the positively-charged carbon, a –CH$_2$CH$_3$ group and a –CH$_3$ group. It is a secondary carbocation.

This carbocation has one R group attached to the positively-charged carbon, a –CH$_2$CH$_2$CH$_3$ group. It is a primary carbocation.

Figure 14.49

There is a greater probability of the secondary carbocation reacting with the bromide ion to form the product. Therefore, 2-bromobutane is the major product.

The Russian chemist, Markovnikov, summarised the reaction by stating, 'When a compound HX is added to an asymmetrical alkene the hydrogen becomes attached to the carbon atom of the double bond which has the most hydrogens directly attached'.

The industrial production of ethanol from ethene

Ethanol is an alcohol. Alcohols are a homologous series with the general formula $C_nH_{2+1}OH$ and the functional group –OH (Chapter 15). Ethanol can be produced by the reaction of ethene with steam using a concentrated strong acid as a catalyst.

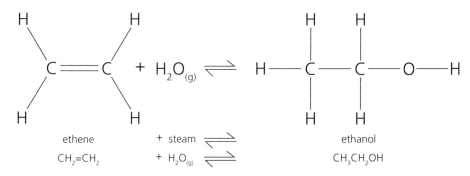

ethene   + steam   ethanol
$CH_2=CH_2$ + $H_2O_{(g)}$ CH_3CH_2OH

Figure 14.50

The addition of a water molecule to a compound is known as **hydration**. This reaction is used industrially to manufacture ethanol from ethene using concentrated phosphoric acid as a catalyst. Ethene and steam are passed over a concentrated phosphoric acid catalyst absorbed on a solid silica surface. The reaction mixture is kept at 60 atm and 600 K. These conditions ensure an almost 100% yield but side reactions such as the formation of methanol and poly(ethene) can reduce the yield a little.

TIP

Hydration and hydrolysis are different.
Hydration is the addition of water to a substance.
Hydrolysis is the decomposition of a chemical compound by reaction with water.

Practice questions

1 a) The following structure shows a section of a polymer.

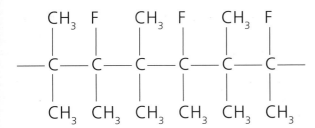

Figure 14.51

  i) Draw the structure of the monomer and give its IUPAC name. *(2)*

  ii) Name the type of polymerisation used to form this polymer. *(1)*

  b) Draw the repeating unit of the polymer poly(2,3-dichloropropene). *(1)*

2 Classify this molecule as an E or Z isomer. *(1)*

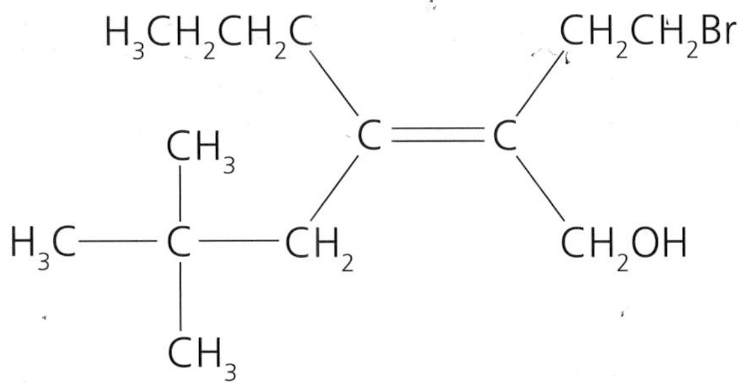

Figure 14.52

3 Alkenes are susceptible to attack by electrophiles.

  a) What is an electrophile? *(1)*

  b) Give an example of an electrophile which is:

    i) a polar molecule *(1)*

    ii) a non-polar molecule. *(1)*

  c) i) Outline a mechanism illustrating the reaction between an alkene and a polar electrophile represented as $H^{\delta+}$—$Br^{\delta-}$. *(3)*

    ii) Name this type of mechanism. *(1)*

4 a) Describe how you would carry out a chemical test used to confirm the presence of an alkene. *(2)*

  b) Outline a mechanism for the reaction described in part a). *(3)*

5 a) This balanced chemical equation describes the reaction between bromine and propene:

$$CH_3CHCH_2 + Br_2 \rightarrow C_3H_6Br_2$$

  i) Is A or B the product of this reaction? *(1)*

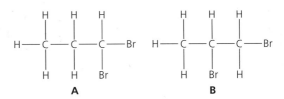

Figure 14.53

  ii) Explain your answer to 6 a) i). *(1)*

  b) Name the mechanism. *(1)*

6 Propan-2-ol is manufactured by the reaction of impure propene with water using concentrated sulfuric acid as a catalyst to form 2-propyl hydrogen sulfate. The 2-propyl hydrogen sulfate is added to water and hydrolyses to form propan-2-ol. The product contains almost no propan-1-ol.

  a) Outline a mechanism illustrating the formation of 2-propyl hydrogen sulfate as described above. *(3)*

  b) Name the type of reactive intermediate and the mechanism. *(2)*

  c) What is hydrolysis? *(1)*

  d) Explain in terms of the relative stability of the reactive intermediate, why the product mixture contains virtually no propan-1-ol. *(2)*

7 a) Compare and contrast the reaction of bromine with ethene with the reaction of bromine with ethane. *(4)*

  b) Explain why 1,1-dibromoethane is not formed in the reaction between bromine and ethene.

8 Which one of the following alkenes, A, B, C or D, will exhibit E–Z isomerism? *(1)*

  A $CH_3CH=CHCH_3$ **B** $(CH_3)_2C=CHCH_3$

  C $C_2H_5(CH_3)C=C(CH_3)_2$ **D** $H_2C=CHF$

9 How many structural isomers have the molecular formula C_5H_{10}?

  A 4   **B** 5   **C** 6   **D** 7 *(1)*

15 Alcohols

TEST YOURSELF ON PRIOR KNOWLEDGE 1

1 State and explain the pattern in boiling points of the Group 6 hydrides: H_2O, H_2S, H_2Se and H_2Te.

2

Figure 15.1

C is an organic compound that is not a hydrocarbon. It can be formed in a series of reactions from compound A.

a) What is the IUPAC name for compound A?

b) State the reaction conditions and name and outline the mechanism for Reaction 1.

Reaction 2 mixes B with steam and passes the mixture over a concentrated phosphoric acid catalyst absorbed on a solid silica surface.

c) Name the homologous series to which the organic product of Reaction 2 belongs.

d) Give two other conditions necessary for this reaction to take place.

3 Write half equations for the following reactions.

a) $Fe^{2+} \rightarrow Fe^{3+}$

b) $MnO_4^- \rightarrow Mn^{2+}$

c) $Cr_2O_7^{2-} \rightarrow Cr^{3+}$

Figure 15.2 The complex mixtures produced by the fermentation process.

Fermentation of carbohydrates to produce wine and beer was one of the earliest chemical reactions exploited by humans. Archaeologists and historians have uncovered evidence of the production of wine and beer in societies which existed 9000 years ago. For the ancient Egyptians (4000 BC), beer was part of their everyday diet. It became and remained popular in Europe as it was often safer to drink than water, as many water supplies were often contaminated.

Ethanol is a member of the homologous series of alcohols. It is present in the complex mixtures produced by the fermentation process. It is commonly referred to as 'alcohol' (Figure 15.2). Alcohols have the functional group –OH and the general formula $C_nH_{2n+1}OH$.

The physical properties of the alcohols

Alcohols have a greater boiling point than alkanes with a similar relative molecular mass. This is due to the presence of hydrogen bonding between alcohol molecules. When a molecular covalent liquid is vaporised, energy is supplied to break the intermolecular forces of attraction between the molecules. Alkane molecules are held together by van der Waals' forces. Both van der Waals' forces and hydrogen bonding are present between alcohol molecules (Figure 15.4). Hydrogen bonding is a stronger intermolecular force than van der Waals' forces.

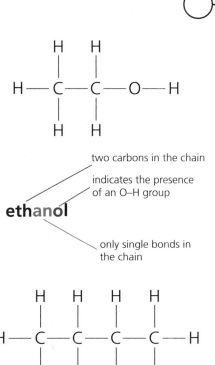

Figure 15.3 Naming alcohols

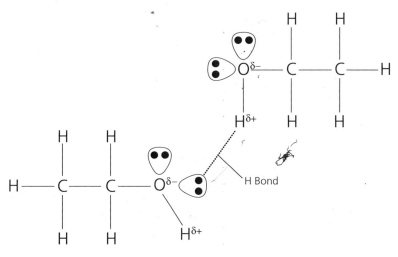

Figure 15.4 Hydrogen bonding between ethanol molecules

For example, propane and ethanol have similar relative molecular masses. The strength of the van der Waals' forces between molecules in a pure sample of propane and between molecules in a pure sample of ethanol should be very similar, leading to similar boiling points. The boiling point of ethanol, however, is 120 °C higher.

In short chain alcohols, such as methanol and ethanol, the –O—H group has a considerable effect on the solubility of the molecule. Both are soluble in all proportions in water while alkanes, which do not possess the –O—H group are insoluble in water.

317

Classification of alcohols

Alcohols are classified as primary (1°), secondary (2°) or tertiary (3°) depending on the position of the –OH group in the carbon chain. The classification can also be derived by considering the number of alkyl groups, R groups, attached to the carbon bonded to the –OH group (Table 15.1).

Table 15.1 Classification of alcohols.

Type of alcohol	General structure	Example	Position of OH group
primary alcohol 1 R group		butan-1-ol	the –OH group is positioned at the end of the chain
secondary alcohol 2 R groups		butan-2-ol	the –OH group is positioned along the length of the chain
tertiary alcohol 3 R groups		2-methylpropan-2-ol	the –OH group is positioned at a branch in the chain

Methanol, CH_3OH, is classified as a primary alcohol as the –OH group is at the end of a chain.

TEST YOURSELF 2

1 Name and classify the following alcohols as primary, secondary or tertiary.
   a) C_2H_5OH
   b) $(C_2H_5)_2C(OH)CH_3$
   c) $CH_3CH(CH_3)CH_2OH$
   d) $CH_3CH(OH)CH_2CH_2CH_3$.

2 Name and draw the displayed formulae of one primary, one secondary and one tertiary alcohol that are isomers of pentan-2-ol.

3 Heptan-1-ol (116.0 M_r) has a boiling point of 180 °C. Octane (114.0 M_r) has a boiling point of 126 °C.

a) Draw the structural formulae for heptan-1-ol and octane.

b) Explain why the boiling point of heptan-1-ol is higher than the boiling point of octane when the M_r of each compound is similar.

4 Ethanol is completely miscible in water while only 0.59 g of hexan-1-ol will saturate 100 g of water at room temperature.

a) Explain why ethanol is so soluble in water.

b) Suggest why hexan-1-ol has an extremely low solubility in water.

Reactions of the alcohols

Oxidation

Primary alcohols are easily oxidised in the presence of oxidising agents ([O]) to produce aldehydes, which can be further oxidised to carboxylic acids.

> **TIP**
> The symbol [O] is used in equations to represent oxygen from an oxidising agent.

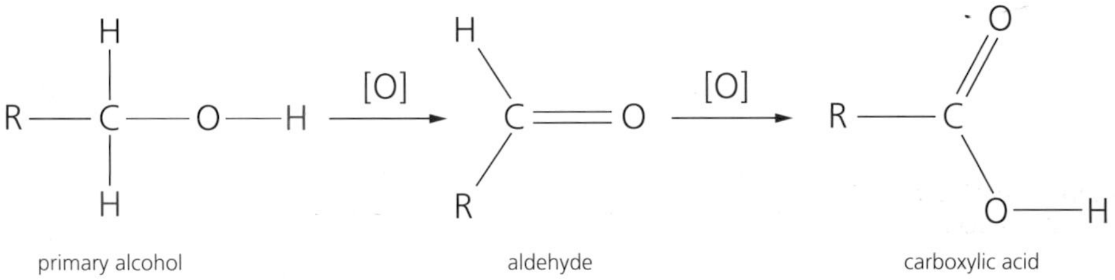

Figure 15.5 Oxidation of primary alcohols.

In the first oxidation shown above, the primary alcohol loses two hydrogen atoms, one from the –O—H group and one from the saturated carbon. In the second oxidation an oxygen atom is added to the remaining hydrogen.

Secondary alcohols can be oxidised to ketones, which do not undergo further oxidation.

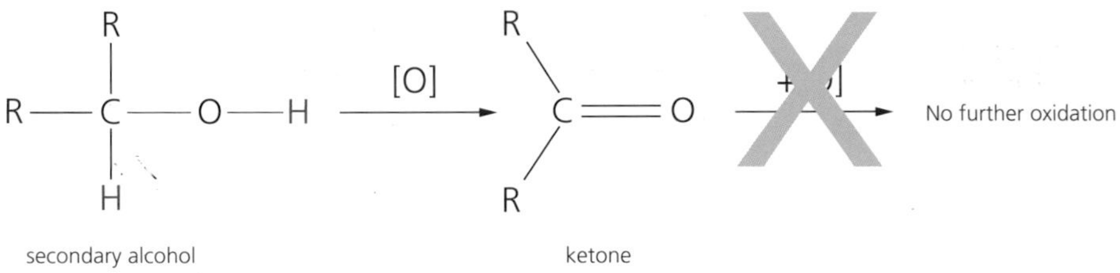

Figure 15.6

Tertiary alcohols are not easily oxidised as they do not have two hydrogen atoms directly attached to the carbon that is bonded to the –O—H group. They can be oxidised using hot concentrated nitric acid. The stringent conditions are required because oxidation of a tertiary alcohol requires the breaking of the strong C—C bonds.

Aldehydes, ketones and carboxylic acids are further examples of organic homologous series. Aldehydes and ketones contain the C=O group which is known as a **carbonyl** group.

> **TIP**
> The carbonyl group, C=O, is a functional group which is present in aldehydes and ketones.

In an aldehyde, the carbonyl group is at the end of the carbon chain, whereas in a ketone, this group is positioned along the chain. Aldehydes are named using the suffix **-al**. Heptanal is an aldehyde with seven carbons. There is no need to number the position of the C=O in an aldehyde as it is always at the end of the chain.

319

Ketones are named with the suffix **–one**. Butanone is a ketone with four carbons. It has no isomers that are ketones as moving the C=O to the end of the chain produces the aldehyde, butanal.

Carboxylic acids contain a –COOH group, positioned at the end of the carbon chain (Figure 15.7). The suffix **–oic acid** denotes a carboxylic acid. Propanoic acid, CH_3CH_2COOH, has three carbons in total.

TIP
When naming a carboxylic acid, remember to count the carbon in the –COOH group. $C_5H_{11}COOH$ is hexanoic acid not pentanoic acid.

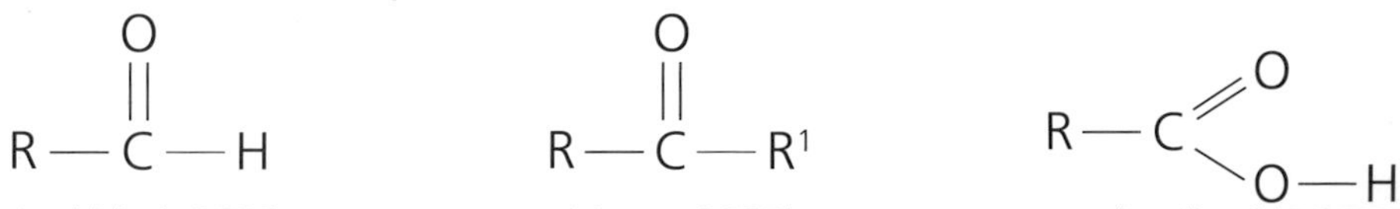

$$\underset{\text{An aldehyde RCHO}}{R - \overset{\displaystyle O}{\overset{\|}{C}} - H} \qquad \underset{\text{A ketone RCOR}^1}{R - \overset{\displaystyle O}{\overset{\|}{C}} - R^1} \qquad \underset{\text{A carboxylic acid RCOOH}}{R - C \overset{\displaystyle O}{\underset{O - H}{\diagdown}}}$$

Figure 15.7 The carbonyl group in different structures.

When oxidising a primary alcohol to produce a **carboxylic acid**, the oxidising agent is added in excess, to encourage the complete oxidation of the alcohol, and the reaction mixture is **refluxed** gently. The equation for oxidation of ethanol is:

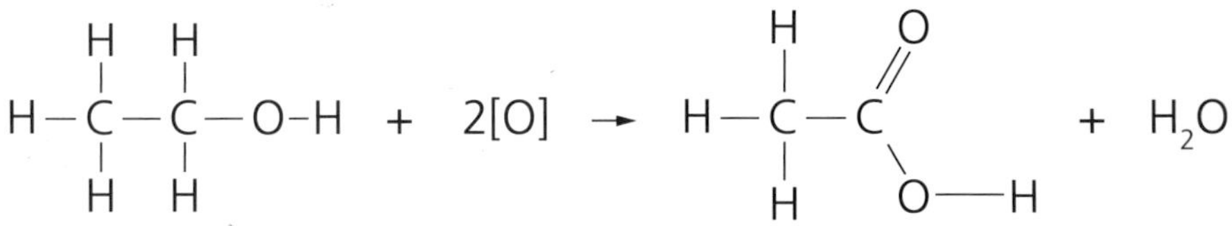

$$H - \overset{\overset{\displaystyle H}{|}}{\underset{\underset{\displaystyle H}{|}}{C}} - \overset{\overset{\displaystyle H}{|}}{\underset{\underset{\displaystyle H}{|}}{C}} - O\text{-}H \; + \; 2[O] \; \rightarrow \; H - \overset{\overset{\displaystyle H}{|}}{\underset{\underset{\displaystyle H}{|}}{C}} - C \overset{\displaystyle O}{\underset{O - H}{\diagup}} \; + \; H_2O$$

$$\underset{\substack{\text{ethanol} \\ \text{(primary alcohol)}}}{CH_3CH_2OH} \; + \; 2[O] \; \rightarrow \; \underset{\substack{\text{ethanoic acid} \\ \text{(carboxylic acid)}}}{CH_3COOH} \; + \; H_2O$$

Figure 15.8 Oxidation of ethanol to ethanoic acid.

The carboxylic acid can be removed from the mixture in the reaction vessel by distillation.

TIP
When a condenser is placed on the reaction vessel the volatile reactants are condensed back into the reaction mixture. This practical technique is known as **refluxing** (page 285). It prevents the volatile reactants evaporating before they have a chance to react.

To produce the **aldehyde**, the primary alcohol must be in excess and the product **distilled off** immediately (Figure 15.9). This is to ensure the oxidation is only partial and the carboxylic acid is not formed.

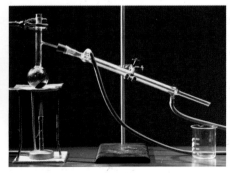

Figure 15.9 Apparatus for oxidation of a primary alcohol to produce an aldehyde.

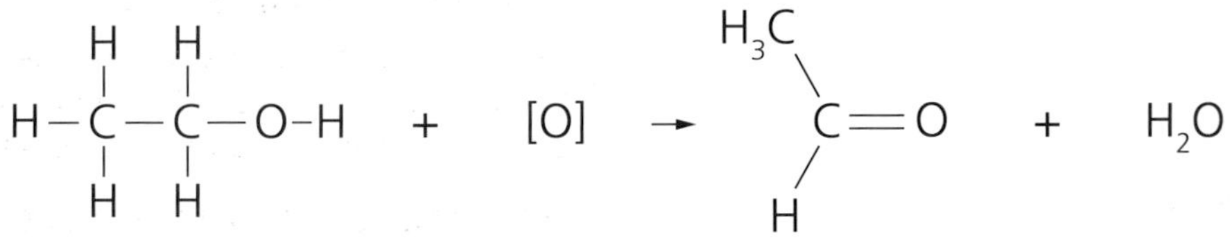

$$H - \overset{\overset{\displaystyle H}{|}}{\underset{\underset{\displaystyle H}{|}}{C}} - \overset{\overset{\displaystyle H}{|}}{\underset{\underset{\displaystyle H}{|}}{C}} - O\text{-}H \; + \; [O] \; \rightarrow \; \overset{\displaystyle H_3C}{\underset{\displaystyle H}{\diagdown \, \diagup}} C = O \; + \; H_2O$$

$$\underset{\substack{\text{ethanol} \\ \text{(primary alcohol)}}}{CH_3CH_2OH} \; + \; [O] \; \rightarrow \; \underset{\substack{\text{ethanal} \\ \text{(aldehyde)}}}{CH_3CHO} \; + \; H_2O$$

Secondary alcohols must be refluxed gently with an excess of the oxidising agent to produce the ketone.

$$CH_3CH(OH)CH_3 \quad + \quad [O] \quad \rightarrow \quad CH_3COCH_3 \quad + \quad H_2O$$

propan-2-ol
(secondary alcohol)

propanone
(ketone)

Acidified potassium or sodium dichromate(VI) solution is a suitable oxidising agent for the oxidation of primary and secondary alcohols. The dichromate(VI) ion, $Cr_2O_7{}^{2-}$, is orange in aqueous solution and is reduced to the green chromium(III) ion, Cr^{3+}, as the alcohol is oxidised.

$$Cr_2O_7^{2-} + 14H^+ + 6e^- \rightarrow 2Cr^{3+} + 7H_2O$$

orange   green

Redox equations similar to the one above are studied in Chapter 8.

TEST YOURSELF 3

1 The following alcohols are oxidised using acidified potassium dichromate(VI) solution. The conditions of the oxidation are also given.
   A pentan-3-ol refluxed gently
   B butan-1-ol warmed gently, immediately distilled
   C hexan-1-ol refluxed gently
   a) Name and give the structural formula of the organic product in each of the oxidation reactions, A, B and C.
   b) Write a half equation to illustrate what happens to the oxidising agent in the oxidation reactions.
2 Refluxing is a laboratory procedure where the reactants are continually evaporated and condensed and returned to the reaction flask. Explain why it is essential to reflux an alcohol/oxidising agent mixture when carrying out the preparation of a carboxylic acid.

REQUIRED PRACTICAL

Ethanal occurs naturally in coffee, bread and ripe fruit, and is produced by plants. It is also produced by the partial oxidation of ethanol and may be a contributing factor to hangovers from alcohol consumption.

Preparation of ethanal (bpt 21 °C)

Ethanal can be prepared in the laboratory using the following method.

- Place 7.5 g of sodium dichromate into a pear-shaped flask.
- Add 15 cm³ of water.

- Set up the apparatus for distillation and slowly add a mixture of $3\,cm^3$ of concentrated sulfuric acid and $6\,cm^3$ of ethanol (density = $0.79\,g\,cm^{-3}$) from a tap funnel into the pear-shaped flask.
- Heat gently until approximately $6\,cm^3$ of distillate is collected.

The distillate will contain a mixture of ethanal, water and ethanol.

1 Calculate the mass of ethanol used.
2 When the reaction is over, the remaining colour in the flask is orange. Explain which reactant is in excess.
3 Ethanal is easily oxidised to compound A.
   a) Name compound A.
   b) Suggest how this experimental procedure prevents the oxidation of ethanal.
4 Why is the distillate collected in a flask surrounded by ice?
5 Suggest how you could obtain ethanal from the mixture collected at the end of the experiment.
6 Write the half equation for the reduction of dichromate(VI) ions to chromium(III) ions.
7 Using the method above and subsequent separation and purification, $1.32\,g$ of ethanal were obtained. Calculate the percentage yield of ethanal.
8 What would be observed when $2\,cm^3$ of the distillate containing ethanol, ethanal and water is added to a beaker of boiling water.

Test for aldehydes and ketones

The fact that aldehydes can be further oxidised and ketones cannot easily be oxidised is used as the basis for a chemical test to distinguish between them.

Tollens' reagent – the silver mirror test

Tollens' reagent, a colourless solution of silver nitrate and dilute ammonia, is also known as ammoniacal silver nitrate. It contains the complex ion $[Ag(NH_3)_2]^+$. It is a mild oxidising agent. When Tollens' reagent is warmed gently in the presence of an aldehyde, the aldehyde is oxidised to a carboxylic acid and the silver ions, Ag^+, in the Tollens' reagent are reduced to silver atoms, $Ag(s)$.

$$RCHO + [O] \rightarrow RCOOH$$

$$Ag^+(aq) + e^- \rightarrow Ag(s)$$

The silver atoms are deposited on the sides of the reaction vessel (Figure 15.10).

The equation for the oxidation of ethanal by warming with Tollens' reagent is:

Figure 15.10 The reaction of Tollens' reagent with an aldehyde – the silver mirror test.

Figure 15.11 Oxidation of ethanal with Tollen's reagent.

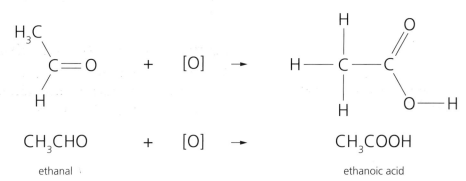

$$CH_3CHO + [O] \rightarrow CH_3COOH$$

ethanal   ethanoic acid

Ketones will not react with Tollens' reagent and the mixture remains colourless and the silver mirror does not form.

Fehling's solution

Fehling's solution contains copper(II) ions, Cu^{2+}. It is also a mild oxidising agent. When warmed gently with an aldehyde, the blue colour gradually disappears and an orange-red precipitate of copper(I) oxide, Cu_2O, forms (Figure 15.12).

$$Cu^{2+} + e^- \rightarrow Cu^+$$

For example, ethanal is oxidised to ethanoic acid by warming with Fehling's solution. This can be represented by the same reaction as in Figure 15.12. The solution remains blue when warmed with a ketone (Table 15.2).

Fehling's solution is freshly prepared in the laboratory. It is made initially as two separate solutions, Fehling's 1 and Fehling's 2, which are added together immediately before use. Fehling's 1 contains the copper(II) ions as aqueous copper sulfate; Fehling's 2 contains sodium hydroxide solution.

Figure 15.12 The reaction of Fehling's solution with an aldehyde.

Table 15.2 Tests for aldehydes and ketones.

Test	Observations with	
	aldehyde	Ketone
warm gently with Tollens' reagent	silver mirror forms on sides of the test-tube	solution remains colourless
warm gently with Fehling's solution	orange-red precipitate forms	solution remains blue

Elimination reactions of the alcohols

> An **elimination** reaction is one where a small molecule is removed (eliminated) from a reactant molecule.

An **elimination** reaction is one where a small molecule is removed (eliminated) from a reactant molecule. A molecule of water can be eliminated from an alcohol to produce an alkene.

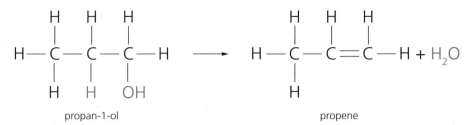

Figure 15.13

propan-1-ol   propene

A **dehydration** reaction is a chemical
reaction where a molecule of water is
eliminated.

This elimination reaction can also be described as a **dehydration reaction**.
A dehydration reaction is a chemical reaction where a molecule of water is
eliminated. Alcohols can be dehydrated using concentrated sulfuric acid as
a catalyst at a temperature of 170°C or by passing the alcohol vapour over a
heated aluminium oxide catalyst, Al_2O_3, at 600°C (Figure 15.14).

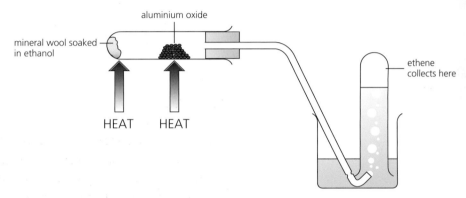

Figure 15.14 Dehydration of an alcohol
to form an alkene using a heated
aluminium oxide catalyst.

The mechanism for the elimination of water from alcohols is shown below.
This mechanism uses sulfuric acid as a catalyst.

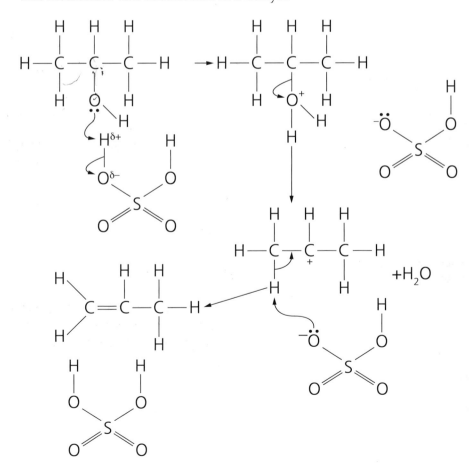

Figure 15.15

15 ALCOHOLS

Initially one of the lone pairs of electrons on oxygen picks up a hydrogen ion from the sulfuric acid and is protonated. A negative hydrogen sulfate ion is produced (HSO_4^-). A water molecule is lost from the protonated alcohol forming a carbocation. Finally a hydrogen sulfate ion removes an H^+ ion from the carbocation and a double bond forms between two carbon atoms.

This **mechanism** can be **simplified** for any acid catalyst by using H^+ instead of the full structure of sulfuric acid.

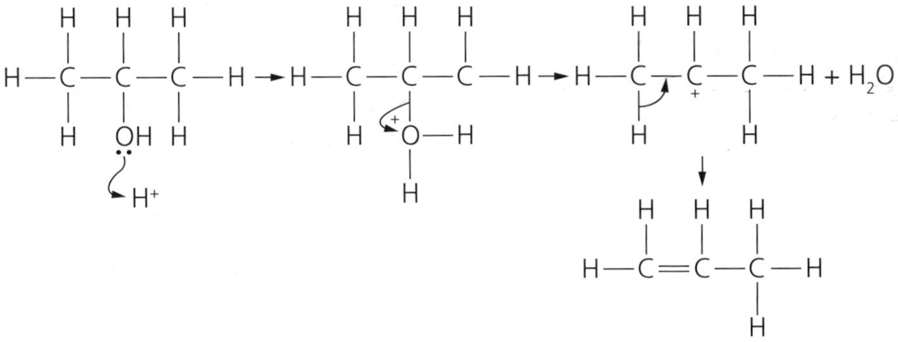

Figure 15.16

Elimination reactions in symmetrical and unsymmetrical alcohols

When propan-2-ol (a symmetrical alcohol) is dehydrated in an elimination reaction the oxygen and hydrogen atoms of the –OH group are eliminated from the molecule with either the hydrogen atom from the carbon on the right, shown in blue in Figure 15.17, or the hydrogen from the carbon on the left, shown in red.

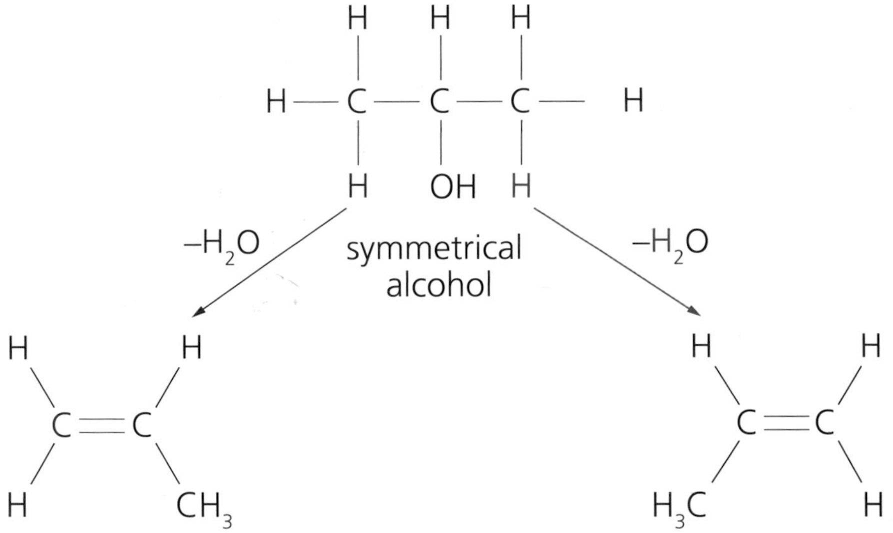

Figure 15.17 The carbonyl group in different structures.

It does not matter which hydrogen atom is eliminated as the product is always propene. However, when a molecule of water is eliminated from an **unsymmetrical** alcohol, a mixture of isomeric products is formed (Figure 15.18).

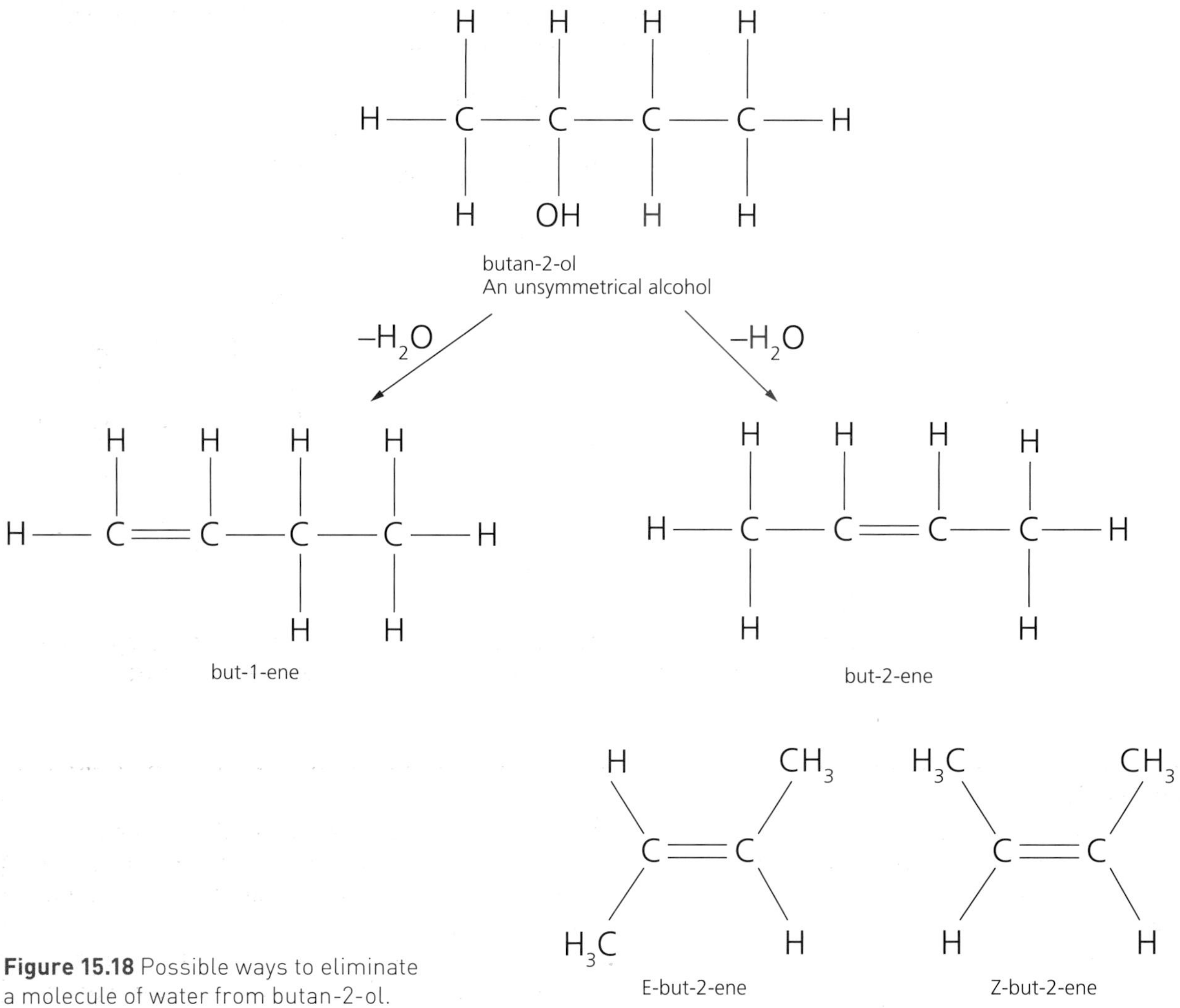

Figure 15.18 Possible ways to eliminate a molecule of water from butan-2-ol.

There are three possible products. But-1-ene is formed when the –OH group and a hydrogen to the left is eliminated (shown in red). But-2-ene is formed when the –O—H group and a hydrogen to the right is eliminated (shown in blue). But-2-ene exists as a pair of geometric isomers or E-Z isomers. The product mixture will contain all three elimination products.

EXAMPLE 1

Outline the mechanism for the elimination of water from butan-2-ol to produce but-1-ene.

Answer

The first stage of the mechanism is:

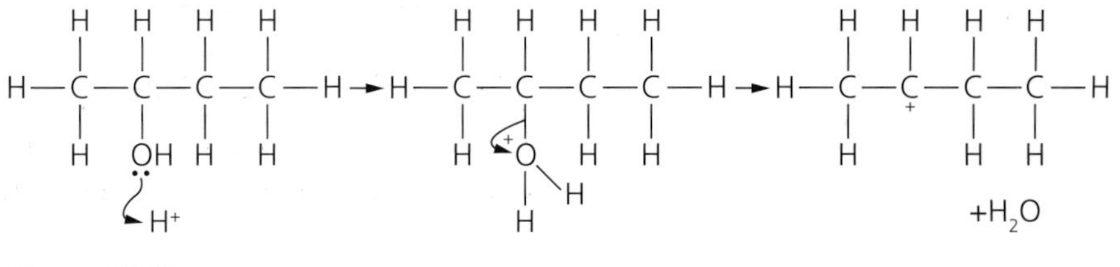

Figure 15.19

Butan-2-ol is an unsymmetrical alcohol so there are different products, depending on which hydrogen is removed.

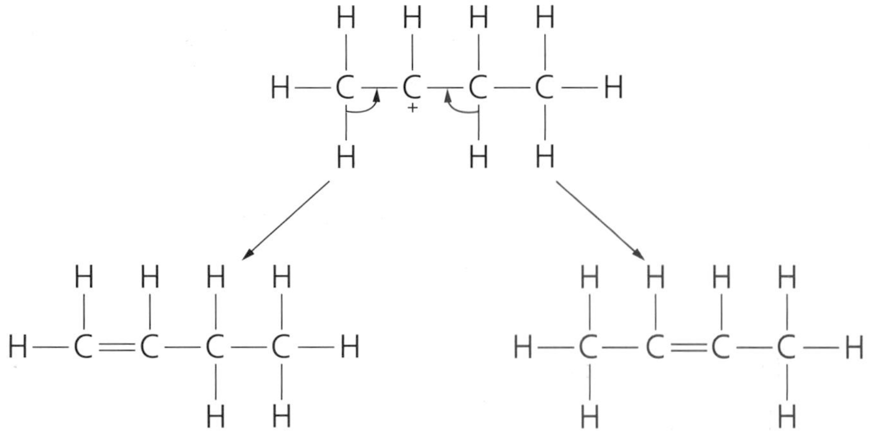

Figure 15.20

But-1-ene is the product required in the question, so the last step of the mechanism which must be shown is the red arrow and red structure shown above.

TEST YOURSELF 4

1 a) Complete the table below to record observations made when Fehling's solution is added to samples of:
   i) propanal
   ii) hexan-2-one.

Name of reagent	Observation with	
	propanal	Hexan-2-one
Fehling's solution		

b) How is Fehling's solution usually prepared in the laboratory?

c) Fehling's solution is a mild oxidising agent. Write a half equation to describe what happens to the copper(II) ions in the solution during the reaction.

d) Describe another chemical test which can be used to distinguish between an aldehyde and a ketone.

2 Name and draw displayed formula for the isomeric alkenes formed when pentan-2-ol is dehydrated in an elimination reaction.

3 a) Name a catalyst which can be used to eliminate water from ethanol.

b) What is the optimum temperature for this reaction?

Industrial production of ethanol

Ethanol is manufactured on an industrial scale by fermentation of carbohydrates and by the reaction of steam with ethene.

Ethanol production by fermentation

Fermentation occurs when yeast and bacteria convert sugars to alcohols, acids and various gases such as carbon dioxide and methane.

By fermentation, yeast converts sugars such as glucose to ethanol and carbon dioxide.

$$C_6H_{12}O_6 \rightarrow 2C_2H_5OH + 2CO_2 \text{ (g)}$$
$$\text{glucose} \rightarrow \text{ethanol} + \text{carbon dioxide}$$

The reaction is carried out:

- in the presence of yeast. Yeast produces enzymes which convert sugars into methanol.
- in the absence of air (anaerobic conditions). This is to prevent the oxidation of ethanol to ethanoic acid (vinegar).
- at a temperature of 35°C. The reaction is too slow below 25°C, while at temperatures above 40°C the enzymes in yeast are denatured, hence a compromise temperature is used.
- in a neutral aqueous solution.

The yeast is killed by concentrations of ethanol above approximately 14%. The ethanol is removed from the reaction mixture by fractional distillation. It can be used to make biofuel.

> **TIP**
> Ethanol can be written C_2H_5OH or CH_3CH_2OH in an equation, but not as C_2H_6O or C_2H_5HO.

Ethanol production by hydration of ethene

$$C_2H_4 + H_2O\text{(g)} \rightleftharpoons C_2H_5OH$$
$$\text{ethene} + \text{steam} \rightleftharpoons \text{ethanol}$$

Conditions

- catalyst of concentrated phosphoric acid absorbed on a solid silica surface
- 60 atm pressure
- 600 K temperature
- excess ethene to give a high yield.

The mechanism for the formation of ethanol by the reaction of steam in the presence of a phosphoric acid (H_3PO_4) catalyst is:

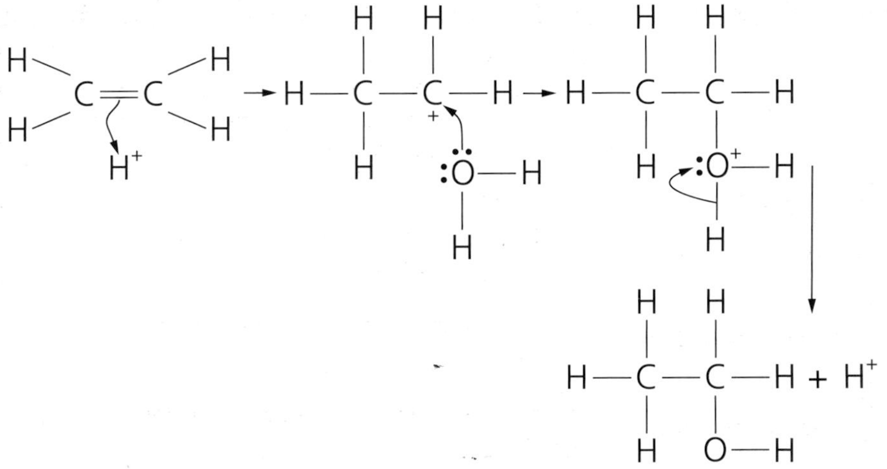

Figure 15.21

The mechanism for the formation of ethanol by the reaction of steam in the presence of a phosphoric acid (H_3PO_4) catalyst is electrophilic addition. In the mechanism the phosphoric acid is often represented as H^+. A carbocation forms and then water adds on to form a protonated alcohol. A proton (H^+) is then removed, regenerating the catalyst and forming the alcohol. Remember that for an unsymmetrical alkene the stability of the carbocation determines the major product.

TEST YOURSELF 5

1 Concentrated phosphoric acid is a catalyst in the hydration of propene to form $CH_3CH(OH)CH_3$. The industrial name for this alcohol is isopropyl alcohol.
   a) State the meaning of the term catalyst.
   b) State the meaning of the term hydration.
   c) Write an equation for the hydration of propene to form isopropyl alcohol.
   d) Give the IUPAC name for isopropyl alcohol.

Figure 15.22 During the months of June and July you are unlikely to get stung by a wasp as they are busy catching insects to feed the wasp larva. However, in autumn the wasps can become more aggressive as they can become intoxicated gorging on a diet of fermented fruit thus making stings more likely!

The environmental and economic advantages and disadvantages of both of the industrial processes used to produce ethanol are summarised in Table 15.3.

Table 15.3

Ethanol production by		Fermentation	Hydration of ethene
Economic	Type of process	A batch process. Everything is put into a container and then left until fermentation is complete. That batch is then cleared out and a new reaction set up. This is inefficient.	A continuous flow process. A stream of reactants is passed continuously over a catalyst. Continuous flow is a more efficient way of carrying out a chemical reaction on an industrial scale than a batch process.
	Rate of reaction	Very slow.	Very fast.
	Quality of product	Produces very impure ethanol which needs further processing.	Produces much purer ethanol.
Environmental	Reaction conditions	Uses gentle temperatures and atmospheric pressure.	Uses high temperatures and pressures, needing lots of energy input.
	Use of resources	Uses renewable resources based on plant material.	Uses finite resources based on crude oil.

Fermentation of carbohydrates produces dilute solutions of ethanol which we know as wine or beer and also **bioethanol**, a biofuel.

Biofuel is any fuel made from living organisms or their waste.

The word biofuel can refer to any fuel made from living organisms or their waste but the term is used in a scientific sense to mean fuels made from crops or waste. Bioethanol is ethanol made from the fermentation of sugar beet followed by fractional distillation.

A carbon-neutral activity is one in which there is no net annual emissions of carbon dioxide into the atmosphere.

Theoretically bioethanol is a **carbon-neutral fuel**. A carbon-neutral activity is one in which there is no net annual emissions of carbon dioxide into the atmosphere. A carbon-neutral fuel is one which uses the same amount of carbon dioxide from the atmosphere in its production as is released into the atmosphere upon its use.

$$CO_{2(emissions)} - CO_{2(uptake)} = 0$$

In theory, the mass of carbon dioxide released into the atmosphere by the production and combustion of ethanol (Equations 15.2 and 15.3) is equal to the mass of carbon dioxide used during photosynthesis (Equation 15.1).

Production of glucose via photosynthesis:

$$6CO_2 + 6H_2O \rightarrow C_6H_{12}O_6 + 6O_2 \quad (15.1)$$
$$\text{glucose}$$

Fermentation of glucose to produce ethanol:

$$C_6H_{12}O_6 \rightarrow 2C_2H_5OH + 2CO_2 \quad (15.2)$$
$$\text{glucose} \qquad \text{ethanol}$$

Combustion of ethanol to produce energy:

$$C_2H_5OH + 3O_2 \rightarrow 2CO_2 + 3H_2O \quad (15.3)$$
$$\text{ethanol}$$

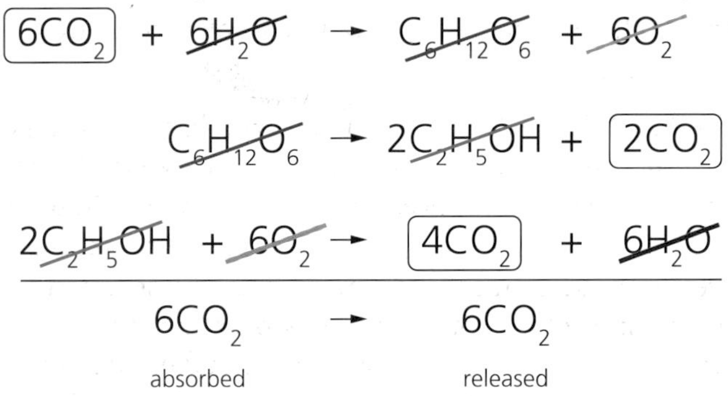

Figure 15.23

Adding Equation 15.1 to Equation 15.2 and (2 × Equation 15.3) to use all of the ethanol produced via the fermentation process gives us the equation in Figure 15.23. This shows that bioethanol can be considered a carbon neutral fuel.

However, the reality is not so neat. Energy is required at each of the stages of production described by Equations 15.1–15.3. This energy at present is almost certainly produced by burning fossil fuels, which of course releases carbon dioxide into the atmosphere.

The plants which are grown to provide the glucose, have to be planted and cared for; fertiliser and pesticides are used. The production of fertilisers and pesticides requires energy. The plant must be harvested and transported to the fermentation facility. The ethanol is extracted from the fermentation mixture by fractional distillation and after this energy intensive separation process, must be transported to the fuel pumps. Hence, in reality bioethanol is not truly a carbon neutral fuel.

The fossil energy balance can, of course, be improved. One way to achieve this would be to use the waste from the plants, for example the stalks, to provide heat to be used during the process.

Figure 15.24 The British Sugar biofuel facility in Norfolk England.

Biofuels account for approximately 3% of global fuel use. The first biofuel plant in England has been built by British Sugar, with support from BP and Dupont, at Wissington in Norfolk (Figure 15.24). It produces a biofuel which is predominately bioethanol with up to 10% biobutanol. Biobutanol, butan-1-ol, is a more efficient fuel than ethanol. It has a longer hydrocarbon chain and can be used in engines designed to run on petrol without engine modification. Locally grown sugar beet (Figure 15.25) previously used to produce sugar for the world market, is now fermented in the presence of yeast to form ethanol.

Apart from carbon neutrality a major advantage of the use of glucose from crops as the raw material for the production of ethanol is that it is sustainable; the crops are a renewable resource. There are however also disadvantages to the use of crops for the production of ethanol. These include:

Figure 15.25 Sugar beet processed at British Sugar biofuel plant in Norfolk is supplied by local growers, ensuring the minimum amount of energy is used in transporting the beet to the facility.

- Our food supply may be depleted, as increasingly land is being used to grow crops for fuel. This is an ethical issue. In some countries, crops that could be used to feed people are used to provide the raw materials for biofuels instead.
- Production of crops is subject to the weather and climate.
- It takes a long time to grow the crops.
- This route leads to the production of a mixture of water and ethanol that requires separation and further processing.

There are more exciting research avenues involving ethanol and butan-1-ol to be pursued. Alcohols will undergo elimination reactions to produce alkenes which are used to make polymers. Polymers have proven to be almost limitless in their use. Most polymers currently produced are derived from crude oil. Imagine if we could further decrease our reliance on crude oil by producing ethene and but-1-ene from the dehydration of ethanol and butan-1-ol produced by fermentation of crops such as sugar beet? Imagine still if we could use bacteria to ferment other crops or waste organic material to produce alcohols with longer carbon chains and subject these to dehydration processes to produce a greater range of alkenes and extend the range of polymers formed from non-crude oil sources. Perhaps you are the generation of chemists who will achieve this.

Figure 15.26 Just some of the uses of polyethene and polybutene. In the future can these be produced from crops such as sugar cane?

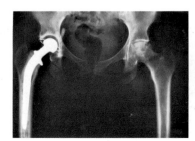

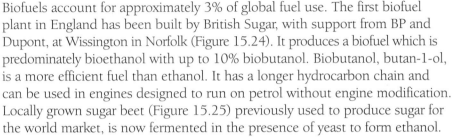

Practice questions

1 Butan-1-ol reacts with acidified potassium dichromate(VI) to produce two oxidation products, A and B. Product A is formed when the butan-1-ol is in excess and is distilled immediately from the reaction mixture.

   a) State the class of alcohols to which butan-1-ol belongs. *(1)*

   b) Draw the displayed formula for both oxidation products, A and B. *(2)*

   c) Butan-1-ol can be converted to an alkene in an elimination reaction.

      i) Write an equation for the elimination of water from butan-1-ol showing the structures of the organic compound. *(2)*

      ii) Identify a catalyst for this process. *(1)*

2 The structures of three isomers with the molecular formula $C_5H_{10}O$ are shown below.

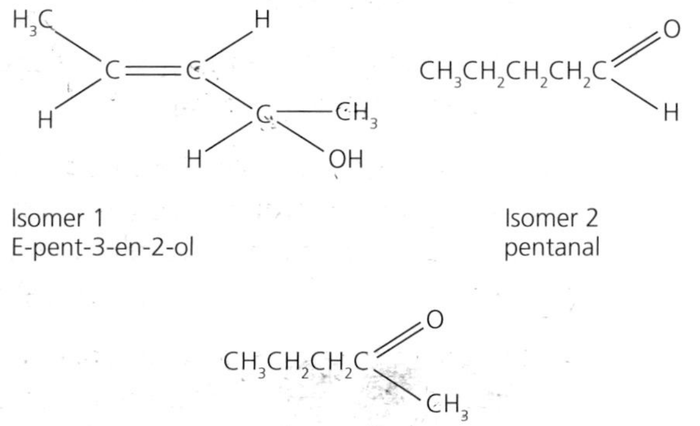

Isomer 1
E-pent-3-en-2-ol

Isomer 2
pentanal

Isomer 3

Figure 15.29

   a) Name isomer 3. *(1)*

   b) Name the type of structural isomerism shown by the three isomers. *(1)*

   c) Draw the structure of Z-pent-3-en-2-ol, the stereoisomer of isomer 1. *(1)*

   d) Explain why isomer 1 exists as a pair of stereoisomers. *(1)*

   e) Describe a chemical test which can be used to distinguish between separate samples of isomer 2 and isomer 3. Include the name of the reagent and observations made with both isomers. *(3)*

3 Compound W is formed from ethene by the following series of reactions.

$$H_2C=CH_2 \xrightarrow[\text{Conc } H_2SO_4]{\text{Reaction 1}} CH_3CH_2OSO_2OH \xrightarrow{\text{Reaction 2}} \underset{W}{CH_3CH_2OH}$$

Figure 15.30

   a) Name and outline a mechanism for reaction 1. *(3)*

   b) Name compound W. *(1)*

   c) State the role of the concentrated sulfuric acid in this process. *(1)*

   d) Outline another two step process which will produce compound W from ethene. Include the conditions for each step. *(6)*

4 The reaction of acidified potassium dichromate(VI) with ethane-1,2-diol produces ethanedioic acid.

   a) Balance the following equation for this reaction.

$$\begin{matrix} CH_2OH \\ | \\ CH_2OH \end{matrix} + [O] \longrightarrow \begin{matrix} COOH \\ | \\ COOH \end{matrix} + H_2O$$

Figure 15.31

   b) An intermediate formed in this reaction is a compound with only aldehyde functional groups and an empirical formula of CHO. Draw the structure of this intermediate. *(1)*

   c) Ethane-1,2-diol can be made from ethene by the following route.

$$H_2C=CH_2 \xrightarrow[\text{Br}_2]{\text{Reaction 1}} BrCH_2CH_2Br \xrightarrow[\text{NaOH}_{(aq)}]{\text{Reaction 2}} \begin{matrix} CH_2OH \\ | \\ CH_2OH \end{matrix}$$

Figure 15.32

   State the type of mechanisms in reaction 1 and reaction 2. *(2)*

5 Nerol is a mono-terpene found in many essential oils such as lemongrass and hops. This colourless liquid is used in perfumery with a stereoisomer known as geraniol.

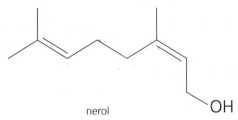

nerol

Figure 15.33

a) Use the skeletal formula above to write the molecular formula and structural formula of nerol. *(2)*

b) Write a balanced symbol equation for the combustion of nerol. *(2)*

c) What is meant by the term stereoisomer? *(1)*

d) State the class of alcohols to which nerol belongs. *(1)*

e) Name the reagent and state the conditions necessary to convert nerol into an aldehyde. *(3)*

6 Which of the following is a tertiary alcohol? *(1)*

A 2-methylbutan-2-ol

B 2-methylbutan-3-ol

C pentan-2-ol

D pentan-3-ol

7 The forces of attraction between ethanol molecules are:

A permanent dipole–dipole attraction only

B covalent bonds and hydrogen bonds

C hydrogen bonds only

D hydrogen bonds and van der Waals' forces *(1)*

8 The catalytic hydration of ethene is described by the equation below.

$$CH_2{=}CH_2(g) + H_2O(g) \rightleftharpoons CH_3CH_2OH(g)$$
$$\Delta H^\ominus = -45\,kJ\,mol^{-1}$$

Ethene and steam are mixed in a molar ratio of 10:6 and continually passed over the catalyst. The gases remain in the catalyst chamber for a relatively short time. Only 5% of the ethene is converted to ethanol each time the gases are passed across and in contact with the catalyst.

a) Explain why a low temperature favours a higher percentage of ethanol in the equilibrium mixture. *(2)*

b) Explain why the temperature actually used is a high temperature. *(1)*

According to the equation the ethene and steam react in equimolar proportions. To encourage the position of equilibrium to shift to the right, an excess of one of the reactants can be used.

c) Suggest why ethene is the reactant in excess despite being more expensive and less abundant than steam. (Consider the nature of the catalyst.) *(1)*

16 Organic analysis

TEST YOURSELF ON PRIOR KNOWLEDGE 1

1 a) Name and draw the structure of the functional group present in an aldehyde and a ketone.
   b) How would you carry out a simple chemical test to distinguish between an aldehyde and a ketone.
   c) Draw the structural formula for pentan-2-one and an isomer of it that is an aldehyde.
2 Describe a simple chemical test which could be used to distinguish between the following pairs of organic compounds.
   a) 2-methylpropan-2-ol and ethanol
   b) cyclohexane and cyclohexene.
3 The mass spectrum of zirconium is shown below.

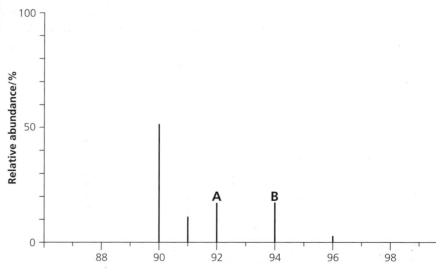

Figure 16.1

a) What value is plotted on the x-axis?
b) Write the symbol for the species which is detected at:
   i) A
   ii) B

Organic analysis

The financial and social benefits of major sporting successes are vast. Unfortunately even before the gladiators of Ancient Rome (Figure 16.2) added herbal extracts to their diets, a minority of athletes had attempted to gain competitive advantage over their rivals by artificially increasing their physical performance. Accompanying the growth in pharmaceutical knowledge in the 19th and 20th centuries was the creation of ever more sophisticated artificial performance enhancing drugs. It was necessary, in the interests of a fair competition to counteract the misuse of drugs by detection and subsequent sanctions. Modern instrumental methods such as mass spectrometry and infrared spectroscopy have made the identification of organic substances much simpler.

Mass spectrometry can identify the chemical composition of a sample based on mass to charge ratio. In addition to its use in detecting banned drugs in sport it can be used to:

- monitor and track pollutants in the air or in water supplies
- detect toxins in food
- locate oil deposits by testing rock samples
- determine the extent of damage to human genes due to the environment
- identify the country of origin of diamonds.

Figure 16.2 Even the gladiators of Ancient Rome used performance enhancing drugs.

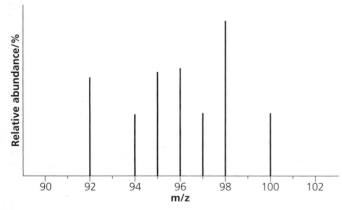

Figure 16.3 Unpolished rough diamonds. Mass spectrometry can be used to identify the country of origin of a diamond and help protect against diamond smuggling.

Mass spectrometry of organic compounds

When a sample of an *element* passes through a mass spectrometer, the spectrum produced consists of several lines. These lines are due to the different isotopes of the element (Figure 16.4) (see also Chapter 1).

When an organic *compound* passes through a mass spectrometer the spectrum produced also consists of several lines. In this case the lines are due to the original molecule and fragments of the molecule. The line with the largest m/z ratio is known as the **molecular ion**. This line has been produced by a molecule which has lost one electron. A simplified version of the mass spectrum of propan-2-one is shown in Figure 16.5. The lines on the spectrum are due to the molecular ion and ions produced by fragments of the molecule. The molecular ion is at 86.

335

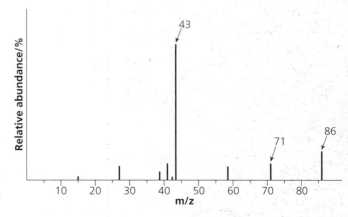

Figure 16.4 The mass spectrum of the metallic element molybdenum. The lines indicate the relative abundance of the different isotopes of the element present in the sample. The M_r can be calculated from this information.

Figure 16.5 The mass spectrum of pentan-2-one.

The mass spectra of compounds containing chlorine or bromine

There are two molecular ion peaks in the mass spectra of compounds containing a single chlorine atom. This is because chlorine exists as two isotopes, ^{35}Cl and ^{37}Cl. The mass spectrum of 2-chloropropane, $CH_3CHClCH_3$, is shown in Figure 16.6.

The peak at m/z ratio 78 is due to the molecular ion $[CH_3CH^{35}ClCH_3]^+$ containing an atom of ^{35}Cl. The peak at m/z ratio 80 is due to the molecular ion $[CH_3CH^{37}ClCH_3]^+$ containing an atom of ^{37}Cl. The ratio of the peaks is 3:1. This ratio reflects the abundance of the chlorine isotopes; $^{35}Cl:^{37}Cl$ 3:1.

There are three molecular ion peaks in the mass spectrum of 2,2-dichloropropane (Figure 16.7).

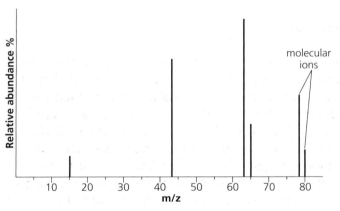

Figure 16.6 The mass spectrum of 2-chloropropane.

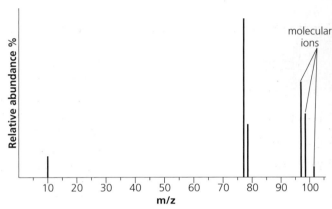

Figure 16.7 The mass spectrum of 2,2-dichloropropane.

They correspond to $[CH_3C^{35}Cl^{35}ClCH_3]^+$, $[CH_3C^{35}Cl^{37}ClCH_3]^+$ and $[CH_3C^{37}Cl^{37}ClCH_3]^+$. The three peaks are in a ratio 9:6:1. This ratio can be understood by working out the probability of a ^{35}Cl or a ^{37}Cl being present in the molecule. Remember there are three out of four (3/4) chances of a chlorine atom being ^{35}Cl and one out of four (1/4) chances of a chlorine atom being ^{37}Cl. In a molecule containing two chlorine atoms the possible combinations are:

	^{35}Cl and ^{35}Cl	^{35}Cl and ^{37}Cl or ^{37}Cl and ^{35}Cl	^{37}Cl and ^{37}Cl
probability	¾ × ¾	(¾ × ¼) + (¼ × ¾)	¼ × ¼
	9/16	3/16 + 3/16	1/16
ratio	9	6	1

This gives a distinctive pattern in a mass spectrum. A pattern of three molecular ions, M, M+2 and M+4, in a ratio of 9:6:1 is an indication that the molecule contains two chlorine atoms.

Bromine exists as two isotopes, ^{79}Br and ^{81}Br, in an almost 1:1 ratio (50.5:49.5). The mass spectrum of 2-bromopropane will show two molecular ions – one at 122 due to $[CH_3CH^{79}BrCH_3]^+$ and at 124 due to $[CH_3CH^{81}BrCH_3]^+$. These molecular ion peaks will be in the ratio 1:1 reflecting the relative abundance of the bromine isotopes.

High resolution mass spectrometry

Consider the three mass spectra, A, B and C in Figure 16.8, which are of three different organic compounds pentane and structural isomers, butanone and butanal. Each spectrum has a molecular ion peak at a mass/charge ratio of 72.

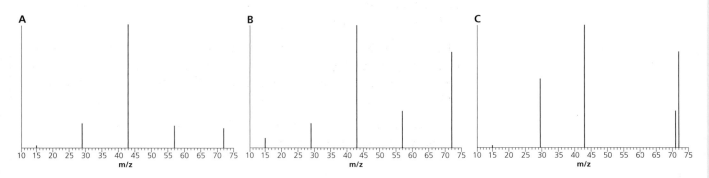

Figure 16.8 The mass spectra of pentane, and isomers butanone and butanal.

$^{1}H = 1.0078$

$^{16}O = 15.9949$

$^{14}N = 14.0071$

$^{12}C = 12.0000$

Figure 16.9

High resolution mass spectrometry can be used to distinguish between these three compounds. It can measure relative atomic masses to 4 decimal places. A more accurate value of relative molecular mass of the molecular ion can establish which compound is pentane. The precise relative atomic mass of some elements is listed in Figure 16.10. The A_r of ^{12}C is exactly 12.0000 by definition. The relative atomic masses of all other atoms are measured relative to the ^{12}C isotope. Using these values the relative molecular mass of pentane is 72.0939 and the relative molecular mass of butanone and butanal is 72.0575. Spectrum A (Figure 16.8) is pentane as the m/z ratio of the molecular ion on the high resolution mass spectra is 72.0939.

EXAMPLE 1

A sample of gas was analysed using a mass spectrometer. The molecular ion was detected at a mass to charge ratio of 28.0.

1 Which two of the following gases could be present in the sample?
   a) ethane
   b) ethene
   c) carbon dioxide
   d) carbon monoxide.
2 Explain how high resolution mass spectrometry can be used to identify the gas in the sample.

Answers

From the information given about the molecular ion you can deduce that the gas has a M_r of 28.0. Several compounds will have a M_r of 28.0, as the value is not measured to a fine degree of accuracy.

1 Using the A_r values given on your Periodic Table, the M_r of both ethene and carbon monoxide is 28.0.

You should include in your answer to Q2 the accurate M_r values for ethene and carbon monoxide. Examination questions will include A_r values to 4 decimal places. High resolution mass spectrometry can measure relative molecular masses to 4 decimal places. This introduces a high degree of accuracy and no two compounds will have exactly the same relative molecular mass when measured this accurately.

2 M_r ethane, C_2H_4 = (2 × 12.0000) + (4 × 1.0078) = 28.0312
M_r carbon monoxide, CO = (12.0000 + 15.9949) = 27.9949
Using high resolution mass spectrometry, the molecular ion of carbon monoxide will be detected at a mass to charge ratio of 27.9949 while the molecular ion of ethene will be detected at a mass to charge ratio of 28.0312.

TEST YOURSELF 6

1 Use the mass spectrum of ethanol (Figure 16.10) to answer the questions which follow.
   a) What label should be placed on the *y*-axis?
   b) What is the mass/charge ratio of the molecular ion?
   c) What is the relative molecular mass of the compound tested?
   d) Explain how you can infer the answer given in part c) from the answer given in part b).

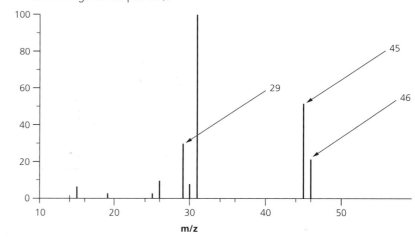

Figure 16.10

2 Caraway seeds are widely used in central and eastern Europe to flavour food. The oils from the caraway seed contain the organic compound carvone. The spectrum of carvone obtained using mass spectrometry has a molecular ion peak at a mass to charge ratio of 150.1041.
Which of the following could be the molecular formula of carvone?
   A $C_9H_{10}O_2$
   B $C_{11}H_{18}$
   C $C_{10}H_{14}O$
   D $C_{10}H_{16}N$.

Infrared spectroscopy

Pairs of atoms joined by a covalent bond continually vibrate. The frequency of vibration is unique to the atom combination of the bond and differs if the bond is single, double or triple. A carbon–carbon single bond vibrates at a different frequency to a carbon–carbon double bond or an oxygen–hydrogen bond. These vibrations have a frequency in the infrared region of the electromagnetic spectrum. This is called the natural frequency of vibration of the bond.

When a beam of infrared radiation is shone onto an organic compound some of the energy is absorbed and the amplitude of vibration of the covalent bond increases. The bond only absorbs radiation that has the same frequency as the natural frequency of the bond. Each type of bond has a natural vibration frequency and the same bond surrounded by different groups of atoms has a different natural frequency of vibration. This knowledge enables chemists to identify groups of atoms in a molecule and the environment surrounding this group. For example, analysis of an infrared spectrum of a compound may indicate that a $C=O$ group is present and will also indicate if it is part of a –CHO group in an aldehyde, or part of a –COOH group in a carboxylic acid.

All organic compounds absorb infrared radiation. Bonds in molecules absorb infrared radiation at characteristic wavenumbers.

Infrared spectra

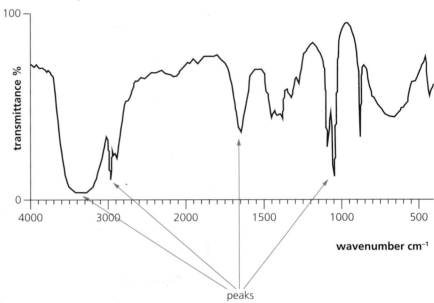

Figure 16.11 An infrared spectrum.

The spectrum begins at the top of the graph and consists of a series of dips, which represent the infrared frequency absorbed by particular bonds. These dips are given the name 'peaks'. This term is a little unusual as normally a dip is known as a trough! Another unusual aspect of an infrared spectrum is the scale. Look carefully at the scale of the x-axis of the infrared spectrum in Figure 16.11 and all of the spectra in this chapter. The scale is different to the right and left of $2000\,cm^{-1}$. It begins at $4000\,cm^{-1}$ and ends at $500\,cm^{-1}$.

An infrared spectrometer does not contain any glass or quartz because these absorb infrared radiation. All internal reflecting and refracting surfaces are made from polished sodium chloride crystals. The samples can easily be prepared. The mass of the sample required is very small; approximately 1 mg is all that is needed.

Analysis of infrared spectra

The functional group or groups of a molecule can be found by analysing infrared spectra in the region between $4000\,cm^{-1}$ and $1500\,cm^{-1}$. Figure 16.12 shows the infrared spectra of ethanol, ethanoic acid and propanone. All these compounds have C—H bonds, two have OH groups and two have $C=O$ groups present in the molecule.

TIP

Table 16.2 shows the characteristic infrared absorptions in organic molecules. A similar table will be given to you to use in your exam. It can be useful to obtain it now to use it to answer the questions given in this chapter.

Table 16.2 Characteristic infrared absorption in organic molecules.

Bond	Wavenumber (cm⁻¹)
N—H (amines)	3300–3500
O—H (alcohols)	3230–3550
C—H	2850–3300
O—H (acids)	2500–3000
C≡N	2220–2260
C=O	1680–1750
C=C	1620–1680
C—O	1000–1300
C—C	750–1100

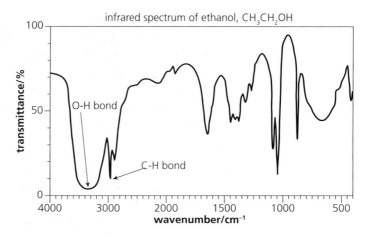

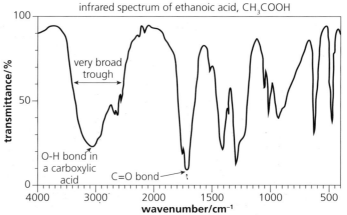

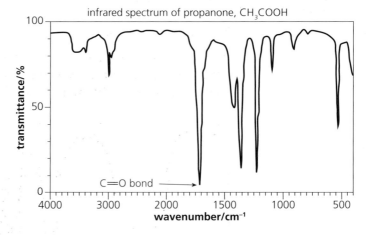

Figure 16.12 The infrared spectra of ethanol, ethanoic acid and propanone.

The C—H bond is present in almost all organic compounds. A peak just under $3000\,\text{cm}^{-1}$ is probably due to C—H bonds.

The absorption of the oxygen–hydrogen bond varies slightly depending on what sort of compound it is in. In carboxylic acids the –OH absorbs IR in the range $2500–3000\,\text{cm}^{-1}$. The –OH bond in an alcohol absorbs at a higher wavenumber than it does in an acid – somewhere between 3230 and $3550\,\text{cm}^{-1}$. It is easily recognised because it produces a broad peak in an alcohol and a very broad peak in an acid.

The C=O produces a peak between $1680\,\text{cm}^{-1}$ and $1750\,\text{cm}^{-1}$. This is present in both the spectra for ethanoic acid and propanone. Notice the absence of the broad OH peak in the spectrum of propanone. The small absorbance at approximately $2900\,\text{cm}^{-1}$ is probably due to C—H bonds.

The fingerprint region

The area of the spectrum below the $1500 \, \text{cm}^{-1}$ is known as the fingerprint region of the spectrum. The absorptions are complex and are caused by the varied and complicated vibrations of the entire molecule. This part of the spectrum is unique to the molecule. IR spectra of many known organic molecules have been recorded and are available in a database. To identify a molecule, the IR spectra is produced and compared to this database.

The infrared spectra of ethanoic acid and butanoic acid are shown in Figure 16.13. There is a broad peak in the range 2500–$3000 \, \text{cm}^{-1}$ present in both spectra indicating the presence of the $-$OH group. The peak present in the range 1680–$1750 \, \text{cm}^{-1}$ indicates the presence of a C=O group. These peaks indicate that both the compounds are carboxylic acids.

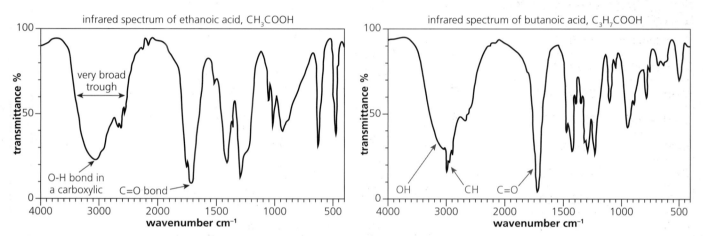

Figure 16.13 The infrared spectra of two different carboxylic acids.

In Figure 16.14, the infrared spectra of ethanoic acid and butanoic acid are superimposed. The region between $1500 \, \text{cm}^{-1}$ and $4000 \, \text{cm}^{-1}$ are very similar but the region below $1500 \, \text{cm}^{-1}$ is unique to each compound.

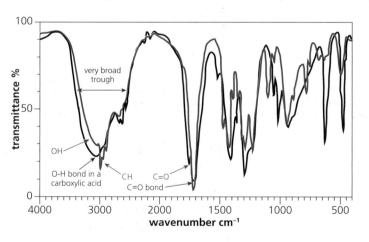

Figure 16.14 The IR spectrum of ethanoic acid superimposed on the IR spectrum of butanoic acid.

Checking purity

Infrared spectra can be used to check the purity of a compound. When the infrared spectrum of a known compound is produced extra peaks can indicate that the compound is not pure.

TEST YOURSELF 3

1 Infrared spectroscopy is a useful instrumental technique.

   a) What happens to the molecule during the absorption of infrared radiation?

   b) Explain how infrared spectroscopy can be used to identify a compound.

   c) To run an infrared spectrum in older spectrometers, the solid sample was ground up with sodium chloride and pressed into a small disc which was placed into the beam of infrared radiation. Suggest why sodium chloride was used to contain the sample.

2 a) Use the information in Figure 16.12 and Table 16.2 to list the expected absorption regions for:

   i) propanoic acid

   ii) butanol.

   b) What labels are placed on the

   i) x-axis

   ii) y-axis of an infrared spectrum?

   c) What is unusual about the scale on the x-axis?

3 The infrared spectra of ethanoic acid, propan-2-ol and propanone were analysed. The major peaks observed in the three spectra are listed below.

   X sharp peak at $1700\,cm^{-1}$, narrow, shallow peak at $3000\,cm^{-1}$

   Y sharp peak at $1700\,cm^{-1}$, deep broad peak in the range $2400–3300\,cm^{-1}$

   Z broad peak in the range $3200–3400\,cm^{-1}$, narrow shallow peak at $3000\,cm^{-1}$

   a) Match the spectra, X, Y and Z, to the compounds.

   b) Spectra X and Z have narrow peaks at $3000\,cm^{-1}$.

   i) Suggest which group gives rise to this peak.

   ii) Why is this peak not noticeable in the spectrum Y?

4 Compound A has the molecular formula $C_3H_6O_2$. The infrared spectrum of A is shown below.

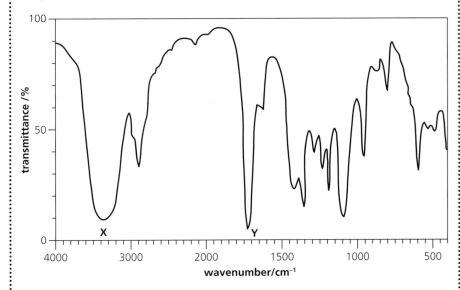

Figure 16.15

   a) Identify the groups which produce the absorptions X and Y.

   b) Draw a possible structure for A.

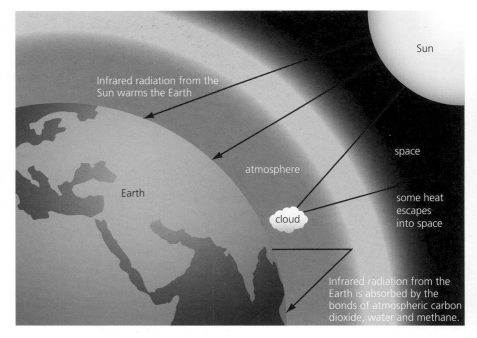

Figure 16.16 Small changes in the balance of greenhouse gases can cause significant temperature changes because they are extremely efficient at absorbing infrared radiation. This leads to the Greenhouse gas effect.

You have already studied the greenhouse effect on p. 272 and Figure 16.16 shows how the greenhouse gases, carbon dioxide, water and methane absorb infrared radiation, leading to the greenhouse effect and global warming. The bonds in these gases are very efficient at absorbing infrared radiation. Infrared spectra of carbon dioxide, methane and water have deep bands indicating their high efficiency in the absorption of infrared radiation. When infrared radiation hits a molecule of these gases it is absorbed and causes the bonds to vibrate. Oxygen and nitrogen, the other major gases in the atmosphere do not have this property. It is the absorption of infrared radiation by greenhouse gases that contributes to global warming.

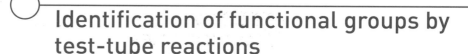

Identification of functional groups by test-tube reactions

Table 16.4 summarises some of the simple chemical tests which can be used to identify the presence of the functional groups you have met during the study of this course.

Table 16.4 Identification tests.

Homologous series	Reagent	Observations	
		Before test	After test
alkenes	bromine water	orange solution	colourless solution
primary and secondary alcohols	acidified potassium dichromate(VI)	orange solution	green solution
tertiary alcohols		orange solution	remains orange
aldehydes	Tollens' reagent	colourless solution	silver precipitate on sides of test-tube known as silver mirror
	Fehling's solution	blue solution	orange-red precipitate formed
carboxylic acids	solid sodium carbonate	white solid	solid disappears, bubbles of gas evolved (effervescence). Gas turns lime water from colourless to cloudy (CO_2)

Most organic compounds do not mix with water however, short chain alcohols and carboxylic acids do.

REQUIRED PRACTICAL

Identification of functional groups

A series of tests were carried out on an organic substance Y which has one functional group, and the observations were recorded in the table below.

Experiment	Observation	Deduction
1 Place 1 cm³ of Y in a test-tube and add 1 cm³ of water. Stopper and shake the mixture.	Two layers are formed	
2 In a fume cupboard add 1 cm³ of Y to a test-tube one quarter full of bromine water and mix well.	Orange colour disappears	
3 Add 6 drops of Y to 1 cm³ of potassium dichromate(VI) solution in a test-tube and acidify by adding 1 cm³ of dilute sulfuric acid. Warm the mixture gently.	No change in colour	

a) Complete the table by inserting appropriate deductions.
b) What colour is observed in the test-tube in experiment 3.
c) Based on the experiments above, suggest a functional group which may be present in Y.
d) Based on the experiments above, suggest a functional group which may be absent from Y.

TEST YOURSELF 4

1 Compounds X and Y both contain five carbon atoms and are insoluble in water. They do not react with either solid sodium carbonate nor with Tollens' reagent. When tested with bromine water, Y causes the orange solution to become colourless. X reacts with acidified potassium dichromate(VI) turning the orange solution to green.
   a) Identify the functional group present in Y.
   b) Which of the following, A, B, C, D or E, could be the structural formula for X?
   A $CH(CH_3)_2CH(OH)CH_3$
   B $CH_3CH_2C(OH)(CH_3)_2$
   C $CH_3CH_2CH_2CH_2CHO$
   D $(CH_3CH_2)_2CO$
   E $CH_3CH_2CHCHCH_2OH$.

2 Give the reagents and observations for test-tube reactions which could be used to distinguish between the following pairs of organic compounds.
   a) propene and propane
   b) ethanoic acid and ethanol
   c) 2-methylpropan-2-ol and propan-2-ol
   d) propanone and propanal.

This chapter has introduced you to some of the instrumental techniques and chemical tests that can be used either separately or in conjunction to obtain information about structure, identity, presence or absence or even concentration of a molecule. You have read how modern instrumental techniques can obstruct diamond smuggling; identify athletes who wish to gain unfair advantage over their rivals by the use of performance enhancing drugs and how these instrumental techniques are providing information to scientists who wish to explain the causes of climate change. It is important to note however that information does not come solely from the use of expensive analytical instruments. It can be gleaned from the most unusual of observations.

Figure 16.17 Even the smallest observations can provide vital information.

In 2013 a fascinating piece of climate change research was initiated using the millions of butterflies in a collection held in The British Museum. For 200 years, peaking in the 1800s, the population of the British Isles was fascinated by butterfly collecting. The collected butterflies were carefully pinned to boards, where they are preserved to this day in perfect condition (Figure 16.17). Among the anatomical and general biological data which were diligently recorded by the butterfly enthusiasts are data on place and time of collection. Due to the competitiveness of the collectors, the collection contains the first butterflies to emerge each year. Butterflies are cold-blooded and as such do not emerge from the chrysalis until the weather is warm enough for them to fly. Digitisation of the amassed data will take 10 years but it will allow statisticians and climatologists to see a pattern of chrysalis emergence of butterflies in specific regions of the British Isles over the past 200 years and thus a pattern of climate change will emerge.

Practice questions

1 Consider the skeletal formula of the five cyclic compounds A, B, C, D and E.

Infrared spectra for four of these cyclic compounds are given below. Match the spectra i, ii, iii or iv with the compounds A, B, C, D or E. **(4)**

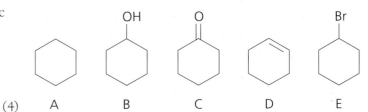

A   B   C   D   E

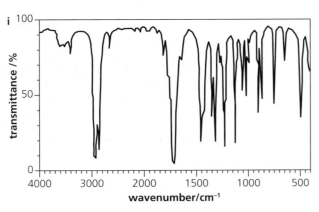

Spectrum 1

Spectrum 2

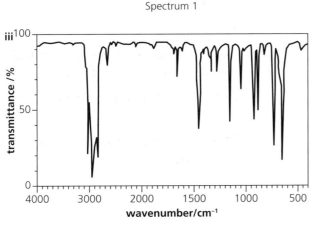

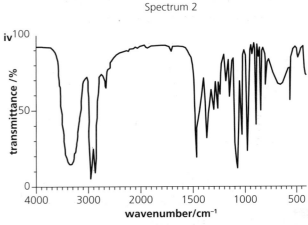

Spectrum 3

Spectrum 4

2 Which one of the following describes how infrared radiation interacts with a Greenhouse gas? **(1)**

A absorption

B emission

C reflection

D transmission

3 The infrared spectrum of a compound with the molecular formula $C_4H_{10}O$ is shown opposite.

a) Identify one feature of the infrared spectrum that supports the fact that this is an alcohol. **(1)**

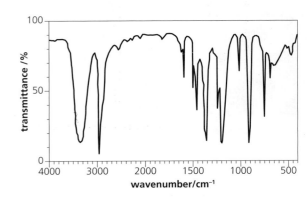

b) Explain how the spectrum can be used to identify the alcohol. **(2)**

4 Chlorine reacts with trichloromethane in UV light to form an organic compound X.

a) Write an equation for the overall reaction of chlorine with trichloromethane in UV light to form X. *(1)*

b) The following infrared spectrum was produced in this reaction. Use this spectrum to explain why it is possible to deduce that this sample of X contains no trichloromethane. *(2)*

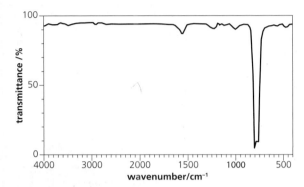

5 The infrared spectra of **water vapour** and **carbon dioxide** are given below.

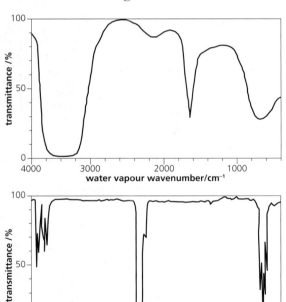

a) Use the information from the infrared spectra to deduce one reason why water vapour is a more effective greenhouse gas than carbon dioxide. *(2)*

b) Use your knowledge of the bonding in carbon dioxide to state why the infrared spectrum of carbon dioxide is not as expected. *(2)*

6 Oxygen has three isotopes ^{16}O, ^{17}O and ^{18}O. A sample of oxygen was analysed in a mass spectrometer and three groups of peaks were obtained:

Group P corresponded to the ion O_2^+
Group Q corresponded to the ion O^{2+}
Group R corresponded to the ion O^+

Which one of the following is the order on the x-axis from left to right of the groups? *(1)*

A PQR

B PRQ

C QPR

D QRP

7 Butan-2-ol can be oxidised by acidified potassium dichromate(VI) solution.

a) Write a balanced symbol equation for this oxidation. *(1)*

b) State the class of alcohol to which butan-2-ol belongs. *(1)*

17

Mathematics for chemistry

Expressing numbers

Significant figures

In calculations you may often get long decimal answers on your calculator display. It is important to round these correctly using significant figures.

The first significant figure of a number is the first digit that is not a zero.

The rules for significant figures are

1 Always count non-zero digits. For example, 21 has 2 significant figures and 8.923 has 4.

2 Never count zeros at the start of a number (leading zeros). For example, 021, 0021 and 0.0021 all have 2 significant figures.

Figure 17.1 A very long decimal answer on a calculator can be rounded to significant figures. What is this number to 3 significant figures?

3 Always count zeros which fall between two non-zero digits. For example, 20.8 has 3 significant figures and 0.00103004 has 6 significant figures.

4 Count zeros at the end of a number (trailing zeros) only if the number contains a decimal point. For example 210 and 21000 both have 2 significant figures while 210.0 has four and 210.00 has five.

Rounding

In calculations you should round the answer to a certain number of significant figures.

The rules for rounding are:

- if the next number is **5** or more, round up.
- if the next number is **4 or less**, do not round up.

For example if you are subtracting:

$$7.799\,g - 6.250\,g$$

your calculation would yield 1.549 g. If the answer is given to 3 significant figures this would be 1.55 g, because the digit '9' is greater than '5'.

When combining measurements with different degrees of accuracy and precision, **the accuracy of the final answer can be no greater than the least accurate measurement**. This means that when measurements are multiplied or divided, the answer can contain no more significant figures than the least accurate measurement.

TIP

A chain is as strong as its weakest link. A calculated answer is as accurate as the least accurate measurement in the calculation.

EXAMPLE 1

Calculate the value of $\dfrac{1.74 \times 4.3}{3.42}$. Give your answer to the appropriate precision.

Answer

Using a calculator the value is 2.187719.

Measurement	Number of significant figures
1.74	3
4.3	2
3.42	3

From the table you can see that the least accurate measurement is 4.3, which has 2 significant figures. Hence your answer should be rounded to 2 significant figures.

2.187719 rounds to 2.2 (to 2 significant figures).

EXAMPLE 2

In a titration 10.5 cm^3 of 0.55 mol dm^{-3} sodium hydroxide solution reacts with 1.5 mol dm^{-3} hydrochloric acid. Calculate the volume of hydrochloric acid required to neutralise the sodium hydroxide solution. Give your answer to the appropriate precision.

Answer

Measurement	Number of significant figures
10.5 cm^3	3
0.55 mol dm^{-3}	2
1.5 mol dm^{-3}	2

The least accurate measurement is to 2 significant figures and only 2 significant figures should be given in your final answer. The answer in full is 3.85 cm^3 so the correctly rounded answer is 3.9 cm^3 to 2 significant figures.

You should check page 56 if you are unsure how to complete this titration calculation.

EXAMPLE 3

A mixture of 4.80 g of ethanoic acid and 0.120 mol of ethanol were allowed to reach equilibrium at 20 °C. A 25.0 cm³ sample of this mixture was titrated with sodium hydroxide added from a burette. The ethanoic acid remaining in the sample reacted exactly with 4.00 cm³ of 0.400 mol dm⁻³ sodium hydroxide solution. Calculate the value for K_c for the reaction of ethanoic acid and ethanol at 20 °C. Give your answer to the appropriate precision.

Answer

Measurement	Number of significant figures
4.80 g	3
0.120 mol	3
25.0 cm³	3
0.400 mol dm⁻³	3
4.00 cm³	3

The precision of the answer (4.57) should be to 3 significant figures.

You should check page 178 if you are unsure how to do this K_c calculation.

TIP

There are many calculations which require you to use an appropriate number of significant figures. Check the calculations in Chapter 2 and enthalpy questions in Chapter 4.

Standard form

Standard form is used to express very large or very small numbers so that they are more easily understood and managed. It is easier to say that a speck of dust weighs 1.2×10^{-6} grams than to say it weighs 0.0000012 grams.

Standard form must always look like this:

'A' must always be between 1 and 10

'n' is the number of places the decimal point moves

$$A \times 10^n$$

Figure 17.2

Figure 17.3 The mass of the earth is 5973600000000000000000000 kg. This is more conveniently written in standard form as 5.9736×10^{24} kg.

EXAMPLE 4

Write 4 600 000 in standard form.

Answer

- Write the non-zero digits with a decimal place after the first number and then write ×10 after it

  4.6×10

- Then count how many places the decimal point has moved to the left and write this value as the n value.

  $4 600 000 = 4.6 \times 10^6$

TIP

Make sure that you are familiar with how standard form is presented on your calculator. Also ensure that you can enter standard form correctly on your calculator.

EXAMPLE 5

Write 0.000345 in standard form:

Answer

- Write the non-zero digits with a decimal place after the first number and then write ×10 after it.

  3.45 × 10

- Then count how many places the decimal point has moved to the right and write this value as the n value – the n is negative because the decimal point has moved to the right instead of the left.

  $0.000345 = 3.45 \times 10^{-4}$

Significant figures and standard form

For numbers in standard form, to find the number of significant figures ignore the exponent (n number) and apply the usual rules.

For example 4.2010×10^{28} has five significant figures.

The same number of significant figures must be kept when converting between ordinary and standard form.

$20.03\,g = 2.003 \times 10^1\,g$ (4 significant figures)

$20.0\,g = 2.00 \times 10^1\,g$ (3 significant figures)

$0.02030\,kg = 2.030 \times 10^{-2}\,kg$ (4 significant figures)

The number 350.99 rounded to:

- 4 significant fig. is 351.0
- 3 significant fig. is 351
- 2 significant fig. is 350
- 1 significant fig. is 400

Using standard form makes it easier to identify significant figures.

In the example above, 351 has been rounded to the 2 significant figure value of 350. However, if seen in isolation, it would be impossible to know whether the final zero in 350 was significant (and the value to 3 significant figures) or insignificant (and the value to 2 significant figures).

Standard form however is unambiguous:

- 3.5×10^2 is to 2 significant figures
- 3.50×10^2 is to 3 significant figures.

TIP

Check out the calculations using the Avogadro constant on page 35 which require use of standard form.

Decimal places

Sometimes in calculations you are asked to present your answer to one or two decimal places.

Rounding a number to one decimal place means there is only one digit after the decimal point.

Rounding a number to two decimal places means there are two digits after the decimal point.

The same rules for rounding apply as for rounding significant figures.

1st decimal place 3rd decimal place

5.743

2nd decimal place

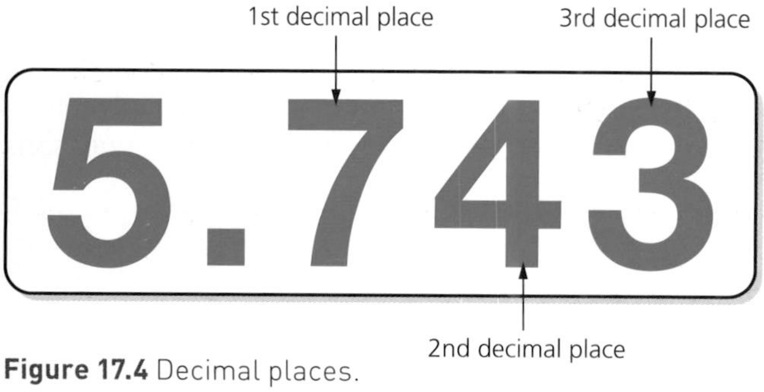

Figure 17.4 Decimal places.

EXAMPLE 6

Round 164.38 to 1 decimal place.

Answer

First underline all the numbers up to 1 number after the decimal point.

164.38

Now look at the number after the last underlined number. Since the number is 8 (above 5) you need to round up.

So the answer is 164.4 (to 1 decimal place).

EXAMPLE 7

Round 4.995 to 2 decimal places.

Answer

First underline all the numbers up to 2 numbers after the decimal point.

4.995

Now look at the number after the last underlined number. Since the number is 5 you need to round up – to round up 9 it becomes 10.

So the answer is 5.00 (to 2 decimal places).

When adding measurements with different degrees of accuracy and precision, **the accuracy of the final answer can be no greater than the least accurate measurement**. This means that when measurements are added or subtracted the answer should have the same number of decimal places as the smallest number of decimal places in any number involved in the calculation.

EXAMPLE 8

Add the following masses of sodium chloride: 40.55g, 3.1g and 10.222g. Give your answer to the appropriate precision.

Answer

Measurement	Number of decimal places
40.55	2
3.1	1
10.22	2

The number with the least decimal places involved is 3.1 (1 decimal place).

The calculation gives 53.872 g and this should be rounded to one decimal place: **53.9 g**.

Units

Many of the calculations used in chemistry will require different units. It is important that you can convert between units.

Volume
Volume is usually measured in cm^3 or dm^3 (decimetre cubed) or m^3.

$$1000\,cm^3 = 1\,dm^3$$

You need to be able to convert between these volume units, particularly for volumetric calculations like those on page 43 and ideal gas equation questions like those on page 63.

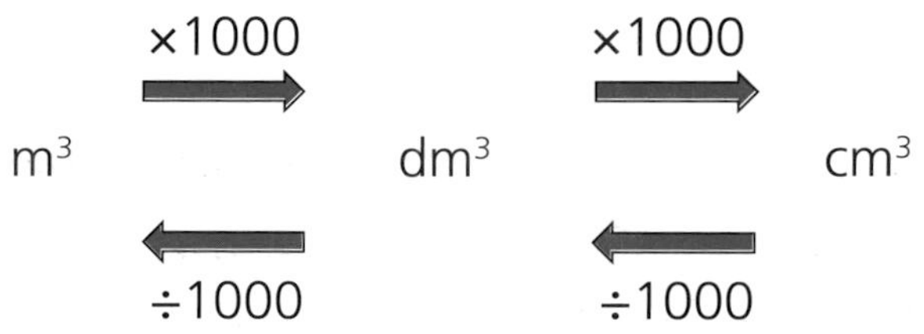

Figure 17.5 Converting between volume units.

$$\xrightarrow{\times 1000} \quad \xrightarrow{\times 1000}$$

m³ $\qquad$ dm³ $\qquad$ cm³

$$\xleftarrow{\div 1000} \quad \xleftarrow{\div 1000}$$

EXAMPLE 9

What is 25.0 cm³ in dm³?

Answer

To convert from cm³ to dm³ you need to divide by 1000.

25.0 cm³ = 0.0250 dm³ (remember to maintain the same number of significant figures).

Temperature

Temperature is measured in kelvin (K) and degrees Celsius (°C). 273K is 0 °C.

Figure 17.6 Conversion between °C and K.

K $\qquad\qquad$ °C

$$\xrightarrow{-273}$$
$$\xleftarrow{+273}$$

EXAMPLE 10

Convert 29 °C into kelvin.

Answer

To convert from °C to kelvin, add 273.

29 + 273 = 302K

Figure 17.7 William Thomson (Lord Kelvin) was a British scientist and mathematician who was born in Belfast in 1824. Knighted by Queen Victoria, for his work on the transatlantic telegraph project, Lord Kelvin is most widely known for determining the correct value of absolute zero as approximately −273.15°C. Absolute temperatures are stated in units of kelvin in his honour.

EXAMPLE 11

Convert 285K to °C.

Answer

To convert from K to °C, take away 273.

285 − 273 = 12 °C

Mass

Mass can be measured in milligrams (mg), grams (g), kilograms (kg) and in tonnes.

353

1 tonne = 1000 kg $\quad$ 1 kilogram = 1000 g $\quad$ 1 gram = 1000 mg

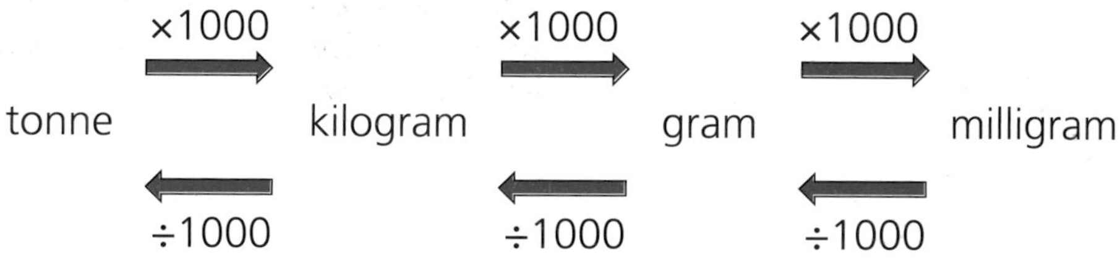

$$\xrightarrow{\times 1000} \quad \xrightarrow{\times 1000} \quad \xrightarrow{\times 1000}$$

tonne $\qquad$ kilogram $\qquad$ gram $\qquad$ milligram

$$\xleftarrow{\div 1000} \quad \xleftarrow{\div 1000} \quad \xleftarrow{\div 1000}$$

TIP

Think logically when converting between units. A kilogram is bigger than a gram so when converting from kilograms to grams you would expect to get a smaller number.

Figure 17.8 Converting between mass units.

Figure 17.9 A top pan balance is used in the laboratory to measure mass.

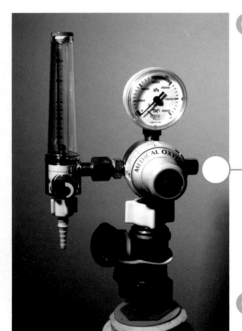

Figure 17.11 The oxygen in a cylinder for use in hospitals for patients with breathing difficulties is stored at a pressure of 13700 kPa.

EXAMPLE 12

Convert 320 mg into kg.

Answer

First you need to covert mg to g by dividing by 1000.

320 ÷ 1000 = 0.32 g

Then convert 0.32 g to kg by dividing by 1000.

0.32 ÷ 1000 = 3.2×10^{-4} kg

Pressure

Pressure can be measured in pascals (Pa) and kilopascals (kPa).

×1000

kPa   Pa

÷1000

Figure 17.10 Converting between pressure units.

EXAMPLE 13

Convert 1200 Pa into kPa.

Answer

To convert Pa to kPa, divide by 1000.

1200 ÷ 1000 = 1.2 kPa

Estimates and means

Arithmetic means

The arithmetic mean is found by adding together all the values and dividing by the total number of values. It may be referred to as the 'average' or simply as the mean.

EXAMPLE 14

The temperature of a solution was measured every 30 seconds for 3 minutes and the results recorded below.

Time/s	0	30	60	90	120	150	180
Temperature/°C	21	22	23	24	24	23	22

Calculate the mean temperature.

Answer

$$\text{Mean} = \frac{21 + 22 + 23 + 24 + 24 + 23 + 22}{7} = 22.7 = 23 \text{ (to 2 significant figures)}$$

Calculating mean titres

To calculate the mean titre, do not include the rough titration value.

A concordant titre is obtained when the titres are within $\pm 0.10\,cm^3$ of each other.

Titres of 23.60 and 23.70 are concordant; titres of 23.60 and 23.85 are not concordant.

To calculate the mean of a set of results **only use concordant values.**

	Rough	Titration 1	Titration 2	Titration 3
Titre/cm³	26.10	25.10	25.45	25.15

$$\text{Mean} = \frac{25.10 + 25.15}{2} = 25.13$$

Titre 2 is not concordant and is not used to calculate the mean titre. The value 25.45 can be referred to as an '**outlier**'.

Algebra

Symbols

You need to be familiar with different symbols which can be used in mathematical equations, or in chemical equations.

Symbol	Meaning
=	is equal to
<<	is much less than
>>	is much greater than
>	is greater than
<	is less than
~	is similar to
∝	is proportional to
⇌	a reversible reaction

Changing the subject of an equation

An equation shows that two things are equal. It will have an equals sign '='. An example of an equation is:

$$\text{moles} = \frac{\text{mass (g)}}{M_r}$$

This means that what is on the left of the equals sign is equal to what is on the right.

The subject of an equation is the single variable (usually on the left of the '=') that everything else is equal to. In the example above, the subject is 'moles'.

One of the very powerful things that can be done with algebra is the rearrangement of an equation so that another variable is the subject and on its own.

EXAMPLE 15

Make x the subject of the equation:

$y = x + 3$

Answer

Switch sides to get the new subject on the left:

$x + 3 = y$

You require x by itself on the left-hand side so you need to subtract 3 from the left side and, to keep the equation true, we need to subtract 3 from the right-hand side as well.

$x + 3 - 3 = y - 3$

Simplify

$x = y - 3$

EXAMPLE 16

Make mass the subject of the equation:

$$moles = \frac{mass\ (g)}{M_r}$$

Answer

Switch sides to get the new subject on the left

$$\frac{mass\ (g)}{M_r} = moles$$

You require mass by itself on the left-hand side so you need to remove M_r by multiplying both sides by M_r and cancelling the M_r on the left.

$$\frac{mass\ (g)}{M_r} \times M_r = moles \times M_r$$

Simplify:

$$mass\ (g) = moles \times M_r$$

Solving algebraic equations

When solving algebraic equations it is important that you change the subject, if necessary and then substitute the values.

EXAMPLE 17

Calculate the amount, in moles, of calcium hydroxide present in $25.0\,cm^3$ of a solution of concentration $0.15\,mol\,dm^{-3}$.

Answer

The subject is the amount in moles, so does not need to be changed. Simply substitute the numerical values and calculate the answer.

$$n = \frac{v \times c}{1000} = \frac{25.0 \times 0.15}{1000} = 0.0038 \text{ to 2 significant figures}$$

If the volume of calcium hydroxide was given in dm^3 then you use the equation $n = v \times c$, and do not divide by 1000.

EXAMPLE 18

Calculate the concentration of calcium hydroxide solution obtained when 0.0038 moles of calcium hydroxide are dissolved in 25.0 cm³ water.

Answer

$$n = \frac{v \times c}{1000}$$

The subject is the amount in moles, n, so it needs to be changed to concentration.

Switch the sides:

$$\frac{v \times c}{1000} = n$$

You require 'c' on its own on the left. To remove the 1000, multiply both sides by 1000 and simplify.

$$\frac{v \times c}{\cancel{1000}} \times \cancel{1000} = n \times 1000$$

$$v \times c = n \times 1000$$

You need to remove 'v' by dividing both sides by 'v' and simplifying the equations.

Simply substitute the numerical values and calculate the answer.

$$\frac{\cancel{v} \times c}{\cancel{v}} = \frac{n \times 1000}{v}$$

$$c = \frac{n \times 1000}{v}$$

Substituting values into the equation above gives:

$$c = \frac{0.0038 \times 1000}{25.0} = 0.15 \text{ mol dm}^{-3} \text{ to 2 significant figures}$$

If the volume of calcium hydroxide was given in dm³ then you use the equation:

$n = v \times c$, and do not divide by 1000.

TIP

You must use the correct units for measurement, when substituting numerical values into an equation. Check back to examples on page 71 on finding gas volumes and on enthalpy changes on page 139 and make sure you can change the units correctly.

Percentages and ratios

Percentages

Per cent means 'out of 100'. If 10 per cent of the population owns a sports car, this means 10 out of every 100 people have one. The symbol % means per cent.

To express values as a percentage, multiply by 100. In AS chemistry, there are several values which are always expressed as percentages.

Percentage yield:

$$\text{Percentage yield} = \frac{\text{actual yield}}{\text{theoretical yield}} \times 100$$

You will find examples of this calculation on page 42.

Atom economy:

$$\% \text{ atom economy} = \frac{\text{molecular mass of desired product}}{\text{sum of molecular masses of all reactants}} \times 100$$

You will find examples of this calculation on page 69.

Figure 17.12 A 15 ml pipette. The error is marked at ±0.03 ml.

Percentage uncertainty:

Uncertainty is an estimate attached to a measurement which characterises the range of values within which the true value is thought to lie. This is normally expressed as a range of values such as 44.0 ± 0.4.

For example, if the mass of an aluminium block is measured with a balance that reads to 0.1 g, the uncertainty is reported as 15.3 ± 0.1 g. This means the 'actual' mass is known to lie between 15.2 and 15.4 g. If the same block is measured on a balance that reads to 0.01 g, the mass might be reported as 15.28 ± 0.01 g. So in this case, the 'actual' mass is known to be between 15.27 and 15.29 g. Thus, in the second mass reading, there is less uncertainty.

When glassware is manufactured there will always be a maximum error. This is usually marked on the glassware. Error is the difference between an individual measurement and the true value of the quantity being measured.

The significance of error in a measurement depends upon how large a quantity is being measured. It is useful to quantify this error as a percentage error.

$$\text{Percentage error} = \frac{\text{error}}{\text{quantity measured}} \times 100$$

EXAMPLE 19

What is the percentage error when a volume of 25.0 cm³ is measured with a burette which has an error of ±0.1 cm³.

Answer

Percentage error $= \dfrac{0.1}{25.0} \times 100 = 0.4\%$

Multiple measurements

For multiple measurements, using a balance with error of 0.005 there will be a maximum error of ±0.005 g for each measurement.

For two mass measurements that give a resultant mass by difference, there are two errors.

EXAMPLE 20

mass of crucible + crystals before heat = 23.45 g (error = 0.005 g)

mass of crucible + crystals after heat = 23.21 g (error = 0.005 g)

What is the percentage error in mass lost?

Answer

overall error = 2 × 0.005 g

mass lost = 0.24 g

Percentage error in mass loss $= \dfrac{2 \times 0.005}{0.24} \times 100 = 4\%$

You must be able to determine the uncertainty when two burette readings are used to calculate a titre value.

EXAMPLE 21

A burette has an error $\pm 0.05\,cm^3$. In a titration the initial burette reading was $0.05\,cm^3$ and the final burette reading was $24.55\,cm^3$. What is the percentage uncertainty in the titre value?

Answer

Initial burette reading = $0.05\,cm^3$

Final burette reading = $24.55\,cm^3$

The overall error in any volume measured in a burette always comes from the two measurements, so the overall error = $2 \times 0.05\,cm^3 = 0.01\,cm^3$

Titre value = $24.50\,cm^3$

Percentage uncertainty = $\dfrac{2 \times 0.05}{24.50} \times 100 = 0.4\%$

When to round off in multi-step calculations

- Rounding off should be left until the very end of the calculation.
- Rounding off after each step, and using this rounded figure as the starting figure for the next step, is likely to make a difference to the final answer as it introduces a rounding error.

EXAMPLE 22

When $6.074\,g$ of a carbonate of M_r 84.3, is reacted with $50.0\,cm^3$ of $2.0\,mol\,dm^{-3}$ HCl(aq) (which is an excess), a temperature rise (ΔT) of $5.5\,°C$ is obtained. The specific heat capacity (c) of the solution is $4.18\,J\,K^{-1}\,g^{-1}$. Calculate the value for the enthalpy change in $kJ\,mol^{-1}$.

Answer

$q = mc\Delta T$

$\Delta T = 5.5\,°C$

$q = mc\Delta T = 50.0 \times 4.18 \times 5.5 = 1149.5\,J = +1.1495\,kJ$

Since the least accurate measurement (the temperature rise) is only to 2 significant figures the answer should also be quoted to 2 significant figures.

Therefore, the enthalpy change = $+1.1\,kJ$

However this figure is to be used subsequently to calculate the enthalpy change per mole so the rounding off should not be applied until the final answer has been obtained.

Number of moles of carbonate = $\dfrac{mass}{M_r} = \dfrac{6.074}{84.3} = 0.07205$

The enthalpy change is $+1.1495\,kJ$ for $0.07205\,mol$

The enthalpy change per mole is:

$1.1495/0.07205 = +15.9541 = +16\,kJ$ to 2 significant figures.

Using the rounded value of $1.1\,kJ$ for the heat produced:

enthalpy per mole = $+15.26671057\,kJ\,mol^{-1}$

Rounding this answer to 2 significant figures gives $+15\,kJ\,mol^{-1}$. Hence there is a rounding error. To avoid this do not round off until the final answer has been obtained.

Figure 17.13 If pastry is 2 parts flour to 1 part fat, then there are 3 parts (2 + 1) altogether. Two thirds of the pastry is flour; one third fat.

TIP
Ratios are used in balancing equations and in finding empirical formulae – check the examples on page 68.

Ratios

A ratio is a way to compare amounts of something. Recipes, for example, are sometimes given as ratios. To make pastry you may need to mix 2 parts flour to 1 part fat. This means the ratio of flour to fat is 2:1. Ratios are written with a colon (:) between the numbers.

In a balanced chemical equation the substances are in a ratio. For example, 2 moles of magnesium react with 1 mole of oxygen to produce 2 moles of magnesium oxide.

$$2Mg + O_2 \rightarrow 2MgO$$

The ratio is 2 moles Mg:1 mole O_2:2 moles MgO.

EXAMPLE 23

In the equation:

$2Al + 6HCl \rightarrow 2AlCl_3 + 3H_2$

The ratio between aluminium and hydrogen is $2Al:3H_2$

The ratio between aluminium and hydrochloric acid is $2Al:6HCl$ which simplifies to $1Al:3HCl$

Ratios are similar to fractions; they can both be simplified by finding common factors. Always try to divide by the highest common factor.

For example in the ratio 12:15 the highest common factor (remember a factor is a number that divides into it exactly) is 3 so the ratio simplifies to 4:5.

Graphs

Plotting graphs

Experiments often involve variables:

- The independent variable is the factor which is changed during the experiment.
- The dependent variable is the factor that is being measured during the experiment.
- The independent variable causes a change in the dependent variable.

A graph is an illustration of how two variables relate to one another. In chemistry a type of graph that you may often be asked to draw is a scatter graph with a best fit curve or line. A scatter graph is used in the situation where both variables are quantitative (numerical) and continuous (any numerical value is possible, not just whole numbers).

General construction

- The graph should have a **title** which summarises the relationship that is being illustrated – this should include the independent variable and the dependent variable, as well as the reaction being studied. For example 'A graph of concentration against time for the reaction between magnesium and hydrochloric acid.'
- The independent variable is placed on the x-axis, while the dependent variable is placed on the y-axis.

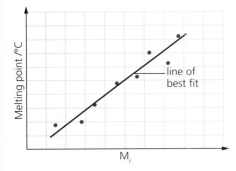

Figure 17.14 Graph of melting point against M_r for some alcohols. Note that this particular graph should not be extrapolated through to zero.

- Appropriate **scales** should be devised for the axis, making the most effective use of the graph paper. You must ensure that the plotted points use at least half of the graph paper provided. The data should be critically examined to establish whether it is necessary to start the scale(s) at zero. For example, if the data for the y-axis is 1002, 1020, 1050, 1060, etc., it may be best to start at 1000 rather than 0. Make sure you show a discontinuity symbol on the axis, if you do this.
- Axes should be **labelled** with the name of the variable followed by a solidus (/) and the unit of measurement. For example the label may be temperature/°C.
- A line of **best fit** should be drawn. When judging the position of the line there should be approximately the same number of data points on each side of the line, resist the temptation to simply connect the first and last points. Ignore any anomalous results.
- It may be important to calculate what is happening beyond the points plotted – extending the best fit line or curve is a process known as extrapolation.

TIP
A line of best fit is added by eye. You should use a transparent plastic ruler or a flexible curve to aid you.

TIP
Check back to Chapter 3 and make sure that you can draw the 3D representation of these shapes using dash and wedge diagrams.

Geometry and trigonometry
2D and 3D shapes
The table shows the difference between 2D and 3D shapes.

3D shapes	2D shapes
Have height, depth and width.	Have length and width.
These figures can be drawn on a sheet of paper using wedged and dashed lines.	These figures can be drawn on a sheet of paper.
3D figures deal with three coordinates: x-coordinate, y-coordinate and z-coordinate.	2D figures deal with two coordinates: x-coordinate and y-coordinate.

The table below shows the 3D shapes that some molecules can have.

Table 17.1 The 3D shapes of some molecules.

3D shape	Name of shape	3D shape	Name of shape
	linear		bent
	trigonal planar		trigonal bipyramid
	tetrahedral		octahedral

361

Index

Free online resources

Answers for the following features found in this book are available online:

- Test yourself questions
- Activities

You'll also find Practical skills sheets and Data sheets. Additionally there is an Extended glossary to help you learn the key terms and formulae you'll need in your exam.

Scan the QR codes below for each chapter.

Alternatively, you can browse through all chapters at www.hoddereducation.co.uk/AQAChemistry1

How to use the QR codes

To use the QR codes you will need a QR code reader for your smartphone/tablet. There are many free readers available, depending on the smartphone/tablet you are using. We have supplied some suggestions below, but this is not an exhaustive list and you should only download software compatible with your device and operating system. We do not endorse any of the third-party products listed below and downloading them is at your own risk.

- for iPhone/iPad, search the App store for Qrafter
- for Android, search the Play store for QR Droid
- for Blackberry, search Blackberry World for QR Scanner Pro
- for Windows/Symbian, search the Store for Upcode

Once you have downloaded a QR code reader, simply open the reader app and use it to take a photo of the code. You will then see a menu of the free resources available for that topic.

1 Atomic structure

4 Energetics

2 Amount of substance

5 Kinetics

3 Bonding

6 Chemical equilibria and le Chatelier's principle

7 Equilibrium constant, Kc, for homogeneous systems

14 Alkenes

8 Oxidation, reduction and redox equations

15 Alcohols

9 Periodicity and Group 2

16 Organic analysis

10 Halogens

17 Maths for Chemistry

11 Introduction to organic chemistry

18 Practical skills

12 Alkanes

19 Preparing for the examination

13 Halogenoalkanes

The periodic table

Key

relative atomic mass
symbol
name
atomic (proton) number

1.0
H
hydrogen
1

(1)	(2)	(3)	(4)	(5)	(6)	(7)	(8)	(9)	(10)	(11)	(12)	(13)	(14)	(15)	(16)	(17)	(18)
1	2											3	4	5	6	7	0
																	4.0 **He** helium 2
6.9 **Li** lithium 3	9.0 **Be** beryllium 4											10.8 **B** boron 5	12.0 **C** carbon 6	14.0 **N** nitrogen 7	16.0 **O** oxygen 8	19.0 **F** fluorine 9	20.2 **Ne** neon 10
23.0 **Na** sodium 11	24.3 **Mg** magnesium 12											27.0 **Al** aluminium 13	28.1 **Si** silicon 14	31.0 **P** phosphorus 15	32.1 **S** sulfur 16	35.5 **Cl** chlorine 17	39.9 **Ar** argon 18
39.1 **K** potassium 19	40.1 **Ca** calcium 20	45.0 **Sc** scandium 21	47.9 **Ti** titanium 22	50.9 **V** vanadium 23	52.0 **Cr** chromium 24	54.9 **Mn** manganese 25	55.8 **Fe** iron 26	58.9 **Co** cobalt 27	58.7 **Ni** nickel 28	63.5 **Cu** copper 29	65.4 **Zn** zinc 30	69.7 **Ga** gallium 31	72.6 **Ge** germanium 32	74.9 **As** arsenic 33	79.0 **Se** selenium 34	79.9 **Br** bromine 35	83.8 **Kr** krypton 36
85.5 **Rb** rubidium 37	87.6 **Sr** strontium 38	88.9 **Y** yttrium 39	91.2 **Zr** zirconium 40	92.9 **Nb** niobium 41	96.0 **Mo** molybdenum 42	[98] **Tc** technetium 43	101.1 **Ru** ruthenium 44	102.9 **Rh** rhodium 45	106.4 **Pd** palladium 46	107.9 **Ag** silver 47	112.4 **Cd** cadmium 48	114.8 **In** indium 49	118.7 **Sn** tin 50	121.8 **Sb** antimony 51	127.6 **Te** tellurium 52	126.9 **I** iodine 53	131.3 **Xe** xenon 54
132.9 **Cs** caesium 55	137.3 **Ba** barium 56	138.9 **La** * lanthanum 57	178.5 **Hf** hafnium 72	180.9 **Ta** tantalum 73	183.8 **W** tungsten 74	186.2 **Re** rhenium 75	190.2 **Os** osmium 76	192.2 **Ir** iridium 77	195.1 **Pt** platinum 78	197.0 **Au** gold 79	200.6 **Hg** mercury 80	204.4 **Tl** thallium 81	207.2 **Pb** lead 82	209.0 **Bi** bismuth 83	[209] **Po** polonium 84	[210] **At** astatine 85	[222] **Rn** radon 86
[223] **Fr** francium 87	[226] **Ra** radium 88	[227] **Ac** † actinium 89	[267] **Rf** rutherfordium 104	[268] **Db** dubnium 105	[271] **Sg** seaborgium 106	[272] **Bh** bohrium 107	[270] **Hs** hassium 108	[276] **Mt** meitnerium 109	[281] **Ds** darmstadtium 110	[280] **Rg** roentgenium 111							

Elements with atomic numbers 112-116 have been reported but not fully authenticated

* 58–71 Lanthanides

140.1 **Ce** cerium 58	140.9 **Pr** praseodymium 59	144.2 **Nd** neodymium 60	[145] **Pm** promethium 61	150.4 **Sm** samarium 62	152.0 **Eu** europium 63	157.3 **Gd** gadolinium 64	158.9 **Tb** terbium 65	162.5 **Dy** dysprosium 66	164.9 **Ho** holmium 67	167.3 **Er** erbium 68	168.9 **Tm** thulium 69	173.1 **Yb** ytterbium 70	175.0 **Lu** lutetium 71

† 90–103 Actinides

232.0 **Th** thorium 90	231.0 **Pa** protactinium 91	238.0 **U** uranium 92	[237] **Np** neptunium 93	[244] **Pu** plutonium 94	[243] **Am** americium 95	[247] **Cm** curium 96	[247] **Bk** berkelium 97	[251] **Cf** californium 98	[252] **Es** einsteinium 99	[257] **Fm** fermium 100	[258] **Md** mendelevium 101	[259] **No** nobelium 102	[262] **Lr** lawrencium 103